THE INTERNATIONAL SERIES OF
MONOGRAPHS ON CHEMISTRY

THE INTERNATIONAL SERIES OF
MONOGRAPHS ON CHEMISTRY

MOLECULAR MOTION IN HIGH POLYMERS

By

R. T. BAILEY
Senior Lecturer in Physical Chemistry

ALASTAIR M. NORTH
Professor of Physical Chemistry

and

RICHARD A. PETHRICK
Senior Lecturer in Physical Chemistry

Department of Pure and Applied Chemistry
University of Strathclyde, Glasgow

CLARENDON PRESS · OXFORD · 1981

Oxford University Press, Walton Street, Oxford OX2 6DP

OXFORD LONDON GLASGOW
NEW YORK TORONTO MELBOURNE WELLINGTON
KUALA LUMPUR SINGAPORE JAKARTA HONG KONG TOKYO
DELHI BOMBAY CALCUTTA MADRAS KARACHI
NAIROBI DAR ES SALAAM CAPE TOWN

© Oxford University Press 1981

Published in the United States by
Oxford University Press, Inc., New York

British Library Cataloguing in Publication Data

Bailey, R. T.
　Molecular motion in the high polymers.—
　(International series of monographs on chemistry).
　1. Molecular dynamics
　2. Polymers and polymerization
　I. Title　II. North, Alastair MacArthur
　III. Pethrick, Richard A.　IV. Series
　547.8'404426　QD381.9.M6　80-41809
　ISBN 0–19–851333–X

Printed in Great Britain by
J. W. Arrowsmith Ltd, Bristol BS3 2NT

PREFACE

Synthetic and naturally occurring macromolecules play such a significant part in all our lives that in many ways we have come to take them for granted. We forget that only in the last two decades or so has it become possible to relate in detail the chemical structure of polymer molecules to their macroscopic physical or mechanical properties. There are several ways of linking the chemistry of plastics and rubbers to their physics, but one of the most useful is by way of an understanding of macromolecular motion, and it is this aspect of molecular physics which forms the content of this book.

The subject of molecular motion in polymers is not an easy one to treat because it involves both an understanding of quite sophisticated theoretical concepts and a description of a very wide range of properties and techniques, each with their own theoretical background. Ideally, of course, the theory underlying any experimental observation or property would relate immediately or obviously to the theory of molecular motion, but, unfortunately, understanding of the subject has not evolved in such a tidy way. As a result, any over-all view requires knowledge of a variety of topics which might be considered quite disparate in other branches of chemistry or physics. The difficulty is compounded by the very size of the molecules involved, which adds a degree of complexity or sophistication to phenomena or theories which present problems enough even when small molecules are under study.

In planning the coverage and subject presentation we have been aware of two factors. First, many readers will be concerned rather more with the observable properties of polymers than with the underlying theoretical molecular physics, and, secondly, the basic theories of many of the techniques as applied to small molecules, are available in standard texts. We have, therefore, divided the book into two sections, the first primarily theoretical in nature and the second phenomenological, and either can be read separately. In the second section we have minimized the basic theories of the several phenomena involved by giving references to the individual standard texts, and by explaining the origin of concepts necessary to extend treatments from small molecules to polymers, rather than deriving all expressions in full detail.

In order to achieve this separation of background theory and observable properties we have constructed the chapters on theory so that they start with the treatment of molecules in an unstressed or unperturbed situation, and then move to the consideration of interaction with an observation probe such as a photon, a phonon, or some other applied field. There is, therefore, a cross-connection between the two sections of the book. Thus Chapter 2 deals first with the resonance vibrational motions of macromolecules, and then with the spectroscopic absorptions arising from these. This chapter requires an elementary familiarity with matrix algebra, and because the treatment is similar in kind, the text advances to scattering from, as well as absorption by, moving molecules. So the theory in the second part of this chapter relates to the photon and neutron absorption and scattering phenomena described in Chapters 11 and 12.

In very much the same way Chapter 3, dealing with damped hydrodynamic modes, establishes a basis for the viscoelastic properties described in Chapter 7. Finally, the statistical mechanical treatments which form the first part of Chapter 4 introduce time-correlation functions for chain movement. This, then, creates a suitable place for the introduction of those time-correlation functions which underlie phenomena such as dipole orientation, photoluminescence, or magnetic spin orientation or decay. These theoretical sections of Chapter 4, therefore, match the theories of chain movement to the properties evidenced in dielectric relaxation (Chapter 5), fluorescence and phosphorescence polarization and decay (Chapter 6), and nuclear and electron spin relaxation (Chapter 9).

Inevitably such an arrangement of material requires abbreviation, which can cause obscurity, and unfamiliar subject organization which, too, might obfuscate rather than elucidate. Nevertheless, we hope that this volume does accomplish the twin objectives of portraying a very wide field as a unified whole, and of elaborating the fascinating science in the several component parts.

Glasgow   R.T.B.
February 1980   A.M.N.
                                                                         R.A.P.

Acknowledgements

Fig. 7.2: reproduced from Johnson, R. M., Schrag, J. L., and Ferry, J. D. *Polymer J.* **1**, (1970) by permission of the publishers.

Fig. 7.3: reproduced with permission from Osaki, K., Mituda, Y., Schrag, J. L., and Ferry J. D. *Macromolecules* **5**, 17 (1972); © 1972 American Chemical Society.

Fig. 7.6: reproduced with permission from Onagi, S., Masuda, I., and Kitagawa, K. *Macromolecules* **3**, 109 (1970); © 1970 American Chemical Society.

Fig. 8.8: reproduced from Nomura, H., Kato, S., and Miyahara, Y. *J. Chem. Soc. Japan* (*Chem. and Ind. Chem.*) 2398 (1973) by permission of the publishers and the authors and from *Mem. Fac. Engng Nagoya Univ.* **17**, 72–125 (1975).

Fig. 8.9: reproduced from Kato, S., Ukondo, H., Fujio, I., Honoran, N., and Miyahara, Y. *J. Chem. Soc. Japan* (*Chem. and Ind. Chem.*) 1981 (1974) by permission of the publishers and the authors.

Fig. 8.10: reproduced from Hawley, S. A. and Dunn, F., *J. Chem. Phys.* **50**, 3523 (1969) by permission of the publishers, American Institute of Physics.

Fig. 9.4: reproduced from Allen, G., Conner, T. M., and Pursey, H. *Trans. Faraday Soc.* **59**, 1525 (1963) by permission of the publishers, Royal Society of Chemistry and the authors.

Fig. 9.5: reproduced from Conner, T. M. *Polymer* **7**, 426 (1966) by permission of the publishers, IPC Business Press Co. Ltd. and the author.

Fig. 9.8: reproduced with permission from Schaefer, J., Stejskal, E. O., and Buchdahl, R. *Macromolecules* **10**, 384 (1977); © 1977 American Chemical Society.

Fig. 9.11: reproduced from Bullock, A. T., Cameron, G. G., and Smith, P. M. *J. Phys. Chem.* **77**, 1635 (1973) by permission of the publishers.

Fig. 9.12: reproduced from Bullock, A. T., Cameron, G. G., and Smith, P. M *J. Chem. Soc. Faraday II* **70**, 1202 (1974) by permission of the publishers, Royal Society of Chemistry.

CONTENTS

PART A INTRODUCTION AND THEORY

PART B PHENOMENOLOGICAL TREATMENT

PART A
INTRODUCTION AND THEORY

THE IMPORTANCE OF MOLECULAR MOTION IN HIGH POLYMERS

1.1. Introduction

An amazing variety of phenomena, familiar in everyday life, depend on the unique properties of large polymeric molecules. For example our very existence results from the replication behaviour of nucleic acids and from the proteins which these help synthesize. Vegetable life stores energy as starch and fabricates skeletal structures from cellulose. However, while the biosynthesis of natural macromolecules evolved slowly over aeons of time, the last fifty years have witnessed the discovery and synthesis of a wide variety of other polymeric materials. Thus modern society has become very dependent on articles constructed from synthetic fibres, plastics, and rubbers. The technological importance of this group of materials lies mainly in their physical properties and the molecular scientist is concerned to ask how the polymer molecular structure is responsible for the observed macroscopic behaviour.

To find an answer to this question we select that aspect of molecular behaviour which seems to exert the greatest influence on the relevant physical properties and then relate this to chemical structure and molecular environment on the one hand and to bulk properties on the other. In the case of fibres, plastics, and rubbers, where we are primarily concerned with mechanical characteristics, we are led to focus attention first on molecular movement,[1] and then on the way this is affected by environmental considerations such as morphology.

1.2. Modes of motion of a polymer molecule

A macromolecule is so large that attempts to characterize its motions in terms of the three spatial coordinates of each atom becomes an almost impossible task. As a result the number of possible modes of motion encountered in such an approach is equally intractable. However, it is possible to introduce a number of simplifying assumptions or to consider various models, so as to divide the possible modes into a number of groups, any one group having a particular relevance to a specific phenomenon exhibited by the polymer system.

It is probably most convenient to start with the very localized vibrational and torsional movements of individual atoms and groups of an isolated chain, considered as if it were *in vacuo*. These are the movements familiar to the vibrational spectroscopist, and are usually discussed in terms of characteristic bond or group frequencies. The allocation of a characteristic frequency to a specific bond or group in a large molecule involves a number of well known assumptions, usually derived from the simple approximation that the rest of the molecule be considered as an inert body of effective infinite mass, motions within which do not couple strongly with the mode of interest. The motion of the atoms within a group is usually described in terms of resonant normal modes. Because of confusion between spectroscopists and rheologists over the term 'normal mode', we have inserted the 'resonance' qualification. This indicates that the mode of motion represents a solution to an equation of motion which has second-order differential (inertial) and linear (elastic restoring force) terms only. In this context a 'normal mode' is most easily visualized as one in which all atoms in the group are in phase, a consideration that allows considerable simplification in the mathematical treatment of the movement. The simplification is that the normal modes can be considered as uncoupled (or independent) harmonic oscillators. This ignores the anharmonicity in vibration, and thus also the coupling between modes, particularly in the higher vibrational states. The neglect of mode coupling is probably of minor importance when polymer properties are being discussed in terms of the relatively energetic vibrations of single bonds or small groups. However it does become significant when consideration is given to very low-frequency modes (such as torsional frequencies) where thermal energies or external stimulation may cause excitation to quite high quantum levels. These problems are well documented in standard texts on infrared (particularly far infrared) and Raman spectroscopy.

This resonance normal-mode picture has been extended from small groups substituted on a polymer chain to include the chain backbone itself. In this context a number of simple modes of motion can be visualized, one of which is the longitudinal or 'accordion vibration' of an extended chain section (Fig. 1.1). This has proved valuable in the determination of extended chain lengths in certain situations, although here, too, caution is necessary in allowing for coupling effects in such low-frequency motions.

When we move to motions of a larger amplitude or less localized character than the spectroscopic 'vibrations' discussed above, it becomes impossible to neglect intermolecular interactions on the moving entity. This is a reflection both of the impossibility of examining completely a polymer molecule *in*

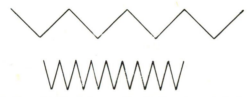

Fig. 1.1. Longitudinal ('accordion') vibration of chain section.

vacuo, and of the fact that in a flexible molecule any particular chain section may experience interactions from other sections of the same chain. The constraints imposed by intermolecular or intergroup interactions can be treated in a variety of ways, but in general the effect is to introduce a retarding force upon, or barrier to, molecular motion. The simplest mathematical accommodation of the effect is to introduce a first-order differential term (viscous drag proportional to particle velocity) into the equation of motion of the moving entity. When this term is larger than the second-order differential inertial term, the equation of motion reduces to a linear first-order differential equation characteristic of relaxation (as opposed to resonance) phenomena. Consequently, as we move to consider the lower-frequency retarded motions of larger chain units, the observed phenomena change from a resonance to a relaxation form.

Of course, there is no sharp discrimination between resonance and relaxation motions, the highest frequency movements of chain segments and side-groups showing characteristics of both, and being describable only by equations of motion containing both inertial and viscous terms. This is especially evident in the quasi-vibrational librations that cause 'collisional' phenomena with neighbouring chains or molecules.

After the torsional or rotational movement of groups substituted on a polymer backbone, probably the most localized motions are those undergone by short segments of the chain backbone. In this context the moving unit consists of a very small number of monomer residues, and the characteristics of the motion depend on the nature and environment of the segment rather than on the molecular weight of the whole chain. The movement can range from a simple torsional libration, to quite large-scale rotational behaviour. This phenomenon generally occurs at frequencies such that in the equation of motion terms arising from intrachain and intermolecular forces are highly significant and give the process a predominantly relaxation character. Indeed relaxation studies of such movements, particularly in dilute solutions in mobile solvents, provide an excellent measurement of chain flexibility. In this case 'flexibility' has a time-dependent definition based on the speed of internal conformation change (segmental rotation), rather than the time-averaged spatial definition used in studies of polymer chain dimensions.

This localized mode of motion is generally viewed as one in which the movement of any one segment is independent of that of any other. In other words it is an incoherent mode of motion with presumably very weak coupling between effective segments. Very recently it has been shown[2] that it is not necessary to assume weak intersegment coupling in order to obtain a high-frequency, molecular-weight-independent mode of motion, and that the choice of an appropriate strong-coupling model also can predict, in theory, modes of motion with these characteristics.

Finally, the largest-scale motions of a polymer molecule are those which involve large sections of the chain, or even the whole macromolecule, moving in a co-operative fashion. When the molecule is rigid, and unable to undergo conformational change, such motions will be simply translational or rotational diffusion. However when the macromolecule is flexible, the translational and rotational character become integrated with the conformational changes or molecular distortion.

In the case of rotational motion the combined rotation and distortion can be analysed using a particular form of the normal mode analysis. In this case, as in the spectroscopic normal modes referred to earlier, the motions involve distortions in which all units in the chain move in phase. The mathematical transformation of the equations of motion of each unit into a set of normal coordinates establishes independent differential equations the solutions of which are the normal modes. This case, however, differs from the spectroscopic analysis in that a space-differential term in the equation of motion arises through the Brownian diffusion process responsible for such gross movements, and the first-order viscous differential completely outweighs any second-order inertial term. As a result the motion has a relaxation character, and the various modes appear with translation as the zeroth, distortion about a single node as the first, and successively more complex distortions about greater numbers of nodes as overtones of the first (Fig. 1.2). It is important to distinguish the relaxation normal modes from the

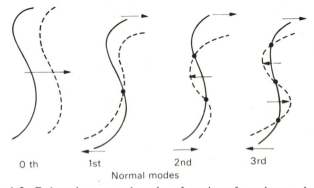

0 th   1st   2nd   3rd

Normal modes

Fig. 1.2. Relaxation normal modes of motion of a polymer chain.

resonance normal modes mentioned earlier, although both result from a similar coordinate transformation in the equations of motion, as a result of which we visualize all particles as moving in phase.

The combination of translation and conformational change presents rather more difficulties to theoretical description. Generally treatments are based upon a wriggling 'worm-like' representation. Some of the more successful consider random walk of the 'worm' in a 'pipe', the shape of which is itself subject to random walk fluctuations.[3] These will be discussed in more detail in Chapter 5.

The translational motion of macromolecules is of enormous importance in a wide variety of diffusion and flow phenomena. Indeed diffusion in polymer systems has been the subject of a number of textbooks,[4] and so generally will be excluded from this volume which will emphasize, instead, intramolecular motions.

1.3. The technological importance of segmental motion

Probably the most important of the modes of motion presented in the preceding section is that one based upon the ability of chemical groups to rotate about single covalent bonds in the chain backbone. The over-all effect is one by which certain segments of the chain move relative to others and is the intramolecular process by which the molecule moves through its numerous possible conformations. It may take place by free rotation about single covalent bonds, or by the summation of partial hindered rotations about a number of bonds.

1.3.1. Molecular movement and mechanical behaviour

Since most of the technological applications of synthetic polymeric materials concern their mechanical behaviour in some way, it is convenient to start with the relationship between molecular motion and macroscopic physical properties.[1,5,6]

In a discussion of the mechanical properties of polymeric materials we are predominantly concerned with some aspect of 'hardness' or elasticity. Consequently it is useful to quantify this as modulus, the stress required to provide a given strain in a sample.

When the moduli of amorphous polymeric materials are examined over a temperature range, they generally conform to the behaviour illustrated in Fig. 1.3. At low temperatures the polymeric solids are hard or glassy, with an elastic modulus close to $10^{10}\,\mathrm{N\,m^{-2}}$, a value characteristic of metals and inorganic glasses. At the other extreme of temperature, thermoplastic materials may flow as viscous liquids. However, where polymeric materials differ from metals and low molecular-weight compounds is in the intermediate region, often designated the 'rubbery plateau region'. At these

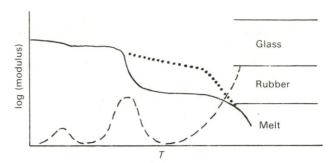

Fig. 1.3. Hardness or energy storage modulus (full curve), energy loss modulus (broken curve) for an ideal amorphous polymer. Storage modulus of a crystalline polymer (dotted curve).

temperatures there is a 'leathery' or 'rubbery' state of matter, (characterized by an elastic modulus of about $10^6 \, \text{N m}^{-2}$), which is intermediate between the normal solid and liquid states.

These three states can be rationalized in terms of segmental and translational motion as follows. In the glassy region neither translational movement nor segmental rotation occur to any appreciable extent; in the intermediate rubber region segmental motion is possible (permitting molecular and solid deformation) but translational motion is very limited; in the viscous flow region both motions occur.

The transition region between the glassy and rubbery states is referred to as the main glass transition. This often occurs over a very short temperature range, so that it is meaningful to refer to a single glass-transition temperature. It is important to realize that this is not the same as a crystal melting temperature (which may also be present in crystalline polymers) or a 'melt–flow transition' temperature (between the rubber and liquid regions). This simple rationalization of the so-called five regions of viscoelastic behaviour (glassy, glass transition, rubbery, melt–flow transition, and viscous liquid) is complicated somewhat when the polymeric chains can crystallize or are cross-linked into a network. In the latter situation translational movement becomes impossible, and the melt–flow transition and viscous liquid regions are not observed. Such behaviour occurs in the wide class of materials known as thermosetting plastics.

In the case of crystalline materials it is necessary to consider the possibility of segmental motion in both the crystalline and amorphous regions of a bulk sample. The intercrystalline amorphous regions may experience some slight onset of segmental motion above the glass transition region, but most polymer chains will be held in the closer packed crystalline domains, called crystallites. Consequently, the sample will exhibit considerable rigidity up to the crystalline melt temperature.

Real polymeric materials also exhibit a number of subsidiary transition phenomena. These are often associated with the onset of a particular mode of molecular motion, and sometimes with order–disorder or other phase transitions. They are responsible for what might be called secondary changes in mechanical behaviour. As an example a transition observed below the main glass transition (and illustrated in Fig. 1.3) is sometimes associated with the change from brittle glass to tough glass, and a transition in the melt is associated with changes in the viscoelastic character of the fluid.

1.3.2. The time dependence of molecular motion

Since the observed transitions in polymers correspond to an onset (or cessation) of molecular motion, there must be a correlation between the speed of the motion and the experimental time available for its observation. In other words, if a polymer is observed over a long time, perhaps put to some use involving a very low-frequency perturbation, a very slow molecular movement will be able to make its effect noticeable. On the other hand, for short times of observation, or in high-frequency uses, only the most rapid molecular motions will be significant. Since molecular motion is a rate process, and generally increases with increasing temperature, the result is that the long-time (low frequency) observation detects the 'onset' of motion ('transition') at a lower temperature than does the short-time (high frequency) experiment.

The fact that the molecular motions are time-dependent means, too, that there is a time dependence in the resulting mechanical behaviour. In other words a finite time is taken for a polymeric material to adopt a strain as a result of a given perturbing stress. As a result of this it is necessary to consider the relative phase of stress and strain.

When stress and strain are in phase, the implication is that polymer molecules can move sufficiently fast to adapt to the applied stress, and the modulus is a real quantity. Since this quantity measures the elastic energy which can be stored and recovered in a sample it is referred to as the 'storage modulus'. On the other hand, if the rate of molecular motion is such that polymer molecules cannot adopt their equilibrium strain in response to the applied stress, the observed strain may lag behind the applied stress. Under these circumstances the strain exhibits a phase lag relative to the stress. This may be represented on an Argand diagram, Fig. 1.4, which shows how the resulting strain can be resolved into two components. Of these, one component is in phase with the stress, whereas one component is 90° out of phase. Consequently the frequency-dependent strain (and so the modulus) is a complex quantity in the mathematical sense.

The component of the modulus which is out of phase with the applied stress is imaginary in the mathematical sense. Since it is a measure of the non-recoverable work done on the system and dissipated as heat, it is termed

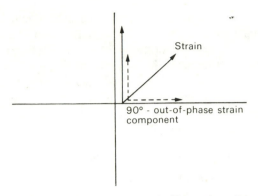

Fig. 1.4. Vector diagram of stress and strain illustrating phase lag of strain.

the 'loss modulus'. The loss modulus of an idealized polymer system is illustrated as a function of temperature in Fig. 1.3, and it can be seen that this passes through a maximum at the transitions corresponding to the onset of some molecular motion. Consequently, just as the apparent 'hardness' of a material may be a function of the observation time or use, so too may be the energy-loss characteristics.

This close interdependence of frequency and temperature implies that it is just as meaningful to consider the complex modulus as a function of observation time as of temperature. Thus, in Fig. 1.5 an apparent modulus for a typical amorphous polymer is plotted against the perturbation frequency which might be used in some technique of observation. It can be

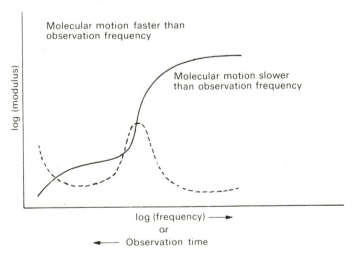

Fig. 1.5. Complex modulus of an amorphous linear polymer. Storage modulus (full curve), loss modulus (broken curve).

seen that the behaviour in the frequency plane is very closely related to that in the temperature plane. High frequencies are equivalent to low temperatures in that molecular motion is slower than the observation frequency, and so is not evidenced. On the other hand at low frequencies or high temperatures molecular motion is faster than the observation frequency and so can be detected.

1.3.3. The 'working range' of plastics

We saw in § 1.3.1. that many polymeric glasses are characterized by at least two transitions. The higher temperature main glass-rubber process is often called the α-transition and the subsidiary low temperature change is named the β-transition.

Many uses of plastics require considerable rigidity without brittleness. This behaviour is often evidenced by polymers in a temperature range below the α-transition but above subsidiary processes such as the β-transition. Furthermore in many uses of plastics, particularly where dielectric properties are concerned, it is desirable that energy-loss phenomena do not vary widely over the temperature 'working range' of the material. These two criteria allow us to discuss a useful working range for any polymer in terms of the temperature difference between the energy-loss maxima (Fig. 1.6). For

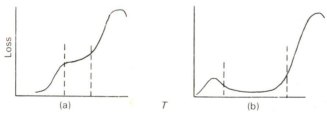

Fig. 1.6. Dielectric or mechanical loss for polymer with transitions: (a) closely; (b) widely separated in temperature. Broken lines indicate possible 'useful working range'.

uses in which energy losses are not too important and a narrow working range is acceptable, polymers of the behaviour illustrated in Fig. 1.6(a) are suitable. Polystyrene is a typical example. On the other hand for specialized uses requiring low losses or a wide working range, polymers of type 1.6(b) must be found. In respect of such requirements polycarbonates, $+OCOR+_n$, are outstanding.

When two modes of molecular motion, which we shall call the α- and β-processes, are responsible for two transitions at different temperatures, the separation of the transition temperatures is frequency-dependent, unless the two molecular motions involved have exactly the same temperature dependence. Generally the process responsible for the higher transition

will be the slower molecular movement at any chosen observation temperature. This conclusion can be appreciated by considering the increase in
energy or temperature needed to speed the molecular movements up to the
observed time scale or frequency. It comes as no surprise to chemists that the
slower process usually has the higher activation energy. This means that
when consideration is given to the joint variation of temperature and
frequency, the α-process has the larger temperature coefficient. Now if we
transform from 'transition temperature observed at a particular measurement frequency', to 'rate of molecular motion at a selected temperature' we
can construct an idealized Arrhenius diagram for the characteristic
behaviour of any amorphous polymer (Fig. 1.7).

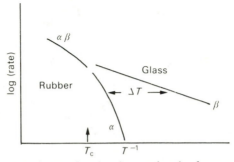

Fig. 1.7. Arrhenius plot of rate of molecular motion (or frequency of observation).
ΔT represents 'useful working range' at a selected frequency. T_c is the coincidence
temperature.

This figure illustrates the important features of the α- and β-processes.
The first is that the α-process has the larger temperature coefficient and that
the 'Arrhenius plot' is non-linear. The second is that the rates of the two
processes must coincide at some high temperature and high frequency.
Above these only a single molecular process or transition is observed, and
this has some of the character of both α- and β-processes. Consequently it is
named the $\alpha\beta$-process. Finally the 'useful working range' can be observed as
the horizontal separation of the α and β curves at any selected frequency.

In the context of the 'useful working range' it is apparent that the
behaviour of polymers will depend on the value of the coincidence
temperature. If this should be below room temperature for a particular
polymer, examination as a function of frequency will show only one transition and that will require a high frequency of observation. For low-
frequency observations the material will be in the rubbery state. Conversely,
if the coincidence temperature is well above room temperature, two widely
spaced transitions will be observed. Thus the working range of the polymer,
as well as the actual position of the main glass transition, depend on the ease,
or 'activation energy' of the molecular motions involved.

1.4. Relaxation properties of polymers

The preceding discussion of the mechanical consequences of molecular motion and the relationship between modulus and frequency may have been familiar to some readers as the phenomenological basis of dynamic mechanical relaxation. In fact relaxation phenomena constitute a wide class of processes, very many of which have their origin in, or depend upon, molecular movement. Consequently it is convenient first to consider relaxation in a general way, and then to assess how specific relaxation processes depend upon the characteristic structure and behaviour of the molecules.

The simplest definition of 'relaxation' is that it is the time-dependent return to equilibrium of some system which has experienced a change in the constraints acting upon it. The constraints may be the familiar thermodynamic variables of temperature, pressure, volume, chemical composition, or may involve an applied field such as mechanical stress, a magnetic field, or an electric field. The relaxation of the system is observed by monitoring the way some property changes with time. In this way we introduce the idea of a molecular response which takes a finite time to adjust to some perturbation.

Very often the rate with which a perturbed system returns to equilibrium is proportional to the distance from equilibrium. This situation is termed *ideal* relaxation, and is familiar in a variety of physical and chemical kinetic situations. If we assign the parameter, $P(t)$, to the difference between the value of an observed property at time, t, and the equilibrium value reached an infinite time after a step function change in constraint (Fig. 1.8) then

$$-\frac{\mathrm{d}P(t)}{\mathrm{d}t} \propto P(t) \qquad (1.1)$$

which has the simple exponential solution with a single time constant,

$$P(t) = P_0 \exp(-t/\tau). \qquad (1.2)$$

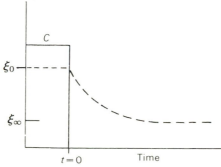

Fig. 1.8. Relaxation after a step change in constraint, C, monitored by observation of property $\xi(t)$.

The constant, τ, is called the relaxation time of the property under consideration, and is the time taken for P to drop to $1/e$ of its original value. The exponential function (1.2) is often called the decay function of the observed property.

In experimental situations the change in constraint can be applied as a step function, but is very often applied as a sinusoidal variation. Under these circumstances the observed response lags in phase behind the applied constraint because of the finite time required by the molecules to make any necessary adjustment (Fig. 1.9). Under these circumstances the observed

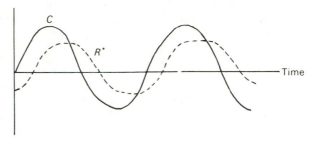

Fig. 1.9. Relaxation in the complex response, R^*, to a sinusoidal variation in constraint C.

response of the system is mathematically complex. If the real response in the time domain is given by the decay function

$$R(t) = \phi(t). \tag{1.3}$$

Then the complex response in the frequency domain is the Laplace transform

$$R^*(\omega) = \mathscr{L}\left(-\frac{\mathrm{d}[\phi(t)]}{\mathrm{d}t}\right). \tag{1.4}$$

The complex response can be resolved into two components in phase with, and 90° out of phase with the constraint. The Laplace transformation of the ideal decay function yields an ideal complex response.

$$\frac{R^*}{R_0 - R_\infty} = \frac{1}{1 + i\omega\tau} \tag{1.5}$$

$$\frac{R'}{R_0 - R_\infty} = \frac{1}{1 + \omega^2\tau^2} \tag{1.6}$$

$$\frac{R''}{R_0 - R_\infty} = \frac{\omega\tau}{1 + \omega^2\tau^2} \tag{1.7}$$

where R' and R'' are the real and imaginary responses, and have been normalized by $R_0 - R_\infty$, where R_0, R_∞ are the values of the response at frequencies much below and much above those in the relaxation region, τ is again the time constant defined by the decay function (1.2), and ω is angular frequency. The functions (1.6) and (1.7) are illustrated in Fig. 1.10.

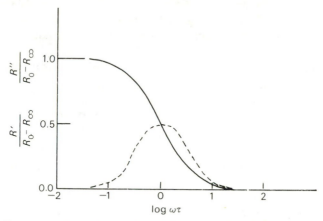

Fig. 1.10. Frequency dependence of real, R' (full line) and imaginary, R'', (broken line) response to an alternating perturbation.

Values of $\omega\tau < 1$, and temperatures greater than T_c (Fig. 1.11) correspond to situations where the speed of molecular change is faster than changes in the applied constraint, and so a large value of the real response is obtained. Conversely, when $\omega\tau > 1$, or at temperatures below T_c, the molecules cannot adjust to the changes in constraint, and so exhibit low values of the real response. The resolved out-of-phase response, R'', is a maximum at the intermediate situation when $\omega\tau = 1$.

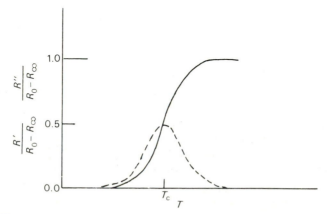

Fig. 1.11. Temperature dependence of real, R' (full line) and imaginary, R'', (broken line) response to an alternating perturbation.

In this way it is possible to correlate a wide variety of apparently quite different relaxation phenomena. Examples (by no means exhaustive) have been listed, in terms of the constraint varied and the response observed, in Table 1.1. In all of the examples selected for this table, it is found that the time required for the molecular response is, or is directly related to, the time taken for some form of molecular motion which, too, is entered in the table. Thus relaxation phenomena are often the direct result of, and have a magnitude controlled by, molecular motion. This can give a molecular understanding of technological properties such as dielectric and mechanical behaviour, and can also give us a means of measuring and characterizing the motion. Consequently the study of relaxation phenomena[6-8] in polymer systems is of both academic and practical importance. Two relaxation phenomena of importance in this way are introduced in a little more detail in §§ 1.4.1. and 1.4.2.

1.4.1. Molecular motion and dielectric relaxation

When a molecular system is placed in an electric field, there is always the tendency for any electrically charged species to migrate along the field in the appropriate direction. If the charged species are completely mobile this results in the conductivity familiar in metals, electrolyte solutions, or semiconductors. However, if the charged entities can move only a certain distance, and then for some reason become localized, the net result is an electric polarization of the sample.

For example, the charge carriers may migrate across a sample, but be unable to cross the boundary between the sample and the electrodes. Under these conditions the trapped charges collect at the surface, causing inter-facial polarization of the sample. At the other extreme on a distance scale, electrons may migrate across the atoms in a molecule but not between molecules. Thus, each molecule in the field suffers a slight distortion of electron distribution, called electronic polarization (forming in the molecules an induced dipole moment). The addition of each molecular dipole along the direction of the field again causes a resultant polarization of the whole sample. A third example of a polarization process occurs when the two opposite charges in a molecular dipole attempt to migrate in the field, thus applying a turning couple to the molecule. This tends to align the dipole in the field. Again the resultant effect is an electric polarization of the sample, called in this case 'orientation polarization'.

It is this electrical polarization that forms the observable property in studies of dielectric relaxation. In this case the polarization charging current can be resolved into two components, the real one of which leads the polarizing voltage by 90°, the other being in phase with it. The real response is the capacitance charging current, and the imaginary component is an ohmic current, or energy loss.

Table 1.1

Some relaxation phenomena associated with molecular movement

Phenomenon	Constraint varied	Property observed	Motion involved
Dynamic mechanical relaxation	Stress, strain	Strain, stress, modulus	Translation and rotation of molecules and chain segments
Viscoelastic relaxation	Shear stress, shear rate	Dynamic viscosity, shear modulus	Translation and rotation of molecules and chain segments
Ultrasonic relaxation	Pressure, temperature	Acoustic absorption, velocity	Any molecular change for which either $\Delta V^0 \neq 0$ or $\Delta H^0 \neq 0$ e.g. conformation change
Dielectric relaxation	Electric field	Electric polarization, capacitance, loss	Limited translation of charges, rotation of dipoles
Fluorescence depolarization	Polarized electromagnetic field	Polarized fluorescence	Rotation of electronic transition dipole moment
Nuclear magnetic and electron spin relaxation	Magnetic field	Nuclear and electron spin magnetic polarization	Rotation of particle spin transition moment

The dielectric characteristics of the system are generally observed as the complex permittivity. In the present situation orientation polarization is of considerable importance, and the principal features follow the generalization illustrated in Fig. 1.5. As the frequency of the applied field is increased from below to above a possible frequency for dipole rotation, the macroscopic permittivity is complex and the real part, ε' (measured as electric capacitance) decreases from a value characteristic of orientation polarization to one in the absence of orientation polarization. The imaginary part ε'' (measured as conductance, dielectric loss, or energy absorption) is zero when orientation is virtually instantaneous or cannot take place, but is a maximum when the frequencies of dipole rotation and electric field change are comparable.

As with other relaxation phenomena, the changes in complex permittivity occur over a rather wide frequency range. In the ideal situation the relaxation is described by a single time constant and the width of the loss curve at half-height is 1.1 decades.

Again, when the permittivity is considered as a function of temperature, the loss component passes through a maximum at the transition temperatures where the dipole movement rate changes from slower than, to faster than, the observation frequency. Just as in the mechanical case, the observed transition temperatures depend on the measurement time or frequency, being highest for high-frequency observations.

The theory of dielectric relaxation is available in several introductory reviews,[9-11] and the consequences in polymer systems are discussed further in Chapter 6.

1.4.2. Dynamic viscosity

A characteristic property of polymer melts and solutions is their high viscosity. For the purpose of the present discussion it is useful to consider viscosity as an energy loss process which occurs when interacting molecules move relative to each other. Consequently any process which causes neighbouring molecules (or parts of molecules) to move must contribute to the macroscopic viscosity of an assembly of molecules. The polymer movements, which we are discussing are no exception to this rule.

The high viscosity of polymer systems may arise through two rather different phenomena. In melts and concentrated solutions there may be interchain entanglements, of which the movement or destruction constitutes the energy absorption process. However, in concentrated and dilute systems, segmental motions, particularly of the co-operative nature described in normal mode analyses, contribute to energy dissipation.

Just as in other relaxation phenomena, the speed of molecular movement is important in observations of viscosity. If an experimental time scale is shorter than the time required for a particular molecular movement, that process will not contribute to the observed viscosity. Instead the system will exhibit elastic characteristics.

These temporal considerations can be observed most easily when viscosity is measured by some periodic process involving an alternating shear field. The observed complex dynamic viscosity may then decrease with increasing shear frequency according to the normal relaxation relationships.

Molecules are subjected to an alternating shear field not only when in contact with an oscillating surface, but also in flow where the effective shear frequency rises with flow or shear rate. Consequently, this decrease in dynamic viscosity appears in a number of technically important situations ranging from polymer melt extrusion to lubrication using solutions of polymers in hydrocarbon oils.

The interrelationship between dynamic viscosity and elasticity[12,13] and the consequent viscoelastic relaxation phenomena, are treated in more detail in Chapter 7.

1.5. Further dynamic properties of technological significance

Although the most familiar dynamic properties of polymers exhibit relaxation characteristics, and so have been introduced in the preceding section, a number of other physico-chemical phenomena are governed by characteristic molecular motions. In this category fall certain optical, electrical conduction, and chemical kinetic characteristics.

The division between optical and dielectric properties is somewhat arbitrary, although in the former case electronic polarization may be more important than atomic movement. A technological example of the close relationship between the two can be seen in the use of polymeric materials for the construction of both dielectric wave guides and 'light pipes' for fibre optics. In both of these uses the important characteristics are the refractive index and absorption coefficient at the relevant frequency. These, in turn, depend on the extent of molecular movement both directly and (through density) indirectly, as will be discussed in later chapters.

Although the limited migration of electrical charge carriers contributes to dielectric polarization, when these are free to be discharged at the electrodes all the phenomena associated with electrical conduction can be observed. Consequently here, too, there is no sharp distinction between the phenomenological basis of dielectric relaxation and electrical conduction. As most organic polymer systems are inherently of low conductivity, a considerable amount of research has been devoted to means of generating charge carriers in specific polymers and, at the same time, providing a structure which permits reasonable carrier mobilities combined with ease of discharge at the electrodes. To this end systems which exhibit some kind of charge transfer interaction in either the ground or photo-excited electronic states, and systems with strong electronic interactions along prolonged chain sections have been most promising. These will be discussed in more detail in Chapter 5.

A process, which, at first sight seems quite distinct from the phenomena presented so far, is the diffusion-controlled reaction of macromolecular species.[14,15]

A diffusion-controlled reaction is one in which the reaction rates depend upon the speed with which two reagents can diffuse together, rather than on the probability of chemical reaction once they have come together. In solutions of low viscosity only very reactive species undergo reactions of this type, which are then confined to free radicals, ions, or electronically excited molecules. However, in very viscous or solid systems even rather slow reactions become diffusion controlled, and such conditions often are characteristic of polymeric materials.

The technological significance of such processes arises in polymer forming processes as well as in degradation or polymer modification reactions. For example the products of a free radical polymerization reaction depend inversely (both in yield and molecular weight) on the efficiency of the radical–radical termination reaction. It has been found that even in mobile solvents this process is diffusion controlled and its rate is governed by the chain mobility in the neighbourhood of the free radical chain-end. Thus the final product depends upon the viscosity conditions and chain flexibility during the growth process.

In the same way many of the photochemical properties of polymers depend upon quenching and energy transfer processes which are intimately dependent upon the ease, or otherwise, of molecular diffusive motions. These too, are discussed in greater detail in Chapter 13.

References

1. North, A. M. Molecular motion in polymers. In *Molecular behaviour and the development of polymeric materials* (ed. A. Ledwith and A. M. North), p. 368. Chapman and Hall, London (1975).
2. McInnes, D. and Pugh, D. *Chem. phys. Lett.* **34**, 139 (1975).
3. Edwards, S. F. and Grant, J. W. V. *J. Phys. A* **6**, 1169 (1973).
4. Crank, J. and Park, G. S. *Diffusion of polymers.* Academic Press, London and New York (1968).
5. North, A. M. The importance of molecular motion in polymers. In *Essays in chemistry*, Vol. 4 (ed. J. N. Bradley, R. F. Hudson, and R. D. Gillard), p. 1. Academic Press, London and New York (1972).
6. McCrum, N. G., Read, B. E. and Williams, G. *Anelastic and dielectric effects in polymeric solids.* Wiley, New York (1967).
7. Bartenev, G. M. and Zelenev, Yu. V. *Relaxation phenomena in polymers.* Wiley, New York (1974).

8. North, A. M. Relaxation phenomena in polymer systems. In *Macromolecules* (ed. C. E. H. Bawn). M. T. P. Physical Chemistry Second Series, Vol. 8, p. 1 (1975).
9. Davies, M. *Quart. Rev.* **8**, 250 (1954).
10. Davies, M. *Some electrical and optical aspects of molecular behaviour.* Pergamon, Oxford and London (1965).
11. Smyth, C. P. *Ann. Rev. phys. Chem.* **17**, 433 (1966).
12. Alfrey, T., Jr. *Mechanical behaviour of high polymers.* Interscience, New York (1948).
13. Ferry, J. D. *Viscoelastic properties of polymers.* (2nd edn.) Wiley, New York (1970).
14. North, A. M. *Quart. Rev.* **20**, 421 (1966).
15. North, A. M. Diffusion control of homogeneous free radical reactions. In *Progress in high polymers* (ed. F. W. Peaker and J. C. Robb), p. 95. Iliffe Books, London (1968).

THE THEORY OF NORMAL VIBRATIONS AND INELASTIC LIGHT SCATTERING

2.1. Resonant normal modes of vibration

In chapter 1 the general principles of polymer chain movement, and the distinction between 'resonant' and 'relaxation' normal modes, were introduced. In this, and the two following chapters, the theoretical treatments of these modes of motion are considered in more detail. We begin, in this chapter, with the resonance vibrational modes, since most readers will be familiar with these as they determine the vibrational behaviour of small molecules.

In order to understand the ensuing treatment of polymer chain vibrations, it is necessary to start from the detailed treatment of small molecules. The complete derivation of the appropriate relationships can be found in a basic reference text,[1] so that here we start only from definitions and expressions, modification of which leads to theories of polymer motion.

Consideration of the vibrational normal modes of a polymer can be developed conveniently in three degrees of approximation, in which the intra- and intermolecular interactions present in a real polymer are introduced progressively. Thus, we begin with the normal vibrations of individual monomer units (small molecules), this is followed by the modes of the perfect infinite polymer chain, and finally we consider the vibrations of a perfectly crystalline, three-dimensional lattice of infinite dimensions. In a real crystalline polymer, of course, the effect of lattice defects, geometrical, and other imperfections should also be considered and these are treated briefly at the end of this chapter.

2.1.1. Normal modes of single isolated molecules

The normal vibrations of a non-linear molecule can be described by $3N - 6$ coordinates chosen so that no linear or angular momentum arises during a vibration. Three of the remaining six coordinates are used to describe the translational motion of the system as a whole and the remaining three account for the rotational motion of the molecule about its centre of mass.

In general, the vibrational motion of any molecular system will be extremely complicated since it will consist of a superposition of a number of different vibrations. However, the motion can always be considered as a superposition of a number of simpler motions, known as normal modes of vibration. In any one normal mode, considered alone, every nucleus performs a simple harmonic motion with the same characteristic frequency ν. All the atoms move in phase with one another, but generally with different amplitudes.

The great significance of the normal coordinates corresponding to the normal vibrations, lies in the simple form they give the expression for the kinetic and potential energy when written in terms of spatial coordinates. The kinetic energy, written in terms of a normal coordinate Q is given by

$$2T = \sum_i \dot{Q}_i^2,$$

whilst the potential energy includes frequency terms, $\lambda_i = 4\pi^2 \nu_i^2$ and then

$$2V = \sum_i \lambda_i Q_i^2$$

The equations of motion also assume correspondingly simple forms in terms of these coordinates.

It is important for the development of theories of polymer motions that the quantities and expressions be cast in matrix notation. So, as a start, we rewrite these familiar relationships as

$$2V = \tilde{\mathbf{Q}} \cdot \mathbf{\Lambda} \cdot \mathbf{Q} \tag{2.1}$$

and

$$2T = \tilde{\dot{\mathbf{Q}}} \cdot \dot{\mathbf{Q}} \tag{2.2}$$

where $\mathbf{Q}$ is the column vector of the normal coordinates, $\dot{\mathbf{Q}}$ the column vector of the corresponding velocities, $\mathbf{\Lambda}$ is the diagonal matrix of elements λ_i, and the symbol $\tilde{}$ represents the transpose of the matrix. In this expression there are no cross terms between Q_i and Q_j.

Cartesian displacement coordinates. If we consider a non-linear molecule having N atoms there will be $3N - 6$ normal vibrational modes. If we represent the displacement of each atom in terms of Cartesian displacement coordinates, x_i, we need $3N$ coordinates to define our system. These can be represented by a single column vector, $\mathbf{X}$, whose elements are the coordinates.

The total kinetic energy of the system, including translational and rotational contributions as well as the vibrational part, can be expressed in terms of the time derivatives $\dot{x}_i$ of the Cartesian coordinates. So, in matrix notation, the *total* kinetic energy, $2\bar{T}$, is

$$2\bar{T} = \tilde{\dot{X}}M\dot{X} \tag{2.3}$$

where $\dot{X}$ is a single column matrix whose elements are the time derivatives of the $3N$ x-coordinates, and M is a diagonal matrix of the nuclear masses.

Internal coordinates. For the purely vibrational problem it is an advantage to work with a set of internal coordinates which describe the relative positions of the nuclei irrespective of translational or rotational motions of the molecule as a whole. For a non-linear molecule, the number of internal coordinates will be $3N - 6$, i.e. the total number of Cartesian coordinates less the three required to define the centre of mass of the system and the three required to define the over-all orientation in space.

The internal coordinates, R, can be generated from the Cartesian coordinates, X, by means of a matrix B which will have $3N - 6$ rows and $3N$ columns,

$$R = BX. \tag{2.4}$$

The purely vibrational kinetic energy, T, may be expressed in terms of internal coordinates using

$$2T = \tilde{\dot{R}}\tilde{A}MA\dot{R} \tag{2.5}$$

where $A = B^{-1}$. This expression assumes that rotation–vibration interaction can be neglected.

Using the G matrix first introduced by Wilson[1] and defined by

$$G = BM^{-1}\tilde{B} \tag{2.6}$$

the vibrational kinetic energy can be expressed as

$$2T = \tilde{\dot{R}}G^{-1}\dot{R}. \tag{2.7}$$

This important relation plays a leading part in the setting up of the secular equation of the vibrational problem. The analogy with the expression for the total kinetic energy, eqn (2.3), should be noted. The matrix G is square and has $3N - 6$ rows and $3N - 6$ columns. Both G and G^{-1} are symmetric.

The potential energy and the F matrix. The potential energy, V, is a function only of the internal coordinates R. Since the nuclear displacements involved are small, it may be expanded in terms of a Taylor series,

$$V(R_1, R_2 \cdots) = V_0 + \left(\frac{\partial V}{\partial R_1}\right)_0 R_1 + \left(\frac{\partial V}{\partial R_2}\right)_0 R_2 + \cdots$$

$$+ \frac{1}{2}\left(\frac{\partial^2 V}{\partial R_1^2}\right) R_1^2 + \frac{1}{2}\left(\frac{\partial^2 V}{\partial R_2^2}\right) R_2^2 + \cdots$$

$$+ \left(\frac{\partial^2 V}{\partial R_1 \partial R_2}\right) R_1 R_2 + \left(\frac{\partial^2 V}{\partial R_2 \partial R_3}\right) R_2 R_3 + \cdots \tag{2.8}$$

where the higher terms can be neglected. V_0 is the potential energy at the equilibrium configuration $(R_1 = R_2 = R_3 = \cdots = 0)$. This may be taken as the zero from which the potential energies are measured. The differentials are taken at the equilibrium configuration so that all the first differential terms will vanish. Thus we are left with the quadratic terms and we arrive at an expression for the potential energy,

$$2V = F_{11}R_1^2 + F_{22}^2 R_2^2 + \cdots + 2F_{12}R_1R_2 + 2F_{23}R_2R_3 + \cdots + \qquad (2.9)$$

or in matrix notation,

$$2V = \tilde{\mathbf{R}}\mathbf{F}\mathbf{R} \qquad (2.10)$$

where **F** is a square matrix whose elements are the force constants F_{ij}. Like **G**, **F** has $3N - 6$ rows and $3N - 6$ columns and, since

$$F_{ij} = \left(\frac{\partial^2 V}{\partial R_i \partial R_j}\right) = F_{ji},$$

it is a symmetric matrix.

The secular equation. The classical treatment uses the Lagrange equation of motion, one for each internal coordinate R_i

$$\frac{\mathrm{d}}{\mathrm{d}t}\left(\frac{\partial T}{\partial \dot{R}_i}\right) + \frac{\partial V}{\partial R_i} = 0 \qquad (2.11)$$

where T and V are expressed by (2.7) and (2.10). This gives rise to $3N - 6$ simultaneous equations the solutions to which are given by a determinantal equation such as

$$\begin{vmatrix} F_{11} - \lambda G_{11}^{-1} & F_{12} - \lambda G_{12}^{-1} & F_{13} - \lambda G_{13}^{-1} \\ F_{21} - \lambda G_{21}^{-1} & F_{22} - \lambda G_{22}^{-1} & F_{23} - \lambda G_{23}^{-1} \\ F_{31} - \lambda G_{31}^{-1} & F_{32} - \lambda G_{32}^{-1} & F_{33} - \lambda G_{33}^{-1} \end{vmatrix} = 0 \qquad (2.12)$$

which is written here for a non-linear triatomic molecule having 3 normal modes. In this equation, again $\lambda = 4\pi^2 \nu^2$, where ν is the vibration frequency. In the general case, the secular equation can be written in matrix notation as

$$|\mathbf{F} - \lambda \mathbf{G}^{-1}| = 0. \qquad (2.13)$$

For a non-linear molecule with N atoms, expansion of the secular determinant gives an equation of order $3N - 6$, the solution of which gives $3N - 6$ normal vibrational frequencies.

This secular equation can be converted into two equivalent forms which are more convenient for practical use. By pre-multiplication of the LHS of (2.13) with **G** we obtain

$$|\mathbf{G}\mathbf{F} - \lambda \mathbf{G}\mathbf{G}^{-1}| = 0 \qquad (2.14)$$

or

$$|\mathbf{GF} - \lambda \mathbf{E}| = 0. \qquad (2.15)$$

Force fields. Calculation of the normal frequencies for a molecule requires a knowledge of the nuclear masses, the geometry of the molecule and the valence force field, which here is assumed to be quadratic. In general, however, the number of quadratic force constants required is greater than the number of observable frequencies, so that there are insufficient data to determine the field completely. In most cases, therefore, it becomes necessary to simplify the force field. One technique is to start off with a general valence force field (GVFF) and then set some of the interaction constants (the off-diagonal elements in the *F*-matrix) to zero. In this way, the number of force constants may be considerably reduced. Sometimes, all the off-diagonal elements of the *F*-matrix of a field of the valence type are set to zero giving a simple valence force field (SVFF). Both these procedures are somewhat arbitrary. An alternative approach is to use a different type of field having fewer parameters. One such field is the Urey–Bradley (U–B) force field. The U–B field uses the principal bond-stretch and angle-change force constants of the SVFF but instead of interaction constants it introduces repulsive forces between non-bonded atoms. This field represents a plausible attempt to take account of van der Waals interaction between non-bonded atoms.

2.1.2. *Vibrations of polymer chains*

The basic concepts discussed in § 2.1.1 for small molecules can be applied to polymer chains with the introduction of a few additional concepts. We shall discuss first as an introduction to the problem, the general case of the vibrations of a chain of coupled harmonic oscillators.

If we take a linear chain made up of equal masses, m, at a distance, d, (repeat distance) and joined by bonds which obey Hooke's law, with a force constant f, we can write down the equation of motion for the nth mass, m_n, as

$$m\ddot{x}_n = f(x_{n+1} + x_{n-1} - 2x_n) \qquad (2.16)$$

where in the interests of clarity, we have restricted discussion to longitudinal vibrational motion along the chain axis x.

Solutions of this equation are of the type

$$x_n = A \, e^{i(\omega t + knd)} \qquad (2.17)$$

which describe periodic motion with a circular frequency ω; n is a running index, and k is the magnitude of the wave vector $\mathbf{k}$. The vector, $\mathbf{k}$, has a modulus equal to $2\pi/\lambda$ where λ is here the wavelength. Differentiation of (2.17) and substitution into (2.16) gives a solution if,

$$\omega = \pm(4f/m)^{\frac{1}{2}} \sin(kd/2). \qquad (2.18)$$

This relationship between ω and the wave vector, $\mathbf{k}$, is called a dispersion relation and the corresponding curve of the function $\omega(\mathbf{k})$ is called a dispersion curve. Physically, eqn (2.18) describes a wave, of length $\lambda = 2\pi/k$ which propagates through the chain. Each particle vibrates with frequency $\omega(\mathbf{k})$. A band of frequencies from $\omega = 0$ to $\omega_{\max} = (4f/m)^{\frac{1}{2}}$, corresponding to $k = 0$ to $k = \pm\pi/d$ can be propagated through the chain. The displacements of two successive and equivalent particles are given by the ratio

$$x_j/x_{j+1} = e^{-ikd}. \tag{2.19}$$

At $k = 0$ all the particles move in phase with one another and the motion corresponds to a rigid translation of the whole chain. At $k = \pm\pi/d$ the particles move completely out-of-phase and there is no longer a travelling wave but a standing wave of wavelength $\lambda = 2d$. In general, the phase difference between two successive particles is given by

$$\phi = kd.$$

The concept of phase difference is basic to our understanding of vibrations in polymer chains. Motions described by values of k outside the limit $k = \pm\pi/d$ simply reproduce the motions already described by values of k within these limits. The region between $+\pi/d$ and $-\pi/d$ is called the first Brillouin zone for a one-dimensional lattice. If large values of k occur, an appropriate integral multiple of π/d should be subtracted to transfer into the Brillouin zone.

If we introduce two kinds of particles of masses M and m into the infinite chain, located alternately at a separation d, two equations of motion can be written, one for each particle of the repeat unit.

$$m\ddot{x}_{2n} = f(x_{2n+1} + x_{2n-1} - 2x_{2n})$$
$$M\ddot{x}_{2n+1} = f(x_{2n+2} + x_{2n} - 2x_{2n+1}). \tag{2.20}$$

Only nearest-neighbour interactions are taken into account and again only longitudinal motions along the chain axis are considered. The solutions of these equations are of the form

$$x_{2n} = A\, e^{i(\omega t + 2nkd)}$$
$$x_{2n+1} = B\, e^{i(\omega t + (2n+1)kd)}. \tag{2.21}$$

Differentiation and substitution into (2.20) leads to the simultaneous equations

$$-\omega^2 mA = 2fB \cos kd - 2fA$$
$$-\omega^2 MB = 2fA \cos kd - 2fB \tag{2.22}$$

whose non-trivial solutions are given by the determinantal equation

$$\begin{vmatrix} 2f - m\omega^2 & -2f\cos kd \\ -2f\cos kd & 2f - M\omega^2 \end{vmatrix} = 0, \tag{2.23}$$

$$\omega^2 = f\left(\frac{1}{m} + \frac{1}{M}\right) \pm f\left\{\left(\frac{1}{m} + \frac{1}{M}\right)^2 - \frac{4\sin^2 kd}{mM}\right\}^{\frac{1}{2}}. \tag{2.24}$$

The values of ω obtained by varying k in this 'dispersion' relation separate into two branches, an acoustic branch ($-$ sign) and an optical branch ($+$ sign).

The origin of these descriptions becomes clear if we investigate the behaviour of the system at a particular k value. The ratio of the amplitudes at a given k is obtained by substituting ω into (2.22). When $k = 0$, $A = B$ for the acoustic branch and $A = -(M/m)B$ for the optical branch. Thus, the acoustic branch particles move together rigidly, giving rise to a translational mode of the whole repeat unit similar to the passage of a sound wave through the medium. For the optical branch, one particle moves in the opposite direction to the other, with amplitudes dependent on the mass of the particles M and m. In this case, the centre of mass of the repeat unit does not move, and if the particles are oppositely charged they can interact with an electromagnetic wave. The vibrations of real polymer chains can be treated in a similar fashion, but the problem becomes very cumbersome and an alternative method using internal coordinates is preferable.

One method is to use the Wilson **GF** technique discussed in § 2.1.1, suitably modified for a polymer chain. If we consider an ideal polymer chain, the whole chain can be constructed geometrically by applying a screw symmetry operation to the starting repeat unit. Consider the repeat unit to be made up of p atoms, and let $\Phi(\theta, l)$ be the screw operation defined by a rotation of an angle θ about the chain axis and a translation of l along the same axis. The repeat distance is obtained after a number of such operations have been performed.

If R_i^n denotes the ith internal coordinate belonging to the nth chemical unit, the vibrational potential energy can be described by a Taylor series expansion truncated to the second order as described in § 2.1.1.

$$2V = \sum_{\substack{n,n' \\ i,j}} (F_R)_{ij}^{nn'} R_i^n R_j^{n'}. \tag{2.25}$$

The periodicity of the chain requires that

$$(F_R)_{ij}^{nn'} = (F_R)_{ij}^{s} \qquad s = |n - n'| \tag{2.26}$$

where s defines the distance of interaction. Thus, we can write

$$2V = \sum_{\substack{n \\ i,j}} (F_R)_{ij}^{0} R_i^n R_j^n + \sum_{\substack{n,s \\ i,j}} \{(F_R)_{ij}^{s} R_i^n R_j^{n+s} + (F_R)_{ji}^{s} R_i^n R_j^{n-s}\}. \tag{2.27}$$

The expression for the kinetic energy of the polymer chain can be written in an analogous way in terms of the momenta P_i^n and the kinetic-energy matrix $\mathbf{G}$

$$2T = \sum_{\substack{n \\ i,j}} (G_R)_{ij}^0 P_i^n P_j^n + \sum_{\substack{n,s \\ i,j}} \{(G_R)_{ij}^s P_i^n P_j^{n+s} + (G_R)_{ji}^s P_i^n P_j^{n-s}\}. \quad (2.28)$$

Substituting the expressions for V and T into the equation of motion leads to an infinite number of second-order differential equations in the unknowns R_i^{n+s}. The solution is given by the plane wave

$$R_i^{n+s} = A_i \exp\{-i(\omega t + s\phi)\}. \quad (2.29)$$

In this equation, ω is the circular frequency, and ϕ is the phase shift between two adjacent equivalent internal coordinates. Substitution of this solution into the system of second-order differential equations leads to a set of $3p$ simultaneous linear equations in the amplitude A_i, where non-trivial solutions are given by the $3p$ values of ω (for each ϕ) for which the determinant vanishes:

$$|\mathbf{G}_R(\phi)\mathbf{F}_R(\phi) - \omega^2(\phi)\mathbf{E}| = 0$$

where

$$\mathbf{G}_R(\phi) = \mathbf{G}_R^0 + \sum_s \{\mathbf{G}_R^s e^{is\phi} + (\tilde{\mathbf{G}}_R)^s e^{-is\phi}\}$$

$$\mathbf{F}_R(\phi) = \mathbf{F}_R^0 + \sum_s \{\mathbf{F}_R^s e^{is\phi} + (\tilde{\mathbf{F}}_R)^s e^{-is\phi}\}. \quad (2.30)$$

Since the determinantal equation 2.30 is of $3p$th degree in ω^2, there are $3p$ characteristic roots $\omega^2(\phi) = 4\pi^2 c^2 \nu^2(\phi)$ for each value of the phase ϕ. The $3p$ functions $\nu_i(\phi)$ can be interpreted as the branches of a multiple-valued function which is the dispersion relation for the polymer chain, analogous to (2.18) and (2.20) for the one-atom and two-atom chains respectively. The function is periodic (period 2π) and one can limit the study to values of ϕ within the first Brillouin zone.

Using this technique to calculate dispersion curves, symmetry-factoring of the secular equation is not required. However, symmetry coordinates are constructed when spectroscopically active fundamental frequencies (i.e. $\mathbf{k} = 0$) are calculated. By suitable computer techniques calculations can be carried out for any phase value provided good force constants, F_{ik}^s, are available and an equilibrium geometry is assumed.

General features of dispersion curves. For a single-chain polymer the dispersion curve shows the following features:

(a) If p is the number of atoms per repeat unit, there will be $3p$ branches;
(b) Two branches always reach zero for $\phi = 0$. These are the acoustic branches which may also reach zero again for values of $\phi \neq 0$;

(c) The shape of the acoustic and optical branches cannot be predicted *a priori* since analytic treatments are too difficult. Numerical techniques are always used. The shape of the branches will depend on the extent of coupling between neighbouring units through the **G**-matrix and also on the extent and distance of the interaction through the **F**-matrix;

(d) When interchain coupling occurs, perturbations of the normal modes, increase of the number of branches, and mixing of normal modes occur.

Symmetry properties and selection rules. Very few of the polymer vibrational frequencies calculated by these techniques are infrared or Raman active. Selection rules must be known in detail for the prediction of the vibrational spectrum of a polymer. Structural information can only be derived from the application of group theoretical methods. These techniques will be discussed in Chapter 11 together with structural analysis of polymers.

Normal coordinate calculations. The main objectives of normal coordinate calculations may be summarized under four headings:

(i) To obtain information on the molecular dynamics of the polymer chain;

(ii) To check the validity of a particular assumed force field;

(iii) To assess the reliability of information obtained from calculations;

(iv) To aid the interpretation of the vibrational spectrum of a polymer (vibrational analysis).

The availability of fast computers has greatly aided the development of normal coordinate calculations in polymers. Until recently, calculations were restricted to $k = 0$ infrared or Raman-active modes, and only very recently have more systematic calculations of dispersion curves of single-chain polymers been performed. For example, calculations have been carried out on three-dimensional crystalline polymers, for a range of k values, on simplified models of polyethylene and poly(methylene oxide)[2,3] and on an actual model of polyethylene.[4]

The problem of the force field, F_{ik}, in polymers is an important one. As pointed out in § 2.3, the force field determined by least square calculations, based on a first-order perturbation of the secular equation, is not unique and often several sets of physically equally reasonable force constants can reproduce the observed vibrational spectrum. This difficulty has been overcome with larger molecules by use of the 'overlay technique' which assumes that the force constants between chemically and structurally similar molecules are the same. In this case, the least squares calculations can include as many molecules as wished provided they are chemically and structurally similar. Striking success with these procedures has been recorded by the calculations made on linear and branched paraffins by Snyder and Schachtschneider.[5] The excellent fit obtained between the calculated and observed frequencies is helped by the similarity of structure of the molecules.

Other force fields such as the U–B field, discussed previously, have been used both alone and modified by the introduction of several valence-type interaction constants with varying degrees of success. In the case of polymers, it is generally necessary to transfer quadratic force constants from model compounds to construct the force field. The accuracy of the force field is determined by the degree of fit between the observed and calculated normal frequencies. For a reliable normal coordinate calculation it is desirable that a particularly good fit is achieved below 1600 cm^{-1} where the influence of vibrational coupling on the modes is most significant, and this imposes the most stringent test on the force field.

Geometric considerations must be taken into account when quadratic force constants are transferred from model molecules. Force constants are calculated for some assumed geometry of the model compound and are thus geometry-dependent. The **G**-matrix is particularly sensitive to bond angles and torsional angles. For example, when the frequencies of polyethylene are calculated using a valence force field transferred from normal paraffins with an assumed tetrahedral geometry,[6] frequency differences arise because the CCC angle is nearer 120° in polyethylene. Calculated frequencies are particularly sensitive to changes in torsional angles, even in the high frequency region, and care must be taken to define the torsional internal coordinate correctly. In general, there is little point in adjusting the force field to improve the fit between calculated and observed frequencies, and normal coordinate calculations should be mainly used as an aid to structural and spectroscopic analysis. By the same token, the precise values of the force constants should not be subject to detailed physical interpretation.

When the calculations for the potentially spectroscopically active $k = 0$ modes are extended to $k \neq 0$ modes, the same general considerations apply. When dispersion curves are calculated, however, longer range interactions, in addition to nearest-neighbour interactions, must be taken into account. This seems to be particularly true for helical polymers where the stability of the chain seems to depend on interactions between repeat units which extend further than second or third neighbours. Calculations on polyethylene, for example, illustrate the effect of longer-range interaction terms on the dispersion curves. The best experimental technique for studying $k \neq 0$ modes is inelastic neutron scattering which is discussed in Chapter 12.

Reliable dispersion curves can also be obtained by studying the vibrational spectra of model compounds of increasing chain length. Linear paraffins (C_2 to C_{19}) have been studied in the solid state where they adopt a planar zig-zag conformation.[5] Each molecule in the crystalline state can be considered as a segment of a one-dimensional crystal if the vibrational perturbation of the CH_3 end group is neglected. If N is the number of CH_2 units in the molecular chain, the corresponding phase values are given by

$$\phi = \pi l / (N + 1) \quad \text{where } l = 1, 2 \cdots N$$

which locate discrete points in the frequency branches. Since each of these frequencies is potentially spectroscopically active, a complete description of each frequency branch can be obtained. The spectroscopically active limiting modes for $\mathbf{k} = 0$ ($\phi = 0$ and $\phi = \pi$) correspond to the infinite chain of polyethylene. This technique is difficult to apply to molecules other than paraffins because the model compounds of different chain lengths are difficult to synthesize. Some work has been carried out on a few model compounds of poly(methylene oxide) and poly(ethyleneglycol).[7]

2.1.3. Perfect crystalline polymers

In polymer crystals, the intermolecular binding forces are relatively weak compared with the intramolecular forces, so that crystalline organic polymers can be treated as molecular crystals. Ordered crystalline forces will generally produce small frequency shifts compared with the free molecule and a multiplication of the number of modes will also occur from either a lowering of the symmetry of the molecular environment or via coupling with neighbouring chains. Low-frequency lattice modes are also expected in the spectrum.

The traditional **GF** technique in terms of internal displacement coordinates is abandoned for polymer crystals, due to the complexity of the problem.

Several techniques are in use, but where phonon dispersion curves are required a technique due to Piseri and Zerbi[8] is suitable. Basically, the method constructs phonon internal displacement coordinates from phonon Cartesian displacement coordinates.

Consider an infinite crystal in which each unit cell is identified by three integers l_1, l_2, and l_3 (which can be zero, positive, or negative) labelled by l. Each unit cell contains q atoms, whose vibrations are described by $3q$ Cartesian coordinates or by $3q$ internal coordinates. Let the Cartesian and internal displacement coordinates of the lth unit cell be represented by the matrix vectors $\mathbf{x}_l$ and $\mathbf{r}_l$ respectively. The kinetic energy of the whole crystal is thus

$$2T = \sum_{l,l'} \tilde{\mathbf{x}}_l \cdot \mathbf{M} \cdot \dot{\mathbf{x}}_{l'} \qquad (l = l'). \tag{2.31}$$

The quadratic potential energy, when no restriction on the distance of interaction is imposed, is

$$2V = \sum_{l,l'} \tilde{\mathbf{r}}_l \cdot \mathbf{F}_{ll'} \cdot \mathbf{r}_{l'} \qquad (2.32)$$

where $\mathbf{M}$ is a diagonal matrix of order $3q$ whose elements are the masses of the atoms in the unit cell; $\mathbf{F}_{ll'}$ is a square matrix of order $3q$ of the harmonic force constants which describe the interaction between the internal coordinates of unit l with those of l'. If this equation is evaluated explicitly, the terms appearing in the expression for the vibrational potential energy can be written in the form

$$2V = \sum (V_j^0 + V_j') + \sum_{ij} V_{ij}. \qquad (2.33)$$

The diagonal blocks in (2.32) (i.e. when $l = l'$) contribute to the $(V_j^0 + V_j')$ terms in (2.33). V_j^0 is the potential energy of the free molecule and V_j' is the contribution from the static field. The V_{ij} terms account for the coupling between internal coordinates of different molecules. The internal and Cartesian phonon coordinates can be written

$$\mathbf{X}(\mathbf{k}) = (\tfrac{1}{2}\pi)^{\frac{1}{2}} \sum_{l=-\infty}^{\infty} \mathbf{x}_l \exp(-i\mathbf{k} \cdot \mathbf{t}(l))$$

$$\mathbf{R}(\mathbf{k}) = (\tfrac{1}{2}\pi)^{\frac{1}{2}} \sum_{l=-\infty}^{\infty} \mathbf{r}_l \exp(-i\mathbf{k} \cdot \mathbf{t}(l)) \qquad (2.34)$$

where $\mathbf{k}$ is the wave vector and $\mathbf{t}(l)$ is the vector which locates the lth unit cell from a suitably chosen origin. Using Wilson's technique the internal coordinate vector $\mathbf{r}_l$ can be expressed as

$$\mathbf{r}_l = \mathbf{B}_{-j}\mathbf{x}_{l-1} + \cdots + \mathbf{B}_1\mathbf{x}_{l-j} + \mathbf{B}_0\mathbf{x}_l + \mathbf{B}_1\mathbf{x}_{l+1} + \cdots + \mathbf{B}_j\mathbf{x}_{l+j}$$

$$= \mathbf{B}_0\mathbf{x}_l + \sum_{j}^{m} (\mathbf{B}_{-j}\mathbf{x}_{l-j} + \mathbf{B}_j\mathbf{x}_{l+j}) \qquad (2.35)$$

where $\mathbf{x}_{l+j}$ is the Cartesian coordinate vector of a unit cell j units from the lth cell; $\mathbf{B}_j$ is the corresponding transformation matrix between $\mathbf{r}_l$ and $\mathbf{x}_{l+j}$. Substitution of (2.35) into (2.34) gives

$$\mathbf{R}(\mathbf{k}) = \sum_{-m}^{m} \mathbf{B}_j \exp(-i\mathbf{k} \cdot \mathbf{t}(j)) \mathbf{X}(\mathbf{k}) \qquad (2.36)$$

from which the transformation matrix from Cartesian to internal phonon coordinates $\mathbf{B}(\mathbf{k})$ can be derived

$$\mathbf{B}(\mathbf{k}) = \sum_{j=-m}^{m} \mathbf{B}_j \exp(-i\mathbf{k} \cdot \mathbf{t}(j)). \qquad (2.37)$$

The potential and kinetic energy matrices can be written in terms of the phonon coordinates defined in (2.34) as

$$2T = \int \tilde{\dot{X}}(k) . M . \dot{X}(k) \, dk \qquad (2.38)$$

and

$$2V = \int \tilde{R}(k) . F_R(k) . R(k) \, dk \qquad (2.39)$$

where the integration is extended over the entire first Brillouin zone. It has been shown (eqn (2.30)) that the matrix $F_R(k)$ takes the form

$$F_R(k) = F_R^0 + \sum_s \{\tilde{F}_R^s \exp(-ik . t(s)) + F_R^s \exp(ik . t(s))\} \qquad (2.40)$$

where F_R^0 is the matrix of the force constants relating coordinates within the same cell, F_R^s is the matrix of the force constants relating the coordinates of two cells s units apart.

We can now express the potential energy in Cartesian coordinates in terms of the potential energy in internal coordinates as

$$F_x(k) = \tilde{B}(k) . F_R(k) . B(k). \qquad (2.41)$$

A further transformation to mass-weighted phonon Cartesian coordinates gives

$$\bar{X}(k) = M^{\frac{1}{2}} X(k) \qquad (2.42)$$

or

$$\bar{X}(k) = (\tfrac{1}{2}\pi)^{\frac{1}{2}} \sum_{l=-\infty}^{\infty} \bar{x}_l \exp(-ik . t(l))$$

where

$$\bar{x}_l = M^{\frac{1}{2}} x_l.$$

This gives the dynamical matrix

$$D(k) = M^{-\frac{1}{2}} . F_x(k) . M^{-\frac{1}{2}} \qquad (2.43)$$

in which M is the diagonal matrix of the masses.

The $3N$ dispersion curves for a given crystal can be obtained numerically by inserting a proper set of k values into the eigenvalue equation

$$|D(k) - \omega^2(k)E| \, L_x(k) = 0 \qquad (2.44)$$

once the force field F_R is known numerically.

If q is the number of atoms per unit cell, there will be $3q$, $k = 0$ vibrations which are potentially infrared and Raman active. However, not all of these will be allowed by symmetry restrictions. In the potential energy expression, eqn (2.33), V_i^0 corresponds to the contribution by molecules in an environment unperturbed by mechanical or electrical interactions. Force constants transferred from model compounds contribute only to V_i^0 terms, the so-called 'oriented gas' model. The V'_j terms, generally unknown for real polymers, arise from the static field on each chain and give rise to frequency shifts relative to the isolated molecule. The V_{ij} terms are responsible for the splitting of accidentally degenerate vibrations of the oriented gas model due to a lowering of the symmetry around the polymer chain. The use of frequency splittings together with lattice vibrations enables the intermolecular interactions in the crystalline polymer to be studied.

The computational problem becomes extremely difficult when phonon dispersion curves for three-dimensional polymer crystals are required and special numerical techniques are necessary. In addition, the number of vibrational degrees of freedom is often reduced by assuming point masses for the groups whose local vibrations occur at fairly high frequencies.

Additional information on the molecular dynamics of polymer systems can be obtained from the study of those physical properties which depend on the frequency distribution $g(\omega)$ or density of vibrational states. Experimentally $g(\omega)$ is obtained using inelastic neutron scattering and will be discussed in Chapter 12.

2.1.4. Real crystalline polymers

So far we have considered the polymer to consist of an ideal, isolated, infinitely long and perfectly regular chain and extended this treatment to polymer chains in an ideal crystalline lattice. However, it is well known that polymers are not ideal, polymer chains are not infinite, and, moreover, bend, fold, and twist, depending on the thermal and mechanical treatment to which they have been subjected. Defects of different types are consequently introduced which perturb the vibrational spectra. Defects in polymers can be grouped broadly under four headings:

(a) Chemical defects—cross-linking, chain branching, etc.;
(b) Conformational defects—rotational isomers, chain folding, etc.;
(c) Steric defects—lack of homogeneous tacticity;
(d) Mass defects.

Defects in polymers generally manifest themselves as additional features in the vibrational spectra. The interpretation of these features using theoretical techniques, such as Green's function methods, is very difficult and numerical techniques are generally employed. When either point defects or geometrical disorder are present the phonon coordinates cannot be defined and the size of the eigenvalue equation to be solved becomes $3Np$, where p is the number of unit cells considered. Even if drastic structural simplifications are introduced the problem still remains intractable by standard numerical procedures.

One technique that can be used, even for complex dynamic systems containing a high concentration of defects, is to compute numerically the number of eigenvalues of a given matrix in a given interval using the negative eigenvalue theorem (NET). This method allows the direct calculation of $g(\omega)$ at any desired accuracy. This numerical technique was first introduced by Dean[9] and used to compute the vibrations of several disordered systems.

The number $n(\omega_2 - \omega_1)$ of eigenvalues of the $3Np \times 3Np$ dynamic matrix $\mathbf{D}$ which lie in the interval ω_1 to ω_2 is given by

$$n(\omega_2 - \omega_1) = \eta(\mathbf{D} - \omega_2\mathbf{I}) - \eta(\mathbf{D} - \omega_1\mathbf{I}) \qquad (2.45)$$

where the frequencies ω_1 and ω_2 are positive real numbers such that $\omega_2 > \omega_1$. $\mathbf{I}$ is the $3Np \times 3Np$ unit matrix and $\eta(\mathbf{D} - \omega_i\mathbf{I})$ is the number of negative eigenvalues of the matrix

$$\mathbf{D}_i = \mathbf{D} - \omega_i\mathbf{I}. \qquad (2.46)$$

The computation of the negative eigenvalues of $\mathbf{D}_i$ is performed by a particular partitioning of $\mathbf{D}_i$ which is applicable to any symmetrical matrix.

The NET states that, given a symmetrical matrix $\mathbf{M}$ of dimensions $r \times r$ partitioned as follows:

$$\mathbf{M} = \begin{bmatrix} \mathbf{A}_1 & \mathbf{B}_2 & & & \mathbf{0} \\ \tilde{\mathbf{B}}_2 & \mathbf{A}_2 & \mathbf{B}_3 & & \\ & & \cdot & \cdot & \\ & & & \cdot & \\ \mathbf{0} & & & \tilde{\mathbf{B}}_k & \cdot \mathbf{A}_k \end{bmatrix} \qquad (2.47)$$

where $\mathbf{A}_i$ has the dimensions of $r_i \times r_i$, $\mathbf{B}_i$ has the dimensions of $r_{i-1} \times r_i$, and $\sum_{i=1}^{k} r_i = r$, then the number $\eta(\mathbf{M} - x\mathbf{I})$ of negative eigenvalues of the matrix $\mathbf{M} - x\mathbf{I}$ is given by

$$\eta(\mathbf{M} - x\mathbf{I}) = \sum_{i=1}^{k} \eta(\mathbf{U}_i) \qquad (2.48)$$

where

$$\mathbf{U}_i = \mathbf{A}_i - x\mathbf{I}_i - \tilde{\mathbf{B}}_i . \mathbf{U}_{i-1}^{-1} . \mathbf{B}_i$$

$$\mathbf{U}_1 = \mathbf{A}_i - x\mathbf{I}_1,$$

the matrix $\mathbf{D}_i$ is then partitioned according to the following procedure. Let $\mathbf{D}_i^{(1)}$ represent the matrix $\mathbf{D}_i$ partitioned as

$$\mathbf{D}_i^{(1)} = \begin{bmatrix} \mathbf{X}_1 & \mathbf{Y}_1 \\ \tilde{\mathbf{Y}}_1 & \mathbf{Z}_1 \end{bmatrix} \qquad (2.49)$$

where $\mathbf{X}_1$ is a 1×1 matrix, $\mathbf{Y}_1$ is a $1 \times (3Np - 1)$ matrix, and $\mathbf{Z}_1$ is a $(3Np - 1)(3Np - 1)$ matrix. From eqn (2.48), it follows that

$$\eta(\mathbf{D}_i) \equiv \eta(\mathbf{D}_i^{(1)}) = \eta(\mathbf{X}_1) + \eta(\mathbf{D}_i^{(2)}) \qquad (2.50)$$

$$\mathbf{D}_i^{(2)} = \mathbf{Z}_1 - \tilde{\mathbf{Y}}_1 . \mathbf{X}_1^{-1} . \mathbf{Y}_1. \qquad (2.51)$$

This process is continued $(3Np - t)$ times until

$$\eta(\mathbf{D}_i) = \sum_{j=1}^{3Np} \eta(\mathbf{X}_j). \qquad (2.52)$$

The intervals ω_1 to ω_2 can be restricted to any desired accuracy. When it is desired to compare the calculated with experimental vibrational spectra, it is sufficiently accurate to use a frequency interval $d\omega = \omega_2 - \omega_1 = 5 \text{ cm}^{-1}$. However, to obtain eigenvectors higher accuracy is required and $d\omega$ in the desired region of the spectrum, is reduced progressively until it contains only one eigenvalue. If $\bar{\omega}_i = \frac{1}{2}(\omega_2 - \omega_1)$ and ω_i is the exact eigenvalue of $\mathbf{D}$ occurring in $d\omega$, the eigenvector $\bar{\mathbf{L}}_i$ associated with $\bar{\omega}_i$ is computed using the 'inverse iteration method'. An arbitrary trial $3Np$ and vector $\mathbf{u}_0$ are chosen and used to construct

$$\boldsymbol{\nu}_i = (\mathbf{D} - \bar{\omega}_i \mathbf{I})^{-1} \mathbf{u}_0$$
$$\mathbf{u}_i = \boldsymbol{\nu}_1 / (\boldsymbol{\nu}_1)_{\max} \qquad (2.53)$$

where $(\boldsymbol{\nu}_1)_{\max}$ represents the element of largest modulus in $\boldsymbol{\nu}_i$. These two iteration steps are repeated s times until convergence is achieved in $\mathbf{U}_s$ and $\mathbf{L}_i$ (the true eigenvector).

This technique has been applied extensively to linear chains[10] and two-dimensional lattices,[11] primarily to test the validity of the procedure. Three-dimensional disordered lattices, silica glasses[12] and ice,[13] have been treated subsequently. These numerical methods have been applied to polymers by Zerbi and co-workers.[14,15] One-dimensional order was assumed, with no interactions between the chains.

Lattice dynamic calculations were first performed on perfect infinite chains. The calculated $\omega(\mathbf{k})$ and $g(\omega)$ were then compared with experimental infrared, Raman, and inelastic neutron scattering data. The vibrational properties of the disordered chains were obtained by introducing the desired type of disorder with preassigned concentration and statistics. The density of states of the disordered chains was calculated using the NET and compared with the experimental spectra of the disordered systems. Extra small features in the $g(\omega)$ were found to occur from the presence of end-groups of finite chain length.

2.2. Inelastic light scattering by macromolecules

In the preceding section we saw how the resonance normal modes of polymers can be resolved into acoustic and optical branches, and that the latter can give rise to infrared and Raman vibrational spectra. While infrared absorption is a resonance process, the Raman phenomenon is one of a number of possible inelastic light scattering processes. Although the phenomenological aspects of these will be discussed in Chapter 11, the underlying theory uses concepts and methods analogous to those required for treatment of molecular vibrations. Since one of the aims of any experiment is to relate observations to theories of molecular behaviour, the means of achieving this often requires the matching of the theory of the experiment to the theory of the appropriate molecular phenomenon. Consequently, it is appropriate at this point to develop the theory of light scattering phenomena to the level where it can be used in conjunction with the foregoing treatments of polymer chain resonance motions, and also of the damped relaxation motions to be discussed in Chapter 4.

There are basically two types of scattering phenomena in liquids and solids. The first occurs when light is scattered elastically from a medium by fluctuations in the local refractive index. For many years now the measurement of the scattered light intensity from polymer solutions as a function of angle, molecular weight, concentration, and temperature, has provided much important information concerning the size, shape, and thermodynamic properties of macromolecules. The theory and uses of such elastic scattering have been widely published in a number of texts.[16,17] However, elastic light scattering experiments do not make full use of the information contained in the scattered beam. By measuring intensities as a function of angle (i.e. momentum transfer), all frequency information is lost. In the second type of scattering process, inelastic scattering, the frequency spectrum of the scattered light is measured and information is obtained about the energy transfer between photons and the medium, and hence about molecular motions.[18,19]

2.2.1. Inelastic scattering phenomena

If a beam of light of frequency, ν_i, is incident on a medium, the oscillating electric field E induces an oscillating electric dipole moment P in a molecule. The oscillating dipole then radiates scattered light of frequency ν_s such that

$$\nu_s = \nu_i \pm \nu$$

where ν are the internal excitations of the molecular system. The frequency shifted light, ν_s, arising from inelastic scattering, provides us directly with the energies of the excitations.

When the light is scattered from periodic, coherent fluctuations in the density of a liquid or solid (i.e. acoustic waves) we have Brillouin scattering. With visible incident light, Brillouin scattering gives frequency shifts of 10^9–10^{11} Hz. When the scattering is from molecular vibrations, we have vibrational Raman scattering where the frequency shifts involved are much larger, ranging up to 10^{14} Hz. Raman scattering today includes scattering from many other types of excitations besides localized molecular vibrations.

These include low-lying electronic states, rotational levels, magnons, plasmons, as well as propagating phonon modes. In fact, the term Raman scattering is now loosely used to cover any form of inelastic light scattering except from acoustic modes and quasi-elastic scattering. This last type of scattering, sometimes called Rayleigh linewidth scattering, arises from neither coherent fluctuations nor localized vibrations, but from random motions usually diffusional in nature. The displaced frequencies, ν, are in the range $10–10^7$ Hz and appear as a broadening of the exciting frequency.

All types of scattering arise from the same source, fluctuations in the local dielectric constant (which at optical frequencies is the square of the refractive index) of the medium, and theoretically the scattering process can be treated similarly whether these fluctuations arise from molecular vibrations, sound waves, or diffusional motion. However, the three frequency regions require quite different experimental techniques, so there is a natural division on this basis.

All scattering observations have a common origin; an incident particle having momentum $\hbar k_i$ and energy $\hbar \omega_i$ is scattered in the medium to emerge with changed momentum and energy, $\hbar k_s$ and $\hbar \omega_s$. An excitation of momentum $\hbar K$ and energy $\hbar \Omega$ is created (or annihilated) in the medium. Conservation of energy and momentum requires that,

$$\omega_i - \omega_s = \pm \Omega \tag{2.54}$$

and

$$k_i - k_s = K \tag{2.55}$$

where the positive and negative signs refer to creation and annihilation of an excitation. These equations show that the range of momentum and energy transfer, i.e. the range of K and Ω, that is measurable will depend on the incident momentum and energy, which is in turn dependent on the nature of the incident particles. The range of energy and momentum parameters covered by a number of scattering experiments is shown in Table 2.1. A

Table 2.1

Energy and momentum parameters for a number of scattering experiments

Radiation	Wavelength (nm)	Momentum (nm)$^{-1}$	Energy (eV)	Energy transfer (eV)
Neutrons	0·02–1	50–1	0·1–0·002	0·1–0·002
Electrons	0·001–10	1000–0·1	5×10^4–0·1	Wide
X-ray	0·1	10	$\sim 10^4$	Wide
Visible				
Raman	5×10^2	2×10^{-3}	3	$0·5–5 \times 10^{-4}$
Brillouin	5×10^2	2×10^{-3}	3	$10^{-3}–10^{-6}$
Rayleigh linewidth	5×10^2	2×10^{-3}	3	$10^{-8}–10^{-14}$

striking feature of this table is the very wide frequency range covered by visible light scattering and the very narrow momentum range. The latter is the most serious limitation of the technique and is due to the fact that $|\mathbf{K}|$ for a visible photon is only of the order of $10^5 \, \text{cm}^{-1}$. Thus, the maximum $|\mathbf{K}|$ possible is $2|\mathbf{k}_i|$ or about $2 \times 10^5 \, \text{cm}^{-1}$. This is very much less than the reciprocal of an interatomic spacing in a solid or liquid, and so light scattering studies are essentially limited to Brillouin zone centres. This limitation does not apply to the neutron, for which momenta of $|\mathbf{k}_i|$ $\sim 10^8 \, \text{cm}^{-1}$ are readily obtainable, so that molecular dynamics can be observed over a wide momentum range.

2.2.2. Time and frequency domains in spectroscopy

Two types of problem can be distinguished in classical spectroscopy: (i) the measurement, and (ii) the interpretation of the information carried by a radiation field $\mathbf{E}(\mathbf{R}, t)$ at a point R and time t. In the case of a scattering process, visible light has a frequency of about $6 \times 10^{14} \, \text{Hz}$ which is much faster than the response of available detectors, so that the time evolution of the field cannot be followed directly.

The best known method of characterizing the electromagnetic field is in the spectral representation—a spectrum of intensity plotted against frequency. Such a form is provided by classical interferometric instruments, e.g. diffraction gratings and Fabry–Perot etalons. These elements serve as variable frequency filters, operating directly on the light field. However, there is a lower bandwidth with such instruments limiting the resolving power to about 10^8. Thus spectral features narrower than a few MHz, the smallest attainable bandwidth, cannot be measured. The limitation is one of size, since these instruments operate by bringing together components of the light field which have been delayed relative to one another. The resolving power is given by the number of wavelengths within the maximum attainable path difference, which is generally a few metres.

In the time domain, the equivalent information is carried by the optical or first-order auto-correlation function, which is the Fourier transform of the spectrum $S(\omega)$. Time correlation functions are discussed in greater detail in Chapter 4. For the present it is adequate to define a correlation function as the decay in correlation of some property between one situation (which may be time or position) and another (which might be later time or different position). An auto-correlation function is when we are concerned with the same property at a point or on the same particle, as opposed to two different properties on the one particle or the same property on two different particles. Such a correlation function is in fact formed in the Michelson interferometer where the full field auto-correlation function, $\langle \mathbf{E}(\mathbf{R}, t)\mathbf{E}^*(\mathbf{R}, t+\tau)\rangle$, is plotted as a function of optical path difference. Instruments of a few metres in size are limited to delay times of less than

10 ns, so that for longer delay times, τ, the field correlation function is not readily formed directly. However, the second-order, or intensity correlation function, is easily measured by post-detection methods, and can be simply related in many cases to the field correlation function, and hence the spectrum.

It is important to appreciate at this point that any spectral feature has a finite bandwidth, requiring that its intensity be fluctuating in time. The relationship between the frequency and time domains, is given by the Wiener–Khinchin theorem, which states that the power spectrum and first-order time correlation function of any stationary ergodic random signal form a Fourier transform pair,

$$S(\omega) = \frac{1}{2\pi} \int_{-\infty}^{+\infty} |\langle \mathbf{E}(t) \cdot \mathbf{E}^*(t+\tau)\rangle| \exp\{i(\omega - \omega_0)\tau\}\, d\tau. \qquad (2.56)$$

Thus, if $|\langle \mathbf{E}(t)\mathbf{E}^*(t+\tau)\rangle|$ is independent of τ (a field of constant intensity), the spectrum is a δ-function, a spike at ω_0 having zero width. More generally, if the field correlation function decays with characteristic time T_c, the frequency width Δf of the spectrum is given by $\Delta f \simeq 1/T_c$.

2.2.3. The field correlation function

In order to treat scattering phenomena, it is necessary to develop the field correlation function as follows. The real field at point R may be expanded as a Fourier series,

$$\mathbf{E}(\mathbf{R}, t) = \sum_\omega a_\omega(\mathbf{R})\, e^{-i\omega t} + \sum_\omega a_\omega^*(\mathbf{R})\, e^{i\omega t}$$
$$= \mathbf{E}(\mathbf{R}, t) + \mathbf{E}^*(\mathbf{R}, t). \qquad (2.57)$$

The two components are called the positive and negative frequency parts of the field, and a_ω represents the field amplitude at frequency ω. The intensity is then given by, $I(\mathbf{R}, t) = \mathbf{E}\mathbf{E}^* = |\mathbf{E}|^2$ which for strong fields is the square of their classical envelope. The first-order field auto-correlation function at a point in space is given by,

$$G^{(1)}(t, t+\tau) = \langle \mathbf{E}(t)\mathbf{E}^*(t+\tau)\rangle \qquad (2.58)$$

where the brackets indicate ensemble (or time), averages. The second-order or intensity correlation function is,

$$G^{(2)}(t, t+\tau) = \langle I(t)I(t+\tau)\rangle. \qquad (2.59)$$

These functions may be written in normalized form as

$$g^{(1)}(t, t+\tau) = \frac{\langle \mathbf{E}(t)\mathbf{E}^*(t+\tau)\rangle}{\langle I\rangle} \qquad (2.60)$$

and

$$g^{(2)}(t, t+\tau) = \frac{\langle I(t)I(t+\tau)\rangle}{\langle I \rangle^2}. \tag{2.61}$$

For $g^{(2)}(t, t+\tau)$; as τ becomes large and the intensities become uncorrelated, its value will approach 1. At the other extreme, as $\tau \to 0$ its value will depend on the exact properties of the field. By contrast, $g^{(1)}(t, t+\tau)$ approaches 1 as $\tau \to 0$ and goes to zero as $\tau \to \infty$. These results arise since the mean value of E is zero, whereas I has a positive mean value.

To specify an arbitrary field completely, all the higher value correlation functions must be known. The most important of these is the second-order, since this is the simplest correlation function that can be constructed directly for low frequency, (≤ 10 MHz), signals. However, the large class of fields having Gaussian statistics, commonly met in random scattering processes, is completely specified by a knowledge of the first-order correlation function. Thus we have the relationship,

$$g^{(2)}(t, t+\tau) = 1 + |g^{(1)}(t, t+\tau)|^2. \tag{2.62}$$

So, by measuring the intensity as a function of time, we can form intensity correlation functions and so determine the first-order field correlation function, which is just the Fourier transform of the observed frequency spectrum.

2.2.4. Scattering from isotropic fluctuations

The interatomic spacing in liquids and solids is much smaller than the wavelength of visible light, so that the medium can be treated as a uniform continuum with optical properties characterized by a dielectric constant, ε. The medium will be considered to be isotropic and non-dispersive.

The electric field associated with a plane monochromatic wave is of the form

$$\mathbf{E}(\mathbf{r}, t) = \mathbf{E}_0 \exp\{i(\mathbf{k} \cdot \mathbf{r} - \omega_0 t)\} \tag{2.63}$$

where E_0 is the amplitude, ω_0 the frequency, $\mathbf{k}$ the wave vector; and the velocity is $\omega_0/|\mathbf{k}|$. Consider the scattering geometry shown in Fig. 2.1, $|\mathbf{k}| = 2\pi n/\lambda_0$ where n is the refractive index of the medium, and λ_0 is the wavelength *in vacuo*. The incident light is assumed to be polarized perpendicular to the scattering plane, the plane of the diagram. Scattered light is detected at the point P, at a distance R from the scatterer, at an angle θ to the incident beam. Consider a volume element dV, small compared with λ^3, at a point $\mathbf{r}$ in the illuminated region. The incident field will induce an instantaneous dipole moment

$$\mathbf{P}(\mathbf{r}, t)\, dV = \mathbf{\alpha}(\mathbf{r}, t) \cdot \mathbf{E}_0(\mathbf{r}, t)\, dV \tag{2.64}$$

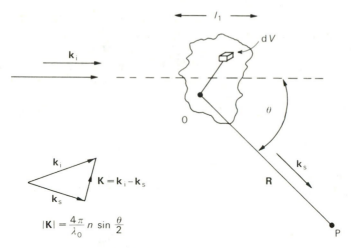

Fig. 2.1. Typical scattering geometry.

where $\mathbf{a}(\mathbf{r}, t)$ is the polarizability tensor. This induced dipole will oscillate at the optical frequency, and will therefore radiate light in various directions. The instantaneous scattered field will be the sum of the electric fields radiated by all the volume elements in the illuminated region. In the far field, $R \equiv |\mathbf{R}| \gg l_1 \gg \lambda$, the instantaneous scattered field is given by

$$\mathbf{E}(\mathbf{R}, t) = \frac{E_0 \omega_0^2}{c^2 R} \exp\{i(\mathbf{k}_s . \mathbf{R} - \omega_0 t)\}$$

$$\times \int dV . \mathbf{a}(\mathbf{r}, t) \exp(i\mathbf{k}_i . \mathbf{r}) \qquad (2.65)$$

where, in the direction of $\mathbf{R}$ in the far field $\mathbf{k}_s$ is a vector of magnitude $k_s \equiv |\mathbf{k}_s|$

$$\mathbf{K} = \mathbf{k}_i - \mathbf{k}_s$$

$$K \equiv |\mathbf{K}| = \frac{4\pi}{\lambda_0} n \sin\frac{\theta}{2}. \qquad (2.66)$$

For simplicity, when we specialize in the case of a scalar polarizability, the scattered light will have the same polarization (perpendicular to $\mathbf{R}$) as the incident light. In general, therefore, the scattered light has the form of a spherical wave of amplitude $\exp\{i(\mathbf{k}_s . \mathbf{R} - \omega_o t)\}/R$, and phase modulated by the integral

$$S(t) \equiv \int dV . \mathbf{a}(\mathbf{r}, t) \exp\{i\mathbf{K} . \mathbf{r}\}. \qquad (2.67)$$

The factor $\exp(i\mathbf{K} . \mathbf{r})$ represents the phase shifts due to the different optical paths taken between the source and detector by light scattered from different parts of the illuminated region. Thus, $S(t)$ may be interpreted as an instantaneous 'interference factor'.

In general, $\mathbf{a}(\mathbf{r}, t)$ will be a fluctuating quantity. Thus, in a pure fluid, the density will undergo spontaneous, thermally induced fluctuations. In a solution of macromolecules of polarizability different from that of the solvent, fluctuations in $\mathbf{a}$ will be caused by local concentration fluctuations. By writing $\mathbf{a}$ in terms of average and fluctuating parts

$$\mathbf{a}(\mathbf{r}, t) = \langle \mathbf{a} \rangle + \delta \mathbf{a}(\mathbf{r}, t) \qquad (2.68)$$

we obtain

$$S(t) \simeq \langle \mathbf{a} \rangle V \delta(\mathbf{K}) + \int dV \, \delta \mathbf{a}(\mathbf{r}, t) \exp(i\mathbf{K} \cdot \mathbf{r}) \qquad (2.69)$$

where $\delta \mathbf{a}$ is the fluctuating increment in $\mathbf{a}$ and $\delta(\mathbf{K})$ is the Dirac delta function. This expresses the well known result that no light is scattered away from the incident direction by a wholly uniform scatterer, or that fluctuations in the polarizability cause the scattering. In the hypothetical case of an optically homogeneous medium, the electric fields from the various volume elements interfere destructively to give a zero sum so that one can always pair two elements having phase shifts differing by $\lambda/2$. In the presence of fluctuations, however, total cancellation generally will not occur between the field scattered by the two elements.

Thus the basic equation for light scattering from systems characterized by scalar susceptibility is, for non-zero scattering angles,

$$\mathbf{E}(\mathbf{R}, t) = \frac{\mathbf{E}_0 \omega_0^2 \exp\{i(\mathbf{k} \cdot \mathbf{R} - \omega_0 t)\}}{c^2 R} \int \delta \mathbf{a} \cdot (\mathbf{r}, t) \, e^{i\mathbf{K} \cdot \mathbf{r}} \, dV. \qquad (2.70)$$

Many important results in light scattering are obtainable directly from this equation.

From (2.69), for $\theta \neq 0$, $S(t)$ is proportional to the spatial Fourier component $\delta \mathbf{a}(\mathbf{K}, t)$ of susceptibility fluctuation with wave vector $\mathbf{K}$. This leads to the well known picture of Bragg scattering from planes of susceptibility fluctuation. The instantaneous field scattered in direction θ is determined by the instantaneous magnitude of $\delta \mathbf{a}(\mathbf{K}, t)$, which in turn is determined by the instantaneous configuration of susceptibility fluctuations throughout the sample, $\{\delta \mathbf{a}(\mathbf{r}, t) \text{ for all } \mathbf{r}\}$. The spatial scale probed by a light scattering experiment is typically the spacing $2\pi/|\mathbf{K}|$ of the Bragg planes, which has a minimum value of about 200 nm. Thus, no direct information concerning atomic positions and motion, etc. can be obtained, a conclusion reached earlier. The susceptibility fluctuation $\delta \mathbf{a}(\mathbf{r}, t)$ can usually be related directly to some more familiar property of the scatterer. For a pure liquid, for example,

$$\delta \mathbf{a}(\mathbf{r}, t) \propto \delta \rho(\mathbf{r}, t)$$

the instantaneous density fluctuation at point $\mathbf{r}$. The intensity and time dependence of the scattered light can then, for this example, be related to the thermodynamic and hydrodynamic properties of the fluid.

From (2.70), the field correlation function is (cf. § 11.2.3),

$$\langle E(R, t), E^*(R, t+\tau) \rangle$$

$$= \frac{E_0^2 \omega_0^4}{c^4 R^2} \int dV \int dV' \langle \delta a(r, t) . \delta a(r', t+\tau) \rangle \exp[iK . (r-r')]$$

$$= \frac{E_0^2 \omega_0^4 V}{c^4 R^2} \int dV \langle \delta a(0, 0) . \delta a(r, \tau) \rangle \exp(iK . r) \tag{2.71}$$

where V is the volume of the illuminated region. Taking the Fourier transform of (2.71) we obtain

$$S(k, \omega) \propto \int dV \int d\tau \langle \delta a(0, 0) . \delta a(r, \tau) \rangle \exp\{i[K . r - (\omega_0 - \omega)\tau]\}, \tag{2.72}$$

i.e. the spectrum of the scattered light is the space–time Fourier transform of the Van Hove susceptibility correlation function.

2.2.5. Rayleigh linewidth spectroscopy of diffusing macromolecules

Consider light scattered from a suspension of non-interacting macromolecules, each small compared with λ. If there are N_s macromolecules in the illuminated region,

$$\delta a(r, t) = a_0 \sum_{i=1}^{N_s} \delta(r - r_i(t)) \tag{2.73}$$

where a_0 is the polarizability difference between a macromolecule and the same volume of solvent and $r_i(t)$ is the position of the macromolecule i at time t. Eqn (2.70) now becomes

$$E(R, t) = \frac{E_0 \omega_0^2}{c^2 R} \exp\{i(k_s . R - \omega_0 t)\} a_0 \sum_{i=1}^{N_s} \exp\{iK . r_i(t)\}. \tag{2.74}$$

Similarly, the electric field correlation function is given by

$$G^{(1)}(\tau) = \langle E(R, t) . E^*(R, t+\tau) \rangle$$

$$= \frac{|E_0|^2 \omega_0^4}{c^4 R^2} \exp(i\omega_0 \tau) . a_0^2 \sum_i \sum_j \exp[iK . \{r_i(t) - r_j(t+\tau)\}]. \tag{2.75}$$

For independent scatterers, the position of particle i will at all times be uncorrelated with the position of particle j. Thus only the terms for $i = j$ survive in the double sum of (2.75).

$$G^{(1)}(\tau) = \frac{E_0^2 \omega_0^4}{c^4 R^2} \exp(i\omega_0 \tau) a^2 \sum_{i=1}^{N_s} \exp[iK . \{r_i(t) - r_i(t+\tau)\}]$$

but

$$\langle\exp\{i\mathbf{K}[\mathbf{r}_i(t) - \mathbf{r}_i(t+\tau)]\}\rangle = \int dV\, G_s(\mathbf{r}, \tau)\exp(i\mathbf{K}.\mathbf{r}) \qquad (2.76)$$

where $G_s(\mathbf{r}, \tau)$ is the probability that, given a particle at point 0 at time zero, it will move to point $\mathbf{r}$ at time τ. For random diffusion under the influence of Brownian motion,

$$g^{(1)}(r) = \exp(i\omega_0\tau)\exp(-D_T K^2\tau) \qquad (2.77)$$

where D_T is the translational diffusion coefficient of the polymer molecule. Light scattered by a macromolecular solution will therefore fluctuate with coherence time $1/D_T K^2$, roughly the time taken by a macromolecule to diffuse a distance $1/|\mathbf{K}|$.

Since Rayleigh linewidth spectroscopy is concerned with the measurement of intensity correlations of the scattered photons, the statistics of the scattered light is of importance. Since, generally, the electric field has Gaussian statistics, all probability distributions can be written in terms of a single quantity, the electric field correlation function. Gaussian statistics are found whenever the scattering volume contains a large number of independent scatterers.

The Gaussian probability distribution of $\mathbf{E}$ is written,

$$P(\mathbf{E}) = \frac{1}{\pi\langle|\mathbf{E}|^2\rangle}\exp(-|\mathbf{E}|^2/\langle|\mathbf{E}|^2\rangle) \qquad (2.78)$$

so that $P(\mathbf{E})$ is completely characterized by the mean intensity

$$\langle|\mathbf{E}|^2\rangle \equiv \lim_{\tau\to\infty} G^{(1)}(\tau) \qquad (2.79)$$

When the number of scatterers is not large, Gaussian statistics will no longer be found, and the analysis of the light scattering experiment becomes considerably more complicated. From the definition of intensity ($I \equiv |\mathbf{E}|^2$), and (2.78), it follows that the intensity probability distribution $P(I)$ of Gaussian light is exponential:

$$P(I) = \frac{1}{\langle I\rangle}\cdot\exp\frac{(-I)}{(\langle I\rangle)} \qquad (2.80)$$

Thus, the most probable intensity is zero and excursions to high intensities, while possible, are rare.

2.2.6. Theory of Brillouin scattering

While incident photons can undergo inelastic scattering from polymer molecules undergoing random diffusive motions, they can also undergo energy exchange with acoustic vibrations in a system of molecules. This can be considered first in terms of fluctuations in the thermodynamic properties of the system and by this means we can evaluate the amplitude of the frequency-shifted spectral peaks. For example, the variations in the dielectric constant with density and temperature can be written as

$$\Delta\varepsilon = \left(\frac{\partial\varepsilon}{\partial\rho}\right)_T \Delta\rho + \left(\frac{\partial\varepsilon}{\partial T}\right)_\rho \Delta T \qquad (2.81)$$

where $\Delta\rho$ and ΔT are the corresponding fluctuations in density and temperature, which are statistically independent. When averages over all molecules are considered, and the quantity $\langle(\Delta\varepsilon)^2\rangle$ is evaluated, cross-terms such as $\langle(\Delta\rho . \Delta T)\rangle$ become zero, and the only terms remaining are $\langle(\Delta\rho)^2\rangle$ and $\langle(\Delta T)^2\rangle$. Fluctuations in entropy and pressure are also statistically independent, so we can write

$$\Delta\varepsilon = \left(\frac{\partial\varepsilon}{\partial S}\right)_P \Delta S + \left(\frac{\partial\varepsilon}{\partial P}\right)_S \Delta P \qquad (2.82)$$

and

$$\langle(\Delta\varepsilon)^2\rangle = \left(\frac{\partial\varepsilon}{\partial S}\right)_P^2 \langle(\Delta S)^2\rangle + \left(\frac{\partial\varepsilon}{\partial P}\right)_S^2 \langle(\Delta P)^2\rangle. \qquad (2.83)$$

The second term represents propagating adiabatic fluctuations, which are the acoustic waves, and the first term represents local isobaric fluctuations which, in normal materials, do not propagate and hence give rise to an unshifted spectral component. Using the thermodynamic relations, $\langle(\Delta P)^2\rangle = kT/V\beta_s$ and $\langle(\Delta S)^2\rangle = kC_p\rho V$, we obtain

$$\langle(\Delta\varepsilon)^2\rangle = \left(\frac{\rho\partial\varepsilon}{\partial\rho}\right)_S^2 \frac{kT\beta_s}{V} + \left(\frac{\partial\varepsilon}{\partial T}\right)_P^2 \frac{kT^2}{C_p\rho V} \qquad (2.84)$$

where C_p is the specific heat, and β_s is the adiabatic compressibility. The relative intensity of the frequency-shifted Brillouin peak compared with the central unshifted line is given by the ratio of the two terms of the right-hand side of 2.84.

Now the fluctuation in dielectric constant can be written in terms of density and temperature changes, i.e.

$$\Delta\varepsilon = \left(\frac{\partial\varepsilon}{\partial\rho}\right)_T \Delta\rho + \left(\frac{\partial\varepsilon}{\partial T}\right)_\rho \Delta T \qquad (2.85)$$

and generally $(\partial\varepsilon/\partial T)_\rho \Delta T \ll (\partial\varepsilon/\partial\rho)_T \Delta\rho$. Thus using the relation

$$\langle(\Delta\rho)^2\rangle_T = \frac{kT\beta_T\rho^2}{V} \qquad (2.86)$$

where β_T is the isothermal compressibility, we can write the total dielectric fluctuation as

$$\langle(\Delta\varepsilon)^2\rangle = \left(\frac{\rho\partial\varepsilon}{\partial\rho}\right)_T^2 \frac{kT\beta_T}{V} \tag{2.87}$$

Keeping in mind that $(\partial\varepsilon/\partial\rho)_T \approx (\partial\varepsilon/\partial\rho)_S$ we can write the expression for the ratio of the central peak intensity I_c to the Brillouin intensity $2I_B$ as

$$\frac{I_c}{2I_B} = \frac{\beta_T - \beta_s}{\beta_s} = \frac{C_p - C_v}{C_v} = \gamma - 1 \tag{2.88}$$

where γ is the ratio of the specific heats C_p/C_v. This ratio $I_c/2I_B$ is called the Landau–Placzek ratio. Eqn (2.88) is obeyed for simple liquids.

Since sound waves experience absorption which is a damping or dissipative process, the Brillouin lines will be broadened and their width will depend on the acoustic absorption coefficient at the corresponding (hypersonic) frequency.

As is discussed in more detail in Chapter 8, energy is lost by a sound wave as it has to overcome frictional forces between molecules. Thermal conduction will also contribute to the loss of energy. The absorption coefficient for a non-relaxing liquid is given by

$$\alpha = \frac{2\pi^2(\nu_s)^2}{\rho v^3}\left\{\tfrac{4}{3}\eta_s + \kappa\left(\frac{1}{C_v} - \frac{1}{C_p}\right)\right\}, \tag{2.89}$$

where η_s is the viscosity, ν_s the phonon frequency, v the phonon velocity, and κ the thermal conductivity.

Generally the thermal conductivity term in eqn (2.89) is negligible compared with the viscosity term. Often, the observed absorption coefficient is larger than that predicted by eqn (2.89) and the 'excess' absorption is due to the volume viscosity η_v. The total viscous absorption is then

$$\alpha_{\text{viscous}} = \frac{2\pi^2\nu_s^2}{\rho v^3}(\tfrac{4}{3}\eta_s + \eta_v) \tag{2.90}$$

At present an ultrasonic or Brillouin scattering measurement is the only way of measuring this volume viscosity.

The phonon lifetime, τ_B, the time taken for the intensity of the sound wave to fall to $1/e$ of its original value, is related to the Brillouin linewidth, Γ_B, by

$$\tau_B = \frac{1}{\Gamma_B}.$$

Since the phonon lifetime is also given by $(\alpha v)^{-1}$, the absorption coefficient can be obtained from the Brillouin linewidth and the wave velocity, both of which can be obtained from a single Brillouin measurement.

Returning now to eqn (2.82) we can calculate the frequency shift. The fluctuation in entropy will decay by some thermal diffusion process towards an equilibrium situation. Thus the fluctuation will have a finite lifetime and be time-dependent. Consequently, the fluctuation in entropy is the mechanism for scattering, and its time dependence, because the fluctuation decays, is the mechanism for line broadening. If a scattering process is not time dependent, the scattered light will show no frequency change. In the case of entropy fluctuations, the linewidth of the scattered light depends on the thermal diffusion coefficient and is related by

$$\Gamma = \frac{\kappa}{\rho C_p} \left\{ \frac{4\pi n}{\lambda_0} \sin \frac{\theta}{2} \right\}^2 \tag{2.91}$$

where κ is the thermal conductivity, n the refractive index, λ_0 the wavelength of the incident light, and θ the angle through which the light is scattered.

As eqn (2.82) shows, fluctuations in pressure also cause light scattering. If the wave vector of the incident light is $\mathbf{k}_i$, that of the scattered light is $\mathbf{k}_s$ and that of the phonon, $\mathbf{K}$, to conserve momentum we have

$$\mathbf{K} = \mathbf{k}_i - \mathbf{k}_s. \tag{2.92}$$

This is illustrated in Fig. 2.1 together with the corresponding vector triangle.

The change in frequency of the scattered light is small (i.e. $|\mathbf{k}_i| \approx |\mathbf{k}_s|$) and generally is about one part in 10^5. To a good approximation therefore

$$|\mathbf{K}| = 2|\mathbf{k}_i| n \sin \theta/2 \tag{2.93}$$

where n is the refractive index of the scattering medium. As

$$|\mathbf{K}| = \frac{2\pi\nu_s}{\upsilon} \quad \text{and} \quad |\mathbf{k}_i| = \frac{2\pi\nu_0}{c}, \qquad \nu_s = 2\left(\frac{\upsilon}{c}\right)\nu_0 . \sin \theta/2, \tag{2.94}$$

we thus expect two shifted lines at $\pm\nu_s$ corresponding to the case where energy is taken from and given to the photon. Classically, the Brillouin line may be considered as Doppler shifted lines scattered from propagating sound waves. From the quantum point of view, the energy change occurring in Brillouin scattering is such that just one phonon is created or annihilated. Little difference is made to the phonon population by the scattering act so the probabilities of creation and annihilation are equal.

2.3. Neutron scattering

The experimental details of neutron scattering will be considered in Chapter 12, but it is appropriate to consider how the theory outlined in this chapter can be extended to neutrons. The principal difference is one of momentum. In the case of light scattering the photon is very light compared with the molecule. Approximations based on a large mass difference are applicable to light scattering but are not necessarily valid for neutron scattering.

Earlier the Born approximation was introduced, and again this is the starting point for discussion. The differential scattering cross-section per unit solid angle and unit interval of energy of the scattered particle (neutron) is given by

$$\frac{d^2\sigma}{d\Omega\,d\varepsilon} = \left| \frac{m^3}{2\pi^2\hbar^6} \frac{k_s}{k_i} W(\mathbf{K}) \sum_i \rho_i \sum_f \left| \left\langle i \left| \sum_{j=1}^N e^{i\mathbf{K}.\mathbf{r}_j} \right| f \right\rangle \right|^2 \right.$$

$$\times \delta\left[\mathbf{k}_s^2 - \mathbf{k}_i^2 + \frac{2m}{\hbar^2}(E_f - E_i) \right] \tag{2.95}$$

where m, $\mathbf{k}_i$ and $\mathbf{k}_s$ are the mass and the initial and final wave vectors of the scattered particle and $\mathbf{K}$ is defined by $\mathbf{K} = \mathbf{k}_i - \mathbf{k}_s$. The operators, $\mathbf{r}_j$, represent the position vectors of the N particles of the scattering system whose initial and final states are labelled by i and f and have energy E_i and E_f. The quantity ρ_i is the statistical probability of the initial state and

$$W(\mathbf{K}) = \left| \int \exp(i\mathbf{K}.\mathbf{r}) \cdot V(\mathbf{r})\,d\mathbf{r} \right|^2 \tag{2.96}$$

where $V(\mathbf{r})$ is the interaction between the neutron and the nucleus of the scattering system. In this expression it has been assumed that all the nuclei interact in the same way.

If, as well as the momentum transfer, $\hbar\mathbf{K}$, we introduce the energy transfer through

$$\hbar\omega = \frac{\hbar^2(\mathbf{k}_i^2 - \mathbf{k}_s^2)}{2m} \tag{2.97}$$

then (2.95) can be written in the form

$$\frac{d^2\sigma}{d\Omega\,d\varepsilon} = AS(\mathbf{K}, \omega) \tag{2.98}$$

where

$$A = \frac{Nm^2}{4\pi^2\hbar^5} \cdot \frac{k_s}{k_i} W(\mathbf{K}) \tag{2.99}$$

and

$$S(\mathbf{K}, \omega) = N^{-1} \sum_i \rho_i \sum_f \left| \left\langle i \left| \sum_{j=1}^N e^{i\mathbf{K}.\mathbf{r}_j} \right| f \right\rangle \right|^2 \delta\left(\omega + \frac{E_i - E_f}{\hbar} \right). \tag{2.100}$$

By introducing the Fourier transform representation of the delta function we obtain

$$S(\mathbf{K}, \omega) = \frac{1}{2\pi N} \int_{-\infty}^{\infty} dt\, e^{-i\omega t} \sum_{i,j} \langle e^{i\mathbf{K}.\mathbf{r}_i(0)}\, e^{i\mathbf{K}.\mathbf{r}_j(t)} \rangle$$

$$= \frac{1}{2\pi} \int_{-\infty}^{\infty} dt\, e^{-i\omega t} F(\mathbf{K}, t) \tag{2.101}$$

where

$$F(\mathbf{K}, t) = \frac{1}{N} \sum_{i,j} \langle e^{-i\mathbf{K}.\mathbf{r}_i(0)} e^{+i\mathbf{K}.\mathbf{r}_j(t)} \rangle. \tag{2.102}$$

The quantity $S(\mathbf{K}, \omega)$ is the scattering function and is a specific form of the general equation presented earlier. We can express $S(\mathbf{K}, \omega)$ in terms of a time-dependent extension of the radial distribution function $G(\mathbf{r}, t)$ which is called the Van Hove space–time correlation function. The reader will see immediately the similarity with the formulation presented for Brillouin scattering. The equations have the form

$$S(\mathbf{K}, \omega) = (2\pi)^{-1} \int \int e^{i(\mathbf{K}.\mathbf{r} - \omega t)} G(\mathbf{r}, t)\, d\mathbf{r}.\, dt \tag{2.103}$$

$$G(\mathbf{r}, t) = (2\pi)^{-3} \int \int e^{-i(\mathbf{K}.\mathbf{r} - \omega t)} S(\mathbf{K}, \omega)\, d\mathbf{K}.\, d\omega. \tag{2.104}$$

Substituting for $S(\mathbf{k}, \omega)$ into eqn (2.104) and integrating over t we obtain

$$G(\mathbf{r}, t) = (2\pi)^{-3} N^{-1} \sum_{i,j} \int d\mathbf{K}\, e^{i\mathbf{K}.\mathbf{r}} \langle e^{-i\mathbf{K}\mathbf{r}_i(0)}\, e^{i\mathbf{K}.\mathbf{r}_j(t)} \rangle. \tag{2.105}$$

It can be seen that the factor $(2\pi)^{-3} N^{-1}$ in this expression makes $G(\mathbf{r}, t)$ independent of N and asymptotically equal to the number density in the thermodynamic limit. The functions $F(\mathbf{K}, t)$ and $G(\mathbf{r}, t)$ are connected through the Fourier relationship

$$G(\mathbf{r}, t) = (2\pi)^{-3} \int d\mathbf{K}\, e^{-i\mathbf{k}.\mathbf{r}} F(\mathbf{K}, t). \tag{2.106}$$

In the classical limit the function $G(\mathbf{r}, t)$ splits naturally into two parts, one describing the correlation of one molecule at different times and the other describing the correlation between distinct particles at different times. Thus we can write

$$G(\mathbf{r}, t) = G_s(\mathbf{r}, t) + G_d(\mathbf{r}, t) \tag{2.107}$$

where

$$G_s(\mathbf{r}, t) = N^{-1} \left\langle \sum_{j=1}^{N} \delta\{\mathbf{r} + \mathbf{r}_j(0) - \mathbf{r}_j(t)\} \right\rangle \tag{2.108}$$

and $G_d(\mathbf{r}, t)$ denotes distinct interactions

$$G_d(\mathbf{r}, t) = N^{-1}\left\langle \sum_{\substack{j \neq i = 1}}^{N} \delta\{\mathbf{r} + \mathbf{r}_i(0) - \mathbf{r}_j(t)\}\right\rangle. \qquad (2.109)$$

For $t = 0$ these reduce to

$$G_s(\mathbf{r}, 0) = \delta(\mathbf{r})$$

$$G_d(\mathbf{r}, 0) = \rho_0 g(\mathbf{r}) \qquad (2.110)$$

where $g(\mathbf{r})$ is the radial distribution function for the system.

If the energy of the incident particle is sufficiently large compared with the energy being transferred (as in the case of X-ray scattering), the momentum transfer for a given scattering angle is independent of the outgoing energy and the differential cross-section per unit solid angle becomes

$$\frac{d\sigma}{d\Omega} = \int \frac{d^2\sigma}{d\Omega\, d\varepsilon} = \hbar A \int S(\mathbf{K}, \omega)\, d\omega$$

$$= \hbar A \int \int e^{i\mathbf{k}\cdot\mathbf{r}}\delta(t)G(\mathbf{r}, t)\, d\mathbf{r}\, dt$$

$$= \left\{1 + \rho_0 \int e^{i\mathbf{K}\cdot\mathbf{r}}g(\mathbf{r})\, d\mathbf{r}\right\}. \qquad (2.111)$$

In practice not all the nuclei interact with the neutrons in the same way. In particular, spin interactions are important. If a_j is an operator that depends upon the spin of the jth particle, we define a spin-dependent space–time correlation function by

$$\Gamma(\mathbf{r}, t) = N^{-1}\left\langle \sum_{i,j=1} \int d\mathbf{r}'\, a_i(0)\delta\{\mathbf{r} + \mathbf{r}_i(0) - \mathbf{r}'\}a_j(t)\delta\{\mathbf{r}' - \mathbf{r}_j(t)\}\right\rangle \qquad (2.112)$$

where

$$a_j(t) = \exp\left(\frac{it\mathcal{H}}{\hbar}\right) a_j \exp\left(\frac{-it\mathcal{H}}{\hbar}\right).$$

For a system of Boltzmann particles with spin-independent Hamiltonian, $\mathcal{H}$ and described by a canonical distribution function $\rho_i = e^{-E_i/Q}$, there is no correlation betweeen spins, nor between spins and positions, and we can write

$$\Gamma(\mathbf{r}, t) = \langle a^2\rangle G_s(\mathbf{r}, t) + \langle a\rangle^2 G_d(\mathbf{r}, t) \qquad (2.113)$$

where a is any of the a_j and $\langle\ \rangle$ denotes an average over the spin states of the corresponding particles. The interaction between a slow neutron and a nucleus of the scattering system is often replaced by the so-called Fermi pseudopotential.

$$V(\mathbf{r}) = \left(\frac{2\pi a \hbar^2}{m}\right)\delta(\mathbf{r}) \tag{2.114}$$

where m is the mass of the neutron and a is called the scattering length of the nucleus (a may depend upon the nuclear spin state).

Thus we write

$$\frac{d^2\sigma}{d\Omega\,d\varepsilon} = \left(\frac{d^2\sigma}{d\Omega\,d\varepsilon}\right)_{coh} + \left(\frac{d^2\sigma}{d\Omega\,d\varepsilon}\right)_{incoh} \tag{2.115}$$

where

$$\left(\frac{d^2\sigma}{d\Omega\,d\varepsilon}\right)_{coh} = \frac{\langle a\rangle^2 N}{2\pi K}\frac{k_s}{k_i}\int e^{i(\mathbf{K}.\mathbf{r}-\omega t)}G(\mathbf{r}, t)\,d\mathbf{r}\,dt. \tag{2.116}$$

$$\left(\frac{d^2\sigma}{d\Omega\,d\varepsilon}\right)_{incoh} = \frac{(\langle a^2\rangle - \langle a\rangle^2)N}{2\pi K}\frac{k_s}{k_i}\int e^{i(\mathbf{K}.\mathbf{r}-\omega t)}G_s(\mathbf{r}, t)\,d\mathbf{r}\,dt. \tag{2.117}$$

Eqns (2.116) and (2.117) are called coherent and incoherent scattering cross-sections, and their extension and use are discussed in § 12.2.2.

When neutron scattering is used to obtain information on the dynamics of polymer molecules, two points must be stressed. First, the time scale of the observation is typically 10^{-10} to 10^{-13} s, and secondly it lacks stress selection in terms of molecular perturbation which influence the scattering of the neutron. This is in contrast to other methods, such as infrared, dielectric, or nuclear magnetic relaxation where the vectorial properties of the probe allow a more precise definition of the motions responsible for the observed dynamic changes. Nevertheless neutron scattering has been applied successfully to certain systems and does provide important information complementary to that from Brillouin scattering.

References

1. Wilson, E. B., Decius, J. C. and Cross, P. C. *Molecular Vibrations*. McGraw-Hill, London, New York, and Toronto (1955).
2. Miyazawa, T. and Kitagawa, T. *J. polym. Sci.* **B2**, 395 (1964).
3. Kitagawa, T. and Miyazawa, T. *Rep. Prog. polym. Phys. Jpn*, **8**, 53 (1965).
4. Tusumi, M. and Krimm, S. *J. chem. Phys.* **46**, 755 (1967).
5. Schachtschneider, J. H. and Snyder, R. G. *Spectrochim. Acta* **19**, 17 (1963); **21**, 169 (1965).
6. Opaskar, G. C. and Krimm, S. *Spectrochim. Acta* **21**, 1165 (1965).
7. Zerbi, G. Molecular vibrations of high polymers. *Appl. Spectrosc. Rev.* **2**, 193 (1969).
8. Piseri, L. and Zerbi, G. *J. mol. Spectrosc.* **26**, 259 (1968).
9. Dean, P. *Proc. Roy. Soc.* **A254**, 507 (1960).

10. Dean, P. *Proc. phys. Soc.* **84**, 727 (1964).
11. Dean, P. *J. phys. Soc. Jpn suppl.* **26**, 20 (1969).
12. Bell, R. J., Bird, N. F. and Dean, P. *J. Phys. C. Solid State Phys.* **1**, 299 (1968).
13. Shawyer, R. E. and Dean, P. *Discuss. Faraday Soc.* **48**, 102 (1969).
14. Tasumi, M. and Zerbi, G. *J. chem. Phys.* **48**, 3813 (1968).
15. Zerbi, G., Piseri, L. and Cabassi, F. *Mol. Phys.* **22**, 241 (1971).
16. Fabelinskii, I. L. *Molecular scattering of light.* Plenum, New York (1968).
17. Fleury, P. A. and Boon, J. P. *Advan. chem. Phys.* **24**, 1 (1973).
18. Berne, B. J. and Pecora, R. *Dynamic light scattering.* Wiley, New York (1976).
19. Chu, B. *Laser light scattering.* Academic Press, London and New York (1974).

DAMPED 'HYDRODYNAMIC' MODES OF MOTION

3.1. Introduction

In the preceding chapter the movement of groups of atoms, and of extended lengths of polymer chains, was presented in terms of resonant normal modes. However, as was pointed out in Chapter 1, many of the molecular motions of interest involve movement through regions such that the various intermolecular forces average out (both in time and direction) as a viscous drag on the moving entity. Such motions come to have more the characteristics of relaxation and diffusion, in that the over-all movement is made up of a large number of relatively small jumps associated with the thermal agitation or Brownian motion of the system. In the equations of motion describing these processes, the viscous and linear terms outweigh the inertial term, so that the equations describe damped modes with a relaxation character.

In the historical development of the theory of such motions there have been two major periods of activity. The first occurred during the 1950s with the development of analytic methods for treating isolated chains in dilute solution. These theories were generally oriented towards a quantitative treatment of solution viscosity, and were based upon a model and mathematical techniques proposed by the late J. G. Kirkwood.

The second period of activity is very recent, and indeed is still gathering momentum at the time of writing. The stimulation has come from the necessity of advancing quantitative theories from solutions to gels and rubbers. Progress has been made by supplementing the statistical mechanics popularized by Kirkwood with procedures developed in other areas of theoretical physics. In particular the application of space and time correlation functions, fluctuation theory, and of Green's functions is now becoming widespread. Techniques based on normal coordinate solutions of the diffusion equation will be discussed in this chapter, and the more modern treatments involving the Langevin equations are presented in Chapter 4.

3.2. The illustration of the elastic dumb-bell

Many of the hydrodynamic properties of polymer molecules can be treated by considering the behaviour of a deformable elastic body immersed in a viscous continuum, which may be subjected to a shearing stress. In order to visualize the most important aspects of molecular behaviour, we consider first the simplest possible model exhibiting both a frictional interaction with the surrounding continuum, and deformation which is recoverable in an elastic fashion. This is the elastic dumb-bell, in which two beads with frictional interaction are connected by an elastic spring with no hydrodynamic interaction (Fig. 3.1(a)). This model has been fully described by Zimm[1] whose treatment is followed here.

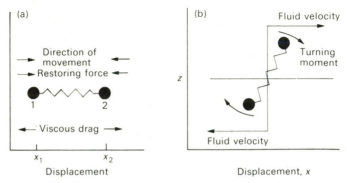

Fig. 3.1. Movement of an extended elastic dumb-bell. (a) Relaxation of extension in a still, viscous environment; (b) rotation and extension in a thermal, shearing liquid.

Consider first the behaviour of the dumb-bell when the beads are displaced from their equilibrium positions in an absolutely static medium. This will illustrate the normal coordinate transformation, after which thermal and shear motions of the fluid can be considered.

The two important forces acting on the extended (or contracted) dumb-bell are the restoring force due to the spring connecting the beads and the viscous drag arising when the beads are pulled through the liquid. The inertial forces can be neglected, since the accelerations involved are small, and so the equations of motion for the displaced system become:

$$\text{for bead 1} \quad \zeta \dot{x}_1 = -g(x_1 - x_2)$$
$$\text{for bead 2} \quad \zeta \dot{x}_2 = -g(x_2 - x_1)$$

(3.1)

where ζ represents the viscous drag (frictional) coefficient, g is the force constant of the spring, x_1, x_2 are the coordinates of the beads, and superscript dots represent derivatives with respect to time. As an example of ideal behaviour the viscous drag is assumed to be proportional to the velocity of the beads, and the Hookean restoring force is assumed to be proportional to the extension (or compression) of the dumb-bell. These are both quite reasonable assumptions.

These simultaneous equations can be solved by adding and subtracting the pair, and changing the positional variables to

$$\xi_0 = \frac{x_1 + x_2}{\sqrt{2}}, \qquad \xi_1 = \frac{x_1 - x_2}{\sqrt{2}} \qquad (3.2)$$

when two new equations arise

$$\dot{\xi}_0 = 0, \qquad \dot{\xi}_1 = -(2g/\zeta)\xi_1. \qquad (3.3)$$

This simple procedure of coordinate transformation has allowed separation of the two new variables into independent differential equations for which the solutions are

$$\xi_0 = \text{constant}, \qquad \xi_1 = \xi_1' \exp(-2gt/\zeta) \qquad (3.4)$$

where ξ_1' is the initial value of ξ.

It is immediately apparent that, neglecting the normalizing factor $1/\sqrt{2}$, ξ_0 is the mean coordinate of the elastic dumb-bell. Obviously since no net force has been applied to move the dumb-bell, the position of its centre of mass cannot change, as designated in (3.4). The quantity ξ_1, again neglecting the normalization factor, is the extension or distortion of the dumb-bell, which relaxes exponentially from its original value ξ_1' towards zero. The two new coordinates, ξ_0 and ξ_1, which permitted this simple solution, are the normal coordinates of the dumb-bell.

It is now possible to introduce the thermal agitations, which in a real system cause the extension of the dumb-bell in the first instance. This is done by recognizing the similarity between the random Brownian motion and diffusion, and so using procedures familiar in the case of the latter phenomenon. At the same time we can recognize the applicability of the treatment to viscoelastic deformations, and introduce the effect of a velocity gradient, or shear field, across the dumb-bell. This is illustrated in Fig. 3.1(b) where shear movement of the surrounding continuum in the xz-plane is represented by a velocity gradient in the z-direction. Assuming again that the shear force on each bead is proportional to the flow velocity, we get the pair of equations

$$\zeta \dot{x}_1 = -kT \frac{d \ln \psi}{dx_1} - g(x_1 - x_2) + Kz_1$$

$$\qquad (3.5)$$

$$\zeta \dot{x}_2 = -kT \frac{d \ln \psi}{dx_2} - g(x_2 - x_1) + Kz_2$$

where K is the rate of shear. The Brownian motion term, $kT (d \ln \psi/dx)$, is introduced as though particles 1 and 2 have average motions in an ensemble of such species. Under these conditions Brownian motion obeys Fick's laws in that the concentration of particles is governed by a diffusive flux from regions of high local concentration. The flux is proportional to the concentration gradient and to kT. In this case ψ is a generalized concentration including both particles 1 and 2, and is introduced into the equations of motion in the standard way.[2]

Once again the coordinates can be transformed to obtain

$$\zeta \dot{\xi}_0 = -kT \frac{\mathrm{d} \ln \psi}{\mathrm{d} \xi_0}$$

$$\zeta \dot{\xi}_1 = -kT \frac{\mathrm{d} \ln \psi}{\mathrm{d} \xi_1} - 2g\xi_1 + K\eta_1 \tag{3.6}$$

where

$$\eta_1 = \frac{z_1 - z_2}{\sqrt{2}}.$$

These equations have now separated ξ_0 and ξ_1. They are solved by setting the net diffusive flux of particles at a point equal to zero, when they become independent second-order differential equations of a type standard in diffusion theory. Of particular significance are the results describing the average (over a molecular ensemble) values of the normal coordinates ξ_1 and η_1

$$\langle \xi_1^2 \rangle = (kT/2g)(1 + K\zeta/g)$$

$$\langle \eta_1^2 \rangle = kT/2g$$

$$\langle \xi_1 \eta_1 \rangle = kTK\zeta/2g^2 \tag{3.7}$$

$$= \frac{kTK}{g}\tau$$

where $\tau = \zeta/2g$.

The qualitative significance of these results is that the mean square displacement along the x-axis increases with temperature and shear rate, and that the cross-term also increases. This means that, on average, the dumb-bell is extended along a 'diagonal' line inclined between the x- and z-axes. In addition the cross-term can be written to include the characteristic time constant τ, descriptive of relaxation of the stretched spring. Consequently the whole motion can be visualized as a rotation (induced by the turning moment of the shear field) in phase with contraction and expansion of the spring, with maximum extension in the $(-x, -z)$- to $(+x, +z)$-direction.

The example of the elastic dumb-bell, although very stylized, is a convenient illustration for several reasons. In the first place it presents the simplest possible system for separation of the coordinates in appropriate equations of motion, and so facilitates understanding of the mathematical techniques used for much more complex systems. It also illustrates the simple result that an elastic body in a shear field is both rotated and deformed in a periodic manner. Finally many treatments of the dynamics of polymer molecules are based upon models in which the chain is visualized as a combination of beads and springs; within such models the dumb-bell, just discussed, forms the elementary unit.

3.3. Models of long 'flexible' chains

3.3.1. Methods and models

The extension of the concepts introduced in the preceding section to long polymer chains has involved two interrelated lines of evolution. On the one hand has been the development of various models for the polymer chain, formulated with the two requirements of mechanical realism and mathematical tractability. The other sequence has been in the sophistication of the mathematical techniques applied to the solution of the diffusion equations derived from the selected molecular model.

A detailed review of the development of theories of dissolved macromolecules has been given by Fixman and Stockmayer,[3] and the most commonly used mathematical derivations reported in full in a textbook by Yamakawa.[4] The basic model from which all these treatments derive was defined by Kirkwood.[5,6] It concerns the method by which diffusion theory is applied to the various basic subunits which are linked to form the polymer chain.

In most recent normal coordinate analyses the chain of infinitesimal point atoms introduced by Kirkwood is altered to a chain of subunits of finite size. The physical picture then begins to resemble that of a chain of macroscopic beads, and indeed such models are often referred to as 'finite bead' or, more romantically, 'necklace' models. The important consequence of this visualization is that the hydrodynamic interaction between polymer and environment (which was solvent in early treatments, but may be other polymer chains) takes place at a finite number of discrete centres along the chain. This permits a mathematical simplification which is judged to outweigh the deficiencies in application to chemical structures which appear to more closely resemble uniform threads or rods.

In the most widely used normal coordinate treatments, based upon analyses by Rouse,[7] Bueche,[8] and Zimm[9] the subunits are assumed to exhibit a frictional drag directly proportional to velocity through the continuum, and a perfectly Hookean elastic restoring force between connected subunits. In molecular terms this implies that each subunit must contain enough covalent backbone bonds that the end-to-end separation within the unit exhibits Gaussian statistics. Under these circumstances the restoring force is the entropic change familiar in rubber elasticity.

The mathematical advantage of such a model is that the potential energy can be written as an explicit quadratic function of displacement, so that the normal coordinate transformation will lead to a series of independent simultaneous quadratic differential equations.

The solution of these diffusion equations represents a task of very considerable mathematical complexity. The original work by Kirkwood applied the Oseen[10] perturbation to the velocity of particles in a fluid. This is based on a solution of the Navier–Stokes equations of hydrodynamics, and describes the velocity perturbation resulting from the application of a force to a particle in terms of the distance between the perturbation and point of application of the force, and the velocities of the particle and of the fluid. A full description of the Oseen velocity perturbation formula, in terms applicable to this problem, has been given in Appendix VIB of ref. 4.

In early treatments, the Oseen interaction was averaged over all the particles in the system, or over all the particles in the polymer chain. Although errors introduced in this way are likely to be less important than those arising elsewhere in the treatment, the validity of the procedure has been questioned. Various arguments have been reviewed by Fixman and Stockmayer,[3] and treatments involving the use of a non-averaged Oseen tensor have been devised. The full mathematical treatment of these, too, is presented in ref. 4.

The formulation of a model to cover the energy-displacement characteristics of a polymer chain must be not only physically realistic and mathematically tractable, but must cover the effect of the chain on the solvent velocity as well as damping of the chain motion. This aspect was ignored in early treatments,[7,8] which models have since been called the 'free-draining approximation'. Zimm[9] first introduced the reduction of solvent velocity in the interior of a coiled chain, a procedure now referred to as the 'non-free-draining model'. Basically the difference between the two arises in the value given to the Oseen perturbation. The free-draining case represents a solution in which the Oseen interaction parameter is given a value of zero, whereas in the simplest non-free draining case it is effectively infinite. This limit corresponds to a very large hydrodynamic interaction between segments, so that much of the behaviour of the molecule is typical of that of a rigid sphere. The most realistic treatments, of course, are those which permit evaluation of the dynamic parameters for all values of the hydrodynamic interaction between these two extreme cases.[5,11–17]

A full mathematical review of these theories of damped hydrodynamic modes of motion, complete with complicating factors such as excluded volume, concentrations, and high-frequency effects, when combined with the relevant phenomenological processes such as viscosity requires[4] almost one hundred pages. In order to cover the field without requiring familiarity with tensor mathematics, we give here a qualitative description of the theoretical approaches, together with the most important resulting equations. The interested reader is referred to ref. 4, for the details of each step in the derivations.

3.3.2. The Kirkwood–Riseman theory of hydrodynamic interaction

In essence, the first step in the theoretical evaluation of hydrodynamic mode parameters is the derivation of an appropriate diffusion equation encompassing a hydrodynamic interaction between chain units. This is introduced by assuming that when the flow of a surrounding fluid is perturbed by the interaction of a polymer segment the effects will be felt in the hydrodynamic force exerted by the fluid on another segment.

This is the point at which the Oseen tensor, $\mathbf{T}$, is introduced. The velocity perturbation, $\mathbf{v}'$, which is experienced in a fluid at a distance r from a point where a force $\mathbf{F}$ is applied, is given by the tensor relationship

$$\mathbf{v}' = \mathbf{T} \cdot \mathbf{F}$$

where

$$(3.8)$$

$$\mathbf{T} = \frac{1}{8\pi\eta_0 r}\left(\mathbf{I} + \frac{\mathbf{rr}}{r^2}\right).$$

In this representation $\mathbf{I}$ is the unit tensor, $\mathbf{r}$ is the vector describing the position of the particle with respect to the point of application of the force as origin, and η_0 is the viscosity coefficient of the fluid continuum. This is then applied to a consideration of the velocity of a segment of a polymer chain of n units, which is then summed over the interactions of all the other $n-1$ segments of the same molecule to yield a set of linear equations for the force on each segment

$$\mathbf{F}_i = \zeta(\mathbf{u}_i - \mathbf{v}_i) - \sum_{\substack{j=0 \\ \neq i}}^{n} \mathbf{T}_{ij}\mathbf{F}_j \qquad (3.9)$$

where ζ is the translational friction coefficient of each subunit in the chain, $\mathbf{u}_i$ and $\mathbf{v}_i$ are respectively the velocity of the ith subunit and the original velocity of the solvent in the absence of a polymer molecule.

Eqn (3.9) is fundamental to the Kirkwood–Riseman[5] theory of the intrinsic viscosity of dilute polymer solutions. For the evaluation of this quantity the micro-Brownian motion of the polymer coil is ignored, and emphasis placed instead on the rotational torque exerted on the molecule. This requires the solution of an integral equation (the integral deriving from the sum over all segments of subunits) which leads to the Kirkwood–Riseman hydrodynamic interaction function

$$F(X) = \frac{6}{\pi^2} \sum_{k=1} \frac{1}{k^2(1 + X/k^{1/2})}$$
(3.10)

where $X/\sqrt{2} = h = \zeta n^{\frac{1}{2}}/\{(12\pi^3)^{\frac{1}{2}}\eta_0 a\}$ (with bonds of length, a, between chain units) is called the hydrodynamic interaction parameter. Values of $XF(X)$, the solution to eqn (3.10), have been tabulated[4] with rather different approximations involved in separate evaluations.

It is interesting to note that the extreme case of $h = 0$ (the free-draining coil) yields for a chain of $n + 1$ units separated by a distance, a,

$$[\eta] = \frac{N_A \zeta n^2 a^2}{36 \eta_0 M}$$
(3.11)

where

$$[\eta] = \lim_{c \to 0} \{(\eta - \eta_0)/\eta_0 c\}$$

is called the intrinsic viscosity of a solution of concentration c. N_A is Avogadro's constant, and M is the chain molecular weight. This can be recast in terms of the mean square radius of gyration of the unperturbed coil $\langle S^2 \rangle_0$,

$$[\eta] = \frac{N_A \zeta_n}{6 \eta_0 M} \langle S^2 \rangle_0.$$
(3.12)

By comparison, when X or h is set equal to infinity, the non-free draining model gives

$$[\eta] = \frac{1 \cdot 3 N_A \pi^{\frac{3}{2}}}{M} \langle S^2 \rangle_0^{\frac{3}{2}}.$$
(3.13)

This last equation, with the hydrodynamic interaction between segments so large, reduces to rigid sphere form, with the intrinsic viscosity depending only on a chain segment molar volume, so eliminating the characteristics of the solvent.

The parameter h, representing the effective change in velocity of solvent (or matrix continuum) flow through the polymer coil, is called the 'draining parameter'. Eqns (3.12) and (3.13) imply that

$$[\eta] \propto M^{1 \cdot 0} \text{ (free-draining) or } M^{0 \cdot 5} \text{ (non-free-draining).}$$
(3.14)

3.3.3. The 'spring and bead' solution of the diffusion equation

The general theory of Kirkwood was originally used to derive a diffusion equation, which was then solved to yield explicit functions for the effective friction coefficient and translational diffusion coefficient. In this instance the intramolecular segmental motion is of greater interest, and this is pursued by the application of a spring and bead model and then by transformation into normal coordinates as introduced in § 3.3.1.

The derivation of the diffusion equation of a chain of n interunit links involves the construction of two $(n+1) \times (n+1)$ matrices, descriptive of the applied force and hydrodynamic mobility. Introduction of the normal coordinate transformation then results in a matrix relationship of which the eigenvalue equation results in a set of eigenvalues, λ_j $(0 \leqslant j \leqslant n)$. These figure prominently in the evaluation of the characteristic time constants for the molecular distortion

$$\tau_j = \frac{\zeta}{2g\lambda_j} = \frac{M\eta_0[\eta]}{RT\lambda_j \left(\sum_{j=1}^{n} \lambda_j^{-1}\right)}. \tag{3.15}$$

In this equation τ_j represents the relaxation time of the jth normal mode as illustrated in Fig. 1.2. The formal similarity between (3.15) and (3.7) illustrates the close similarity of the deformable body models. This result was derived on the additional assumption that the hydrodynamic interaction is independent of the rate of shear, and so of the shear stress. Consequently it is strictly applicable only to situations in which the molecule is subjected to an infinitely weak distorting force.

The mathematically complex intrinsic viscosity, $[\eta^*]$, associated with this solution is

$$[\eta^*] = \frac{RT}{M\eta_0} \sum_{j=1}^{n} \frac{\tau_j}{1 + i\omega\tau_j}. \tag{3.16}$$

Since the complex shear modulus is related to the complex viscosity by $G^* = i\omega\eta^*$ it is possible to define a complex shear intrinsic modulus as

$$[G^*] = [G'] + i[G''] = \lim_{c \to 0} \frac{G^* - i\omega\eta_0}{c} \tag{3.17}$$

$$[G^*] = i\omega\eta_0[\eta^*]$$

$$[G'] = \omega\eta_0[\eta''] \tag{3.18}$$

$$[G''] = \omega\eta_0[\eta']$$

leading to

$$[G'] = \frac{RT}{M} \sum_{j=1}^{n} \frac{\omega^2 \tau_j^2}{1 + \omega^2 \tau_j^2}$$

$$[G''] = \frac{RT}{M} \sum_{j=1}^{n} \frac{\omega \tau_j}{1 + \omega^2 \tau_j^2}. \tag{3.19}$$

3.3.4. Evaluation of the eigenvalues, λ_j, and relaxation times, τ_j

The matrix relationship describing the interaction of one bead on another through the hydrodynamic interaction with the solvent and through the elastic connecting links is an integrodifferential equation. The integral enters through the summation over all subunits, and of course the differential exists through the diffusion and velocity terms. The eigenvalues resulting from this equation depend on the value of the parameter, h, describing the hydrodynamic interaction.

When the interaction term is set equal to zero, the Rouse eigenvalues[7] are obtained

$$\lambda_j = \pi^2 j^2 / n^2 \qquad (j = 0, 1, \ldots, n) \tag{3.20}$$

and so

$$\tau_j = \frac{6 M \eta_0 [\eta]}{\pi^2 R T j^2} \qquad (h = 0, n \gg 1). \tag{3.21}$$

The non-free-draining coil matrix equation is more difficult to solve because the matrix is not symmetric. However the eigenvalues can be recast[9,17] in the form

$$\lambda_j' = (n^2 / 4h) \lambda_j \tag{3.22}$$

whence

$$\tau_j = \frac{M \eta_0 [\eta]}{0 \cdot 586 R T \lambda_j} \qquad (h = \infty). \tag{3.23}$$

These eigenvalues can be expressed in terms of Fresnel integrals as follows

$$\lambda_j' = \pi j^{\frac{1}{2}} \{ \pi j C(\pi j) - \tfrac{1}{2} S(\pi j) \} \tag{3.24}$$

where $C(x)$ and $S(x)$ are the Fresnel integrals

$$C(x) = (2\pi)^{-\frac{1}{2}} \int_0^x t^{-\frac{1}{2}} \cos t \, dt$$
$$S(x) = (2\pi)^{-\frac{1}{2}} \int_0^x t^{-\frac{1}{2}} \sin t \, dt. \tag{3.25}$$

Calculated[9] values of the reduced eigenvalues for $h = \infty$ are given in Table 3.1.

Hearst[13] has used the Rouse eigenfunctions to calculate eigenvalues for different values of the draining or interaction parameter, h. The use of the Rouse eigenfunctions is effectively an approximation neglecting off-diagonal elements in the matrix equation. In this approximation the eigenvalues, λ_j are given by the sum of free-draining and non-free-draining terms

Table 3.1

Reduced eigenvalues, λ'_j, for a non-free-draining coil $(h = \infty)$[9]

j	λ'_j
0	0
1	4·04
2	12·79
3	24·2
4	37·9
5	53·5
6	70·7
7	89·4
$j \gg 1$	$\sim \frac{1}{2}\pi^2 j^{\frac{3}{2}}(1 - 1/2\pi j)$

((3.20) and (3.22)),

$$\lambda_j = \frac{\pi^2}{n^2}\left(j^2 + \frac{4h\lambda'_j}{\pi^2}\right). \tag{3.26}$$

The relaxation times of the jth modes can then be evaluated by combining (3.21) and (3.26).

More recently less approximate solutions of the matrix equations have been evaluated. Fixman and Pyun[15] used the Rouse solution (considering only diagonal elements in the matrix equation) as a zeroth approximation, and introduced the off-diagonal terms using perturbation theory. A similar procedure has been adopted by Thurston and Morrison,[16] who removed the requirement that the number of subunits should be very large.

This last consideration is important, because many polymers encountered in practice have chain lengths such that the number of Gaussian subunits will be a finite number between 1 and 100. In the treatment of Thurston and Morrison the interaction parameter is redefined[18] so as to remove a molecular weight dependence,

$$h^* = h/n^{\frac{1}{2}(1+\varepsilon)} \tag{3.27}$$

where $1 + \varepsilon$ is the term in the exponent relating the end-to-end separation of the polymer coil to the molecular weight of, or number of subunits in, the coil

$$\langle r^2 \rangle \propto n^{1+\varepsilon}. \tag{3.28}$$

The eigenvalues for the free draining case, $(h^* = 0)$ are given exactly by

$$\lambda_j = 4 \sin^2\{\pi j/2(n + 1)\}. \tag{3.29}$$

Thurston and Morrison have tabulated exact eigenvalues for h^* of 0·01, 0·1,

0·3, and 0·4 for integral values of n from 1 to 15. Sufficient data are printed to allow interpolation. In addition approximate eigenvalues for n ranging between 20 and 100 are presented in graphical form.

The foregoing theories, with their emphasis on intrinsic viscosity, effectively relate the hydrodynamic modes of motion to polymer coil dimensions. In so doing they ignore first the effect of excluded volume in causing real dimensions to be greater than calculated using simple random flight models, and second that steric and other restrictions to internal motion can give rise to what is called 'internal viscosity'.[19]

The many calculations[20–5] of the excluded volume effects on dynamic properties have been oriented towards calculation of chain dimensions and intrinsic viscosity. However it is possible to obtain an expression in terms of Fresnel integrals which corresponds to (3.24) with the insertion of the index, ε, defined in (3.28)

$$\lambda_j' = (2\pi)^{\frac{1}{2}} 2^{-\delta} (\pi j)^\delta \{\pi j C_\delta(\pi j) - \delta S_\delta(\pi j)\} \tag{3.30}$$

where $\delta = \frac{1}{2}(1 + \varepsilon)$, and the Fresnel integrals are modified to

$$C_\delta(x) = (2\pi)^{-\frac{1}{2}} \int_0^x t^{-\delta} \cos t \, dt$$

$$S_\delta(x) = (2\pi)^{-\frac{1}{2}} \int_0^x t^{-\delta} \sin t \, dt. \tag{3.31}$$

It will be noted that eqns (3.30) and (3.31) reduce to eqns (3.24) and (3.25) respectively for $\varepsilon = 0$.

The neglect of internal viscosity has been considered in detail by Peterlin[26–8], but in general the error introduced in this way does not seem of major importance at low (or zero) shear rates.

It must be emphasized that the foregoing considerations apply only to isolated polymer chains, and so dilutions which are effectively infinite. Mathematical theories have not yet been totally successful in predicting the Huggins constant, k', in the viscosity-concentration equation,

$$\eta = \eta_0(1 + [\eta]c + k'[\eta]^2 c^2 + \cdots) \tag{3.32}$$

and the corresponding eigenvalues have not been used in the discussion of hydrodynamic mode relaxation times. In such cases the equations derived for infinite dilution are applied, with concentration entering the expression for the relaxation time only through the replacement of intrinsic viscosity by solvent and solution viscosities (first two terms of (3.32)). This constitutes quite a severe limitation in the comparison of theory and experiment since results are usually obtained from solutions of such concentration that the third term of (3.32) is significant.

3.4. Models of 'stiff' chains

In the preceding section the underlying assumption to all the mathematical considerations was that the chain was sufficiently flexible to form a number of subunits within which the end-to-end distance obeyed Gaussian statistics, and a sufficient number of conformations were available to create an entropic, perfectly elastic (energy quadratic in displacement) restoring force. For many real polymer molecules this is rather a severe requirement, and we now examine certain situations where it will not hold good.

3.4.1. The rigid rod

The extreme case of a non-flexible chain is provided by the perfectly straight rigid rod. The general diffusion equation[5,6] was first applied to this system by Kirkwood and Auer.[29] These authors retained the picture of a bead-like array, but disallowed all modes of motion except rotational degrees of freedom of the rod.

The insertion of a Brownian motion term and solution of the resulting integral equation yields a relaxation time for the end-over-end rotational motion in three dimensions,

$$\tau_{\text{rot}} = \pi\eta_0 L^3 / 18kT \ln(L/b) \qquad (3.33)$$

where η_0 is again solvent (continuum) viscosity, L is the rod length, and b is the spacing of the subgroups along the rod. For an array of contiguous beads, b is the bead diameter, and so can be considered as the rod diameter.

A rather similar expression has been obtained by Broersma,[30] who inserted a correction term, γ, to account for end-effects, and where b is now specifically a rod diameter or ellipsoid minor diameter. The unidimensional end-over-end motion is given by

$$\tau_{\text{e-o-e}} = \pi\eta_0 L^3 / 6kT\{2L/b) - \gamma\} \qquad (3.34)$$

$$\gamma = 1 \cdot 57 - 7 \left\{ \frac{1}{\ln(2L/b)} - 0 \cdot 28 \right\}^2 : \quad \text{cylinder}$$

$$\gamma = 0 \cdot 50: \quad \text{ellipsoid.}$$

Both of these derivations require the rod length to be considerably greater than the diameter. This will be so for chains of high molecular weight, but then these long chains will be beginning to exhibit curvature so that the equations for rods are again inapplicable, although consideration as a very prolate ellipsoid is a possibility.

When $L/b \gg 2$, these expressions differ only by a factor of 3. This has its origin in the way the coordinates of diffusion are introduced. In the derivation of (3.33) the rod is considered as rotating at random through the three dimensions of a coordinate system fixed in space–laboratory coordinates. On the other hand (3.34) is based upon rotation in a plane based on the rod major and minor axes–molecular coordinates. Consequently this latter equation represents a one-dimensional projection of the former.

The rotation of a prolate ellipsoid has also been investigated by Simha[31] and by Cerf and Scheraga,[32] who also suggest that the rotational relaxation time varies as somewhat less than the cube of the major diameter.

3.4.2. The 'broken rod' and 'worm-like' chains

There have been two theoretical approaches to the problem of very 'stiff' chains. The first has been to start with theories of perfect rods, and then to 'break' the rods into a small number of subrods connected by freely rotating links. The second has been to start at the other extreme of the flexible coil, and introduce factors that inhibit tight coiling of the chain, or which introduce a 'persistence' in the direction of the successive chain units.

The problem of the once-broken rod was first treated by Yu and Stockmayer,[33] who adapted the Kirkwood–Auer treatment for a single rigid rod to the case of two rods of equal length connected by a freely rotating universal joint. Their treatment was directed, principally, at an evaluation of intrinsic viscosity, but of course included an evaluation of the rotational diffusion coefficient, and so led to the rotational relaxation time and the real and imaginary components of the shear modulus. In fact it transpires that the rotational time is half that of the unbroken rod of same total length, while quantities such as the intrinsic viscosity are reduced by a factor of 0·85 and the mean radius of gyration by 0·63 of the values in the absence of the central joint. These two latter quantities agree well with experiments[34] made on a helical polymer grown from a bifunctional initiating molecule.

The model of a chain as a number of rods connected by 'universal joints' has been considered further by Gotlib,[35] with a view to formulating models for segmental rotation, as will be discussed later.

Treatments of the 'worm-like' chain have, in general, been directed more towards an understanding of those properties which depend on the time-averaged molecular dimensions than towards a consideration of damped internal modes of motion. Thus the parameters evaluated are usually end-to-end dimensions (and derived quantities such as dipole moments), whole molecule diffusion constants (both translational and rotational), intrinsic viscosities, and sedimentation coefficients. The calculations start with the concept of chain stiffness, and persistence in vector properties, as presented by Kratky and Porod[36] and Kuhn and Kuhn,[37] and have been reviewed in ref. 4.

3.5. Normal-mode models at high frequencies

An essential feature of the analyses presented in § 3.3 is the division of the chain into a number of Gaussian subunits. A corresponding assumption in the analyses of stiff chains in § 3.4 is that the effective major axis of the molecular shape should be much longer than the minor axis. The normal modes of motion of the flexible coil then define fundamental and overtone motions in which progressively smaller sections of the chain move in phase in the same direction. Obviously the credibility and applicability of this picture reaches a limit as the mode number approaches the number of subunits, and the section of the chain moving as an entity becomes smaller than the necessary subunit. The highest normal modes have the shortest relaxation times, and consequently are those of importance in observations made at high frequencies. Consequently, at these high frequencies, the relevant forces can no longer be described in terms of perfectly elastic beads constructed from freely rotating chain units, and the rates of motion become dependent upon chemical structural effects.

It can be seen that the problem is formally similar to that encountered[16] in short 'stiff' chains. In such macromolecules it may be necessary to include so many backbone atoms in the subunit in order to obtain Gaussian behaviour, that the required chain length may be greater than that of the actual molecule.

The physical significance of these high-frequency characteristics will be that the number of normal hydrodynamic modes effective at the observation frequency will be less than predicted by theory, so that there will not be the expected contributions to the viscoelastic behaviour. On the other hand physical observations will begin to detect contributions from the very localized, and structure-sensitive, movements which have been defined as segmental motion.

3.5.1. The 'damped torsional oscillator' model

One approach to this difficulty of matching the high frequency normal mode to the truly localized segmental behaviour is to revise the molecular model so as to base it upon a much smaller subunit. This is the technique adopted by Tobolsky,[38,39] in what he has called the 'damped torsional oscillator' model. In this representation the requirement that the subunit show a quadratic dependence of energy on a linear displacement is replaced by a quadratic dependence on torsional movement. Since such a potential can, in principle, be provided by the bonding and steric interactions in quite a short-chain section, the necessity of building a large number of bonds into a Gaussian unit is relaxed.

The mathematical derivation then follows an identical course to that used for the free-draining Rouse–Bueche model, and yields a comparable set of eigenvalues and relaxation times

$$\tau_j = \frac{\zeta z^2}{\pi^2 C j^2} \tag{3.35}$$

where j is again the mode number. In this instance the parameters differ from the bead-spring model in that:

(a) C is a torsional force constant which may be an order of magnitude or more greater than the entropic force constant.

(b) ζ is a friction factor with an origin in the barrier opposing rotation. It is thus totally intramolecular, as opposed to the totally hydrodynamic origin of the bead friction factor. In this case it is given the value $8\pi^2 m \nu_t \exp(H/RT)$ where H is the barrier height and ν_t is the torsional frequency at the bottom of the quadratic well. This results in a factor about a thousand times greater than that in the spring-bead representation.

(c) z is the number of subunits. This may be approximated to the number of discernible monomer units, in distinction to the bead-spring model where it is indeterminate and so has to be eliminated from equations for use in practical situations.

The relaxation time for the first normal mode of such a chain of damped torsional oscillators can then can be written as

$$\tau_1 = (2z^2/\pi^2 \nu_t) \exp(H/RT). \tag{3.36}$$

Tobolsky and DuPré have suggested[39] that consideration of experiments which observe such relaxation times should allow comparison of the relative advantages of the two representations. However, this is unlikely to be the case, since the Rouse–Zimm analysis is most meaningful in the low-frequency limit, whereas the damped torsional oscillator model, with its neglect of intermolecular viscous drag, can only be meaningful at higher mode numbers or for short chains. In this respect the two approaches are complementary rather than competitive. Although the authors are not aware of any detailed test of the damped torsional oscillator model, it does seem to be applicable to the motions causing acoustic absorption[40] in C_9 to C_{14} n-alkanes.

3.5.2. Further implications of internal viscosity

The basic equations of the Rouse–Bueche–Zimm theories predict that the dynamic intrinsic viscosity should become zero at infinite frequency,

$$[\eta'] = \frac{RT}{M\eta_s} \sum_i \{\tau_i/(1 + \omega^2 \tau_j^2)\} \tag{3.37}$$

$$[\eta''] = \frac{RT}{M\eta_s} \sum_j \{\tau_j^2/(1+\omega^2\tau_j^2)\}$$ (3.38)

where we now represent solvent viscosity by η_s and retain subscript zero to represent a value at zero frequency. These functions are illustrated graphically in Figs. 3.2 and 3.3.

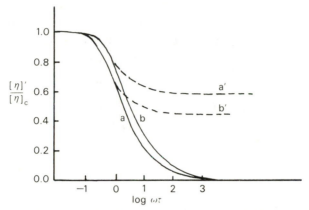

Fig. 3.2. Real intrinsic viscosity of chains without internal viscosity. Full lines: a, free-draining; b, non-free-draining. With internal viscosity, $\alpha = 0.5$. Broken lines: a', free-draining; b', non-free-draining.

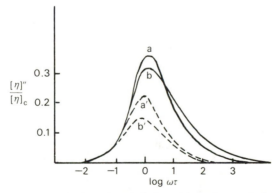

Fig. 3.3. Imaginary intrinsic viscosity of chains without internal viscosity. Full lines: a, free-draining; b, non-free-draining. With internal viscosity, $\alpha = 0.5$. Broken lines: a', free-draining; b', non-free-draining.

A first view of the effect of a finite number of beads will be that the dynamic viscosity decreases more rapidly than expected because of the absence of highest frequency modes, but that the effectively rigid collection of spheres will then exhibit non-relaxing viscosity[41,42] as is observed in suspensions of solid materials. However Peterlin[43] has suggested that energy dissipation in the non-flexible chain could be more important than the space-filling solid sphere contribution. In order to pursue this, the necklace model of Cerf[44] is used in the diffusion equations.

In the Cerf model the internal viscosity is assumed to be proportional to the rate of deformation of the chain rather than to the actual rate of movement of any particle. In other words, the internal viscosity introduces a new first-order differential term, which is the difference between the total viscosity and its rotational component. Solution[43] of the normal-mode matrices in the usual way yields for the complex intrinsic viscosity

$$[\eta^*] = \frac{RT}{M\eta_s} \sum_j \left\{ \tau_j \frac{1 + i\omega(\tau_j' - \tau_j)}{1 + i\omega\tau_j'} \right\} \qquad (3.39)$$

where the consideration of internal viscosity has introduced the new time constant, τ_j'. The real dynamic viscosity is given by

$$[\eta] - [\eta_\infty'] = \frac{RT}{M\eta_s} \sum_j \{\tau_j^2/\tau_j'(1 + \omega^2\tau_j'^2)\}. \qquad (3.40)$$

which differs from (3.37) (where $[\eta_\infty']$ is zero) formally by the factor τ_j/τ_j' and replacement of $\omega\tau_j$ by $\omega\tau_j'$. The imaginary dynamic viscosity is given by

$$[\eta''] = \frac{RT}{M\eta_s} \sum_j \{\omega\tau_j^2/(1 + \omega^2\tau_j'^2)\}. \qquad (3.41)$$

Importantly the infinite-frequency real intrinsic viscosity, $[\eta_\infty']$ is not zero

$$[\eta_\infty'] = \frac{RT}{M\eta_s} \sum_j \{\tau_j(\tau_j' - \tau_j)/\tau_j'\}. \qquad (3.42)$$

The time constants, τ_j and τ_j', are related through the expression

$$\tau_j' = (1 + \nu_j\phi_j/\zeta)\tau_j \qquad (3.43)$$

where ϕ_j are the diagonal elements of the internal viscosity tensor yielding eventually, eigenvalues ν_i. ζ is again the external friction coefficient.

Peterlin has listed the eigenvalues λ_j, ν_j for the free-draining and non-free-draining model, and evaluated the relaxing dynamic viscosity for various values of the internal/external friction factor, $\alpha = \phi/n\zeta$ where n is again the number of beads. The example of $\alpha = 0.5$ is illustrated in Figs. 3.2 and 3.3.

The treatment of Peterlin still requires the mode number, j, to be less than the number of subunits, n, and so will not be applicable to the highest frequencies. In addition the retention of the normal coordinate transformation in this situation has been questioned,[3] but no easily applied alternative transformation seems to be available.

A slightly different approach to the introduction of internal viscosity terms has been adopted by Caroli, Saint-James, and Jannink.[45] In this case the starting point is again the flexible chain of beads and springs, but instead of applying the Kirkwood diffusion equation the equations of motion are written in terms of coupled Langevin equations as described in the next chapter. This has the advantage that the internal damping coefficient can be introduced as a correlation in the forces acting on any bead. This correlation is assumed to decrease with distance from the reference unit, and so is essentially a short range effect. The mathematical approach involves writing[46] the Langevin equations of motion in Fourier space and then determining average properties by the use of the fluctuation–dissipation theorem. In this calculation the Fourier transform of the mean square displacement is evaluated in order to make comparison with neutron scattering measurements. The conclusion is that the frequency dependence of the scattering cross-section (see Chapter 12) is modified in the vicinity of the highest-mode frequency.

A related calculation with a most significant conclusion has been published by Pugh and MacInnes.[47] These authors start with the assumption that the covalent bonds and steric forces which give rise to 'internal viscosity' will cause strong coupling between local segments so that the motion of the chain must be a co-operative phenomenon. They then model the polymer by a continuous elastic space curve (so eliminating any bead-spring considerations) acted upon by the normal random, entropic and frictional forces and the internal viscosity opposing changes in curvature. Hydrodynamic interactions are neglected. Treatment of the Langevin equations of motion by Fourier transformation and application[48] of the fluctuation–dissipation theorem yields an expression for the number of modes in unit frequency range—effectively a density-of-states function.

The mode density function has a familiar Rouse–Zimm form at low frequencies, but has a unique characteristic at high frequencies when the internal viscosity term outweighs the external viscous drag. Under these circumstances the density function, $n(\omega)$, has a sharp peak whose shape and location are almost independent of molecular weight but depend on the ratio of the elastic and internal force constants. The authors suggest that it is these modes which are observed in dielectric and acoustic observations of localized segmental motion.

The importance of this calculation does not lie in the exact value of the result, since the model selected and the inclusion of hydrodynamic interactions can obviously be made more sophisticated. However it does illustrate that it is possible to have a molecular weight-independent peak in the mode spectrum of strongly coupled units, whereas hitherto the molecular weight independence had been assumed to indicate weak coupling of independent segmental motions.

References

1. Zimm, B. H. In *Rheology* (ed. F. R. Eirich), Vol. 3, p. 1. Academic Press, London and New York (1960).
2. Crank, J. *Mathematics of diffusion*. Oxford University Press, London (1956).
3. Fixman, M. and Stockmayer, W. H. *Ann. Rev. phys. Chem.* **21**, 407 (1970).
4. Yamakawa, H. *Modern Theory of Polymer Solutions*, Chapter 6. Harper and Row, New York, Evanston, San Francisco and London, (1971).
5. Kirkwood, J. G. and Riseman, J. *J. chem. Phys.* **16**, 565 (1948).
6. Kirkwood, J. G. *J. polym. Sci.* **12**, 1 (1954).
7. Rouse, P. E. *J. chem. Phys.* **21**, 1272 (1953).
8. Bueche, F. *J. chem. Phys.* **22**, 603 (1954).
9. Zimm, B. H. *J. chem. Phys.* **24**, 269 (1956).
10. Oseen, C. W. *Hydrodynamik*. Akademische Verlagsgesellschaft, Leipzig (1927).
11. Kirkwood, J. G., Zwanzig, R. W. and Plock, R. J. *J. chem. Phys.* **23**, 213 (1955).
12. Auer, P. L. and Gardner, C. S. *J. chem. Phys.* **23**, 1545 (1955).
13. Hearst, J. E. *J. chem. Phys.* **37**, 2547 (1962).
14. Tschoegl, N. W. *J. chem. Phys.* **39**, 149 (1963).
15. Pyun, C. W. and Fixman, M. *J. chem. Phys.* **42**, 3838 (1965); **44**, 2107 (1966).
16. Thurston, G. B. and Morrison, J. D. *Polymer* **10**, 421 (1969).
17. Zimm, B. H., Roe, G. M. and Epstein, L. F. *J. chem. Phys.* **24**, 279 (1956).
18. Thurston, G. B. and Peterlin, A. *J. chem. Phys.* **46**, 4881 (1967).
19. Kuhn, W. and Kuhn, H. *Helv. Chim. Acta* **28**, 97, 1533 (1945); **29**, 72 (1946).
20. Kurata, M. and Yamakawa, H. *J. chem. Phys.* **29**, 311 (1958).
21. Peterlin, A. *J. chem. Phys.* **23**, 2464 (1955).
22. Ptitsyn, O. B. and Eizner, Yu. E. *Zh. fiz. Khim.*, **32**, 2464 (1958).
23. Tschoegl, N. W. *J. chem. Phys.* **40**, 473 (1964).
24. Bloomfield, V. A. and Zimm, B. H. *J. chem. Phys.* **44**, 315 (1966).
25. Fixman, M. *J. chem. Phys.* **45**, 785, 793 (1966).
26. Peterlin, A. and Čopič, M. *J. appl. Phys.* **27**, 434 (1956).
27. Peterlin, A. *J. chem. Phys.* **33**, 1799 (1960).
28. Reinhold, C. and Peterlin, A. *J. chem. Phys.* **44**, 4333 (1966).
29. Kirkwood, J. G. and Auer, P. L. *J. chem. Phys.* **19**, 281 (1951).
30. Broersma, S. *J. chem. Phys.* **32**, 1626 (1960).
31. Simha, R. *J. phys. Chem.* **44**, 22 (1940).
32. Cerf, R. *Compt. Rend.*, **234**, 1549 (1952).
33. Yu, H. and Stockmayer, W. H. *J. chem. Phys.* **47**, 1369 (1967).
34. Teramoto, A., Yamashita, T. and Fujita, H. *J. chem. Phys.* **46**, 1919 (1967).
35. Gotlib, Yu. Ya. In *Relaxation phenomena in polymers* (ed. G. M. Bartenev and Yu. V. Zelenev), p. 242. Wiley, New York and Toronto (1974).
36. Kratky, O. and Porod, G. *Rec. Trav. Chim.*, **68**, 1106 (1949).
37. Kuhn, W. and Kuhn, H. *J. polym. Sci.* **5**, 519 (1950).
38. Tobolsky, A. V. and Aklonis, J. J. *J. phys. Chem.* **68**, 1970 (1964).
39. Tobolsky, A. V. and DuPré D. B. *J. polym. Sci.* **A-2**(6), 1177 (1968); *Advan. polym. Sci.* **6**, 103 (1969).
40. Cochran, M. A., Jones, P. B., North, A. M. and Pethrick, R. A. *J. chem. Soc., Faraday Trans. II* **68**, 1719 (1972).
41. Lamb, J. and Matheson, A. J. *Proc. Roy. Soc. (Lond.)* **A281**, 207 (1964).
42. Ferry, J. D., Holmes, L. A., Lamb, J. and Matheson, A. J. *J. chem. Phys.* **70**, 1685 (1966).

43. Peterlin, A. *J. polym. Sci. A-2* **5**, 179 (1967).
44. Cerf, R. *Fortschr. Hochpolymer. Forsch.*, **1**, 382 (1959).
45. Caroli, C., Saint-James, D. and Jannink, G. *J. polym. Sci., A-2* **11**, 2467 (1973).
46. See, for example, Whittaker, E. T. *Analytical dynamics*. Cambridge University Press, London (1937).
47. Pugh, D. and MacInnes, D. A. *Chem. phys. Lett.* **34**, 139 (1975).
48. Edwards, S. F. and Freed, K. F. *J. chem. Phys.* **61**, 1189 (1974).

STATISTICAL THEORIES OF POLYMER MOLECULAR MOTION

4.1. Introduction

This discussion of the statistical treatments of polymer molecular motion divides naturally into two sections. The first section concerns development of theories of the molecular movement of a polymer chain or chains and the second gives an exposition of theoretical expressions, particularly time-correlation functions, descriptive of the behaviour of molecules subject to the perturbation or probe of some observation technique. Since the objective of many experimental techniques is to relate macroscopic observations to molecular behaviour, the most sophisticated procedure is to search for ways in which these two theoretical approaches can be linked, so that evaluation of the one leads to information on the other. In this respect, this chapter is the link between the two sections, mainly theoretical and mainly phenomenological, of the book.

Spherically symmetric organic molecules can exhibit a plastic mesophase, one characteristic of which is whole molecule rotation. However, polymer molecules, although they adopt an approximately spherical outline overall are usually so entangled that such whole-molecule rotational motion is virtually non-existent in the solid phase. Also, in the melt whole-molecule rotational motion makes a very small contribution to the relaxation spectrum, although it does become increasingly more important as the concentration is decreased.

Thus the most important motions of polymers have their origin in the conformation changes of the chain segments, as we saw in Chapter 1. The principal difference between the relaxation spectrum of an isolated polymer molecule and that exhibited by simple, substituted ethane molecules involves co-operative motion.[1,2] It is the theoretical description of this which is the ultimate objective of the analyses presented in this chapter. As we have seen earlier, the local conformational changes, which in small molecules appear as internal rotation,[3-8] in a polymer give rise to 'local' and 'normal' mode motions. Whereas the former are sensitive to the nature of the intramolecular potential energy profile, the latter are independent of chemical structure and are purely a function of the molecular weight.[9,10] The role of co-operative motion can be expressed in the comparison that normal-mode motions are *phase-coherent* distortions of the whole polymer molecule (Fig. 4.1) whereas the so-called segmental or local motions are *phase-incoherent* distortions of small elements of the polymer backbone.

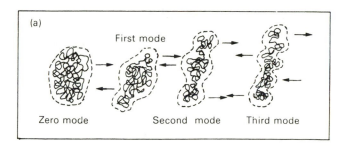

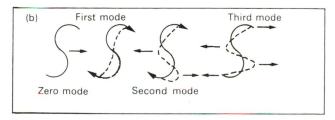

Fig. 4.1. Normal-mode relaxation of a polymer molecule. The zeroth mode corresponds to simple diffusion of the whole molecule, higher modes correspond to a co-operative diffusion of one part of the molecule relative to the remainder. For the first mode there is one node produced, higher order modes lead to a corresponding increase in the number of nodes.

The long-range normal-mode motions can be described in terms of simple bead and spring models and they can be modelled mathematically using the concepts of continuum mechanics as introduced in the last chapter. However, it is also possible to start from the rotational isomerism of each segment and build in the co-operative phase coherence, and it is this approach that we consider next.

4.2. Rotational isomerism in polymeric materials

The energy changes associated with rotational isomerism are best understood by considering the case of *gauche–trans* isomerism in 1,2-disubstituted ethane (Fig. 4.2). At thermal equilibrium the populations of the *gauche* and *trans* isomers are defined in terms of a Boltzmann

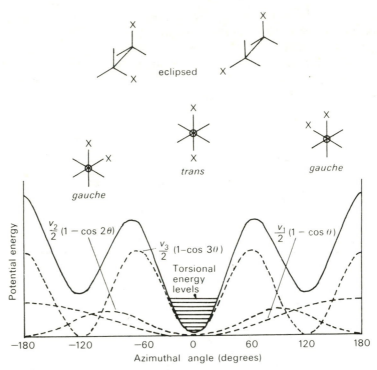

Fig. 4.2. Rotational isomerism in 1,2-disubstituted ethane. The azimuthal angle is defined as the angle subtended between the projection of the one C—X bond in the plane of the other C—X bond. For the *trans* isomer the angle equals zero; for the *gauche* isomer it equals 120 degrees. The dotted line corresponds to the component terms which when summed describe the total potential energy profile associated with the rotational isomerization process.

distribution function, the energy difference between states being determined by a combination of non-bonding van der Waals and dipolar interactions. The activation energy associated with the rotational isomeric process is determined by the magnitude of the non-bonding interactions in the eclipsed states, due allowance being made for the effects of distortion on the molecular structure in achieving such transition states.[5–7] Polymers with the general structure: $-CXY-CH_2-$ can exhibit configurational as well as conformational isomerism (Fig. 4.3). Studies of stereo-specific polymers indicate the dramatic effects which changes in the precise stereochemistry can have on the conformational energetics and rates of internal rotation (Table 4.1).

The rotational isomeric approach has had its greatest success in predicting the thermodynamic properties of semi-flexible polymers in dilute solu-

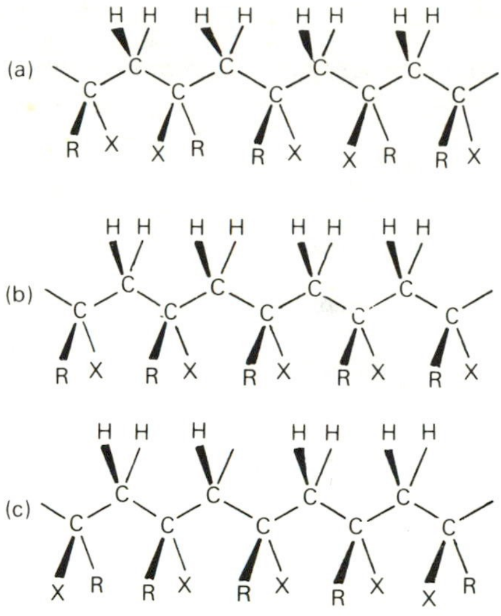

R–CH_3, X–CO_2CH_3 : Poly (methylmethacrylate)

R–CH_3, X–C_6H_5 : Poly (α-methylstyrene)

Fig. 4.3. Structural isomerism in vinyl polymers. —CH_2—CXY— tacticity in poly(methylmethacrylate). (a) Syndiotactic; (b) isotactic; (c) atactic.

tion.[1,2] Attempts to model the total dynamic spectrum of a polymer molecule starting from rotational isomeric models have been less successful, although they do provide a useful point from which to discuss the time-dependent properties of polymers.

4.2.1. Mechanism of internal rotation

At equilibrium a polymer molecule will experience a continuous bombardment from solvent molecules. Collisions may be either elastic or inelastic. Elastic collisions occur when solvent molecules collide with a polymer and momentum is conserved. These will produce either whole-molecule rotation, normal-mode motion, or translation (Fig. 4.4). Inelastic collisions involve conversion of part of the translational energy into internal energy of the polymer either as increased vibrational energy or as population of higher conformational states. Exchange of energy between internal vibrational states and the translational energy reservoir gives rise to vibrational relaxation. Preferential activation of the lowest vibrational states,

Table 4.1

Activation energy of rotational isomerism[2]

Polymer	Solvent	Activation energy (kJ mol^{-1})		
		Effective	Viscous	Internal
p-chlorostyrene	toluene	22·6	8·7	13.4
methylmethacrylate	toluene	28·0	8.7	19·4
	dioxane	30·1	13·0	20·0
	tetraline	47·6	15·9	21·7
methylacrylate	toluene	26·5	8·7	21·8
butylmethacrylate	toluene	22·6	8·7	14·4
	dioxan	20·9	13·8	7·1
	chloroform	15·0	6·7	8·3
	1,2-dichloroethane	17·6	9·2	8·4
butylmethacrylate (isotactic)	toluene	22·5	8·7	13·8
(syndiotactic)		24·6	8·7	15·9
ethylmethacrylate	toluene	24·2	8·7	15·4
ethylene				4–6
ethyleneterephthalate				4–6
oxymethylene				6
methylmethacrylate (isotactic)				17
(syndiotactic)				20
methylacrylate (isotactic)				7
(syndiotactic)				3
propylene (isotactic)				4–5
(syndiotactic)				3·4

which in the case of a polymer are torsional oscillations, provides the mechanism for rotational isomerism.

4.2.2. *Internal rotation and the over-all dynamics of a polymer chain*

Two types of relaxation can be identified and associated with the local and normal mode motions. Whether or not both are observed depends upon the relative magnitude of the intramolecular rotational potential and the molecular weight. Also the precise definition of a 'local' mode depends on these parameters, and a wide variation in the number of monomer units involved in the local conformational change is possible. For example, less distortion and coupling is to be expected in rotation about the Si—O bond in poly(dimethylsiloxane) than in the shorter and more hindered C—C bonds of polystyrene.[11] This concept of co-operative 'local' relaxation implies that in polystyrene even segmental motion may involve up to about ten monomer units. Of course, the extreme limit of co-operative behaviour is that of a rigid rod, the intra-segment barrier to rotation being so high that conformational change is very unlikely and the dynamic spectrum is reduced to relaxation via whole molecule motion.

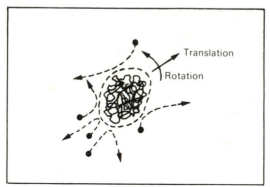

Elastic collisions

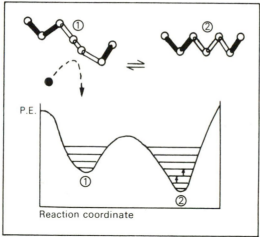

Inelastic collisions

Fig. 4.4. Inelastic and elastic collisions between solvent molecules and polymer. The dotted lines in the above diagram indicate the trajectories followed by the solvent molecules during their collisions with the polymer. The heavy arrows in the upper diagram indicate the directions of over-all rotation and translation of the polymer. The lower diagram depicts a conformational change accompanying an inelastic collision between solvent and a section of the polymer chain. The arrows between energy levels for conformation (2) depict torsional transitions of the chain backbone and correspond to the element units of energy changed in the collisional process.

4.2.3. Theoretical models connecting normal-mode and internal rotational relaxation[12-14]

In order to demonstrate the intimate connection between normal-mode motion and conformational change in a macromolecule we shall consider the behaviour of a polymer constructed from $N + 1$ beads, joined by N links (bonds or groups of monomers) each of length b. The direction of the bonds from bead $i - 1$ to bead i is given by the unit vector $\mathbf{a}_i$. The pattern of interaction between the ith and $(i + s)$th bond in the chain changes firstly discretely with increasing s, and then becomes independent of it. If the spatial position vector of the zeroth bead is $\mathbf{r}_0$, the location and conformation of the chain is then specified by the set of $N + 1$ vectors $\mathbf{r}_0$, $\mathbf{a}_1$, $\mathbf{a}_2$, $\mathbf{a}_3$, ..., $\mathbf{a}_N$ (Fig. 4.5).

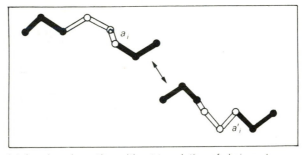

(a) Local mode motion without translation of chain ends

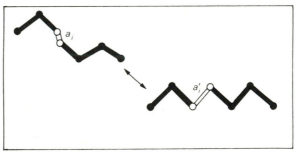

(b) Local mode motion with translation of chain ends

Fig. 4.5. Two possible forms of local backbone motion of a polymer chain. Motion in each case is achieved by a series of bond flips. The usual constraint applied to such conformational changes is that the motion involves the interchange of closely connected vectors maintaining constant the vector sum, as in (a).

The chain conformation will vary as a function of time due to the random activation of high-energy conformations, the beads being considered to move one at a time. Realistic chain or rotational relaxation models are generally too complicated to use in the development of theories and certain of the inadequacies which arise from the use of an ideal chain are attributed to the neglect of excluded volume effects. The simplest model of the motion of interior beads ($i \neq 0$ or N) consists of a jump or 'flip' whereby the vectors $\mathbf{a}_i$ and $\mathbf{a}_{i+1}$ exchange values (Fig. 4.5). In order to differentiate between the bond vectors before and after flip, primes are used for the latter, and the process corresponds to conservation of $(\mathbf{a}_i + \mathbf{a}_{i+1})$. It must be emphasized that such an approximation is only valid for the ideally flexible polymer chain, and real chains may involve simultaneous activation of a number of chain units. In this simplified model the chain is assumed to be sufficiently long for the effects of end-groups to be neglected; then the probability that the chain has a conformation $\{\mathbf{a}_1, \mathbf{a}_2, \mathbf{a}_3, \ldots, \mathbf{a}_N\} = \{\mathbf{a}^N\}$ may be designated as $p\{\mathbf{a}^N, t\}$. The dynamic experiment is concerned with the probability, $w_i dt$, that bead i will execute a flip during the time interval dt to form a new conformation which may be considered in terms of a master equation.[12-14]

$$dp(\mathbf{a}^N, t)/dt = p(\mathbf{a}^N, t)_i w_i + w_i p (\cdots \mathbf{a}_{i+1}, \mathbf{a}_i, \ldots, t). \tag{4.1}$$

If the end-groups were involved the above equation would require a somewhat more complex form. For a freely jointed chain the flip probability may be written as

$$w_i = (1 - g \mathbf{a}_i \cdot \mathbf{a}_{i+1}) \tag{4.2}$$

where $|g| \leq 1$, which permits weighting the mobility of a bead according to its instantaneous bond angle. The over-all effect of such weighting on a three-dimensional chain would be simply to increase the rates of all the flips by a factor $1 + (g/3)$. The chain may also be expected to show a weak correlation due to the effects of intramolecular potential on the nearest link motions. This correlation may be given the form

$$\langle \mathbf{a}_i \cdot \mathbf{a}_{i+1} \rangle = \beta \tag{4.3}$$

and is related to the equilibrium mean square length of the chain by

$$\langle r^2 \rangle_{eq} = Nb^2 (1 + \beta)/(1 - \beta). \tag{4.4}$$

The above correlation is a first step towards the consideration of a real polymer where specific angular geometries may be expected to be favoured, and allows particular distortions of groups of bonds to be identified with the observation of correlated motion.

In application of the above formulism it is assumed that β is small, restricting the analysis to semi-flexible chains. Using (4.3), and including a flip frequency†, α, the flip probability for a bead may be written as

$$w_i = \alpha (1 + 3\beta \mathbf{a}_{i-1} \cdot \mathbf{a}_{i+1} + 3\beta \mathbf{a}_i \cdot \mathbf{a}_{i+2}). \tag{4.5}$$

---

† The flip frequency in this type of derivation is not given any absolute physical significance. For the simpler polymers it may be associated with the frequency of segmental conformational change.

In an experimental observation the motion of a polymer chain is observed as time-averaged changes in the conformation represented by an average value of one bond vector, $\mathbf{q}_i(t)$. To obtain this we define[13] the probability $p(\mathbf{a}^N, t)$ that at time t the chain has the conformation $\{\mathbf{a}_1, \mathbf{a}_2, \ldots, \mathbf{a}_N\} = \{\mathbf{a}^N\}$. Then the time-averaged changes are expressed in terms of the reference bond and its two neighbours as

$$\mathbf{q}_i(t) = \langle \mathbf{a}_i(t) \rangle = \int \ldots \int_{\text{all } \{\mathbf{a}^N\}} \mathbf{a}_i p(\mathbf{a}^N, t) \, \mathrm{d}\{\mathbf{a}^N\} \tag{4.6}$$

$$\mathrm{d}\mathbf{q}_i/\mathrm{d}t = \langle \mathbf{a}_{i-1} w(\mathbf{a}_{i-1} . \mathbf{a}_i) + \mathbf{a}_{i+1} w(\mathbf{a}_i . \mathbf{a}_{i+1}) - \mathbf{a}_i w(\mathbf{a}_{i-1} . \mathbf{a}_i)$$
$$- \mathbf{a}_i w(\mathbf{a}_i . \mathbf{a}_{i+1}) \rangle \tag{4.7}$$

where the averages are over the non-equilibrium† ensemble of N bond vectors. Introducing the flip probability yields

$$\mathrm{d}\mathbf{q}_i/\mathrm{d}t = -2\mathbf{a}_i + \mathbf{a}_{i-1} + 3[\langle \mathbf{a}_{i-1}(\mathbf{a}_{i-2} . \mathbf{a}_i) \rangle + \langle \mathbf{a}_{i-1}(\mathbf{a}_{i-1} \cdot \mathbf{a}_{i+1}) \rangle$$
$$+ \langle \mathbf{a}_{i+1}(\mathbf{a}_{i-1} . \mathbf{a}_{i+1}) \rangle + \langle \mathbf{a}_{i-1}(\mathbf{a}_i . \mathbf{a}_{i+2}) \rangle$$
$$- \langle \mathbf{a}_i(\mathbf{a}_{i-2} . \mathbf{a}_i) \rangle - \langle \mathbf{a}_i(\mathbf{a}_i . \mathbf{a}_{i+2}) \rangle - 2\langle \mathbf{a}_i(\mathbf{a}_{i-1} . \mathbf{a}_{i+1}) \rangle]. \tag{4.8}$$

If all terms in $O(\beta^2)$ are neglected, the triple correlations in (4.8) are those of a freely jointed chain and may be evaluated[13] in terms of linear averages. The final result, obtained by Stockmayer, is

$$\alpha^{-i} \, \mathrm{d}\mathbf{q}_i/\mathrm{d}t = -2\mathbf{q}_i + (1+\beta)(\mathbf{q}_{i-1} + \mathbf{q}_{i+1}) - \beta(\mathbf{q}_{i-2} + \mathbf{q}_{i+2}) \tag{4.9}$$

which can be written as

$$\alpha^{-1} \, \mathrm{d}\mathbf{q}/\mathrm{d}t = -\mathbf{q}^{-1} . \mathbf{B} . \mathbf{q} \tag{4.10}$$

where $\mathbf{q}$ is a column vector whose elements are all the $\mathbf{q}_i$, and $\mathbf{B}$ is a square $N \times N$ matrix, of which the elements

$$\mathbf{B}_{jj} = 2\mathbf{B}_{ij} = -(1+\beta) \quad \text{for } i = j-1$$
$$= \beta \quad \text{for } i = j+2 \tag{4.11}$$
$$= 0 \quad \text{otherwise.}$$

The matrix, $\mathbf{B}$ is diagonalized by the transformation $\mathbf{Q}^{-1} . \mathbf{B} . \mathbf{Q} = \lambda_p$, where

$$\mathbf{Q}_{ip} = (2/N)^{\frac{1}{2}} \sin(ip\pi/N) \tag{4.12}$$

and the eigenvalues of $\mathbf{B}$ are

$$\lambda_p = 4(1 - 3\beta) \sin^2(p\pi/2N) + 16 \sin^4(p\pi/2N) \tag{4.13}$$

---

† The choice of initial states need not correspond to an equilibrium conformation of the molecule. Experiments such as viscoelastic relaxation induce states which deviate appreciably from those of an equilibrium conformation.

assuming that the value of N is large and that any distinction between N and $N+1$ can be ignored. It is usual to consider the motion of the chain in terms of the normal coordinates $X_p(t)$ which are related to the average chain coordinates and form a column vector A defined by the orthogonal transformation

$$A = Q \cdot q. \tag{4.14}$$

The time dependence, or relaxation, of such modes may be assumed to have the form

$$A_p(t) = A_p(0) \exp(-t/\tau'_p) \tag{4.15}$$

where

$$\tau'_p = \tau_p/\alpha. \tag{4.16}$$

The above analysis indicates that for the first few (slowest) normal modes ($p \ll N$) the relaxation spectrum of the freely jointed chain possesses values of τ_p which are proportional to p^2; however deviations from this relationship will occur for the faster modes. The ratio of the longest to the shortest relaxation time is given by

$$\tau_1/N^2\tau_N = 4(1+4\beta)/\pi^2 \tag{4.17}$$

and the over-all relaxation spectrum is broadened as the energetic preference for the extended conformation and thus the correlation, β, are increased.

In this approach the time dependence of r_0, translational diffusion, has been neglected. The mean square displacement per unit of the centre of mass as a result of the random perturbation of the beads may be described in terms of the translational diffusion coefficient D,

$$6D = \sum_{i=0}^{N} \frac{w_i(\Delta r'_i)^2}{(N+1)^2} \tag{4.18}$$

in which the displacement of bead i at flip is

$$\Delta r_i = b(a'_i - a_i) = b(a_{i+1} - a'_{i+1}) \tag{4.19}$$

which can be shown to equal

$$D = b^2(1 - \beta/N) = kT/N\xi \tag{4.20}$$

where the last quantity illustrates the connection between the basic flip frequency, α, and the effective friction coefficient, ξ, of a bead in the chain. It leads to the result for slow relaxation that

$$1/\tau'_p = 3\pi^2 p^2 D/\langle r^2 \rangle. \tag{4.21}$$

This indicates that the normal-mode and segmental processes are intimately connected, although it predicts the existence of only one relaxation process. Deviations at high frequency from the predictions of a simple bead and spring model are usually attributed to the failure of random flight statistics. Inclusion of these deviations appears to require the introduction of a specific chemical structure or localized torsional effects. The effects of torsional vibration and internal rotation are often lumped together and termed 'internal viscosity'.[15–18]

Experimentally it is found that the dynamic spectrum of a polymer in solution is composed of two distinct regions:

(i) long-range or normal-mode relaxation—defined purely in terms of the molecular weight;

(ii) local or segmental relaxation, sensitive to changes in chemical structure.

Whether or not both of these processes are detected depends upon the relative magnitude of the barrier to internal rotation and the molecular weight of the polymer. For example, in polystyrene (M_n 100 000) these processes partially overlap one another (Fig. 4.6) whereas in poly(dimethylsiloxane) of the same molecular weight the processes are completely resolved.

It has been shown recently[19] that two relaxations can be predicted from a modified form of the bead and spring model by considering the spectrum to be generated from the simple model plus torsional and conformational excited states. These latter mix with the over-all mode spectrum and give rise to specific branches. In the simple case of a semi-flexible polymer the relaxation spectrum exhibits two branches. The detail of this approach may be expected to demonstrate the effects of bond distortion in the production of coupling of conformational transitions.

4.2.4. Relationship between viscosity and effective barrier to internal rotation

Local motions of segments of the polymer chain can be described in terms of a combination of torsional-vibrational motions and rotational conformation change. The effect of solvent viscosity on these two processes is somewhat different and provides a means of characterizing the observed relaxation process. The relaxation time for torsional-vibrational motion is proportional to τ_0, the rotational diffusion time of a chain link. The effective relaxation time for such motions will depend on the local viscosity of the medium, η_{loc}, and a factor $(\kappa/kT)^a$ where κ is the torsional force constant. 'a' varies between 1 and 3 depending upon the model employed, being between 1 and 3 for pure transverse motion and equal to 3 for longitudinal motion. The effective activation energy for such a relaxation process is given

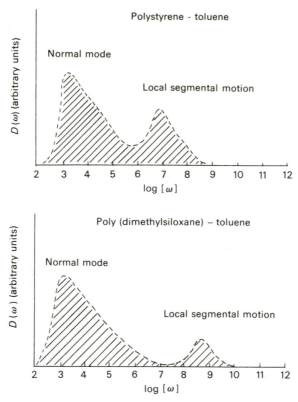

Fig. 4.6. Distribution of relaxation processes for a flexible and semi-flexible polymer. The normal-mode process is observed as a viscoelastic loss and can be described by a superposition of a series of relaxation processes, the amplitude of each decreasing as the mode number-frequency is increased. The segmental processes usually appear as a single relaxation process although the detailed conformational changes involved may not be identical for every part of the polymer chain.

by

$$U_{\text{eff}} = U_{\eta} + (1 + a)RT. \tag{4.22}$$

This model does not include a term which specifically describes the internal rotational process and so does not lead to the prediction of a specific resolved high-frequency relaxation process.[20,21] It has, however, been used to provide a realistic description of the way in which the local viscosity and the internal restrictions to free motion modify the high-frequency tail of the normal-mode process.

Alternatively, approaching the segmental motion as a conformational change, the relaxation may be described by a modified rate equation[22-5]

$$\tau \propto \tau_0 \exp(U_{\text{int}}/kT)g(T, U) \tag{4.23}$$

where $g(T, U)$ depends on the detail of the mechanism of internal rotation and involves consideration of terms involving the shape of the barrier, the number of degrees of freedom in which energy is stored and frequently has the form $(U/kT)^n$. A more detailed discussion of the form of this function is to be found elsewhere,[8,26,27]

It is found experimentally (Table 4.1) that U_{eff} is independent of temperature for a number of polymers. It is determined by a combination of an internal contribution and the local viscosity

$$U_{eff} = U_\eta + U_{int}. \tag{4.24}$$

Experimental values of U_{int} for many systems vary by less than 10 per cent for a wide variety of solvents (Table 4.1).

For polymers which do not contain a CH_3 group in the alpha position, a comparison of this theory with experiment indicates that the most probable mechanism for relaxation is a jump over a single energy barrier. The geometric constraints give rise to a minimum size for the rotational unit of five or six bonds. It is further observed that in polymers with side groups the rotation of the backbone is often weakly coupled to the motion of the side chain.

4.3. Correlation theory approach to molecular relaxation

The work of Green[28] and Kubo[29] in the 1950s began the development of a non-equilibrium statistical description of transport processes and other time-dependent phenomena. Using this approach phenomenological coefficients can be formulated. These are called time-correlation functions, and their integrals over time describe the transport and dynamics of the ensemble. These correlation functions play a similar role in non-equilibrium statistical mechanics to that played by partition functions in equilibrium statistical mechanics.[30]

Before considering the application of this approach to specific dynamic observations, it is desirable to consider briefly a time-correlation function in its most general form. A particle can have its energy and position defined in terms of two functions, momentum, $\mathbf{p}(t)$, and spatial coordinate, $\mathbf{q}(t)$. At some initial time, $t = 0$, the value of the phase-space coordinates will be $\mathbf{p} = \mathbf{p}(0)$ and $\mathbf{q} = \mathbf{q}(0)$. The values of $\mathbf{p}(t)$ and $\mathbf{q}(t)$ are related to the values of $\mathbf{p}$ and $\mathbf{q}$ through the equations of motion of the system. To emphasize this, we write

$$\mathbf{p}(t) = \mathbf{p}(\mathbf{p}, \mathbf{q}; t)$$
$$\mathbf{q}(t) = \mathbf{q}(\mathbf{p}, \mathbf{q}; t). \tag{4.25}$$

Let $A\{\mathbf{p}(t), \mathbf{q}(t)\}$ be some function of the phase-space coordinates, then

$$A\{\mathbf{p}(t), \mathbf{q}(t)\} = A(\mathbf{p}, \mathbf{q}; t) = A(t) \tag{4.26}$$

and define a classical time auto-correlation function of $A(t)$ by

$$C(t) = \langle A(0) \cdot A(t) \rangle = \int \cdots \int A(\mathbf{p}, \mathbf{q}; 0) \cdot A(\mathbf{p}, \mathbf{q}; t) f(\mathbf{p}, \mathbf{q}) \, d\mathbf{p} \, d\mathbf{q}$$

(4.27)

where $f(\mathbf{p}, \mathbf{q})$ is the equilibrium phase-space distribution function and where $d\mathbf{p}d\mathbf{q}$ stands for $d\mathbf{p}_1, d\mathbf{p}_2, d\mathbf{p}_3 \cdots d\mathbf{p}_N, d\mathbf{q}_1, d\mathbf{q}_2, d\mathbf{q}_3, \cdots d\mathbf{q}_N$. For simplicity, it is usual to consider the motion of a particular molecule or group relative to its own position, as self- or auto-correlation, and its motion relative to its neighbours as cross-correlation. Thus the appropriate auto-correlation function for the velocity of a single molecule would have the form

$$C(t) = \langle \mathbf{v}(0) \cdot \mathbf{v}(t) \rangle$$

(4.28)

where $\mathbf{v}(0)$ and $\mathbf{v}(t)$ are respectively the initial and instantaneous values of the velocity after time t. It must be noted, of course, that for a polymer, since $\mathbf{v}(t)$ depends upon the momentum and position of all the bonds, an explicit calculation of this function is obviously very difficult.

For most processes $C(t) \to 0$ as $t \to \infty$, and the correlation time τ_0 can be defined as[31]

$$\tau' = dt/d \ln(C(t)/C(0)).$$

(4.29)

For the special case of an exponential correlation function $C(t) = C(0) \exp(-t/\tau')$ and $\tau_0 = \tau'$. For a distribution of correlation times $p(\tau')$,

$$C(t) = C(0) \int p(\tau') \exp(t/\tau') \quad \text{and} \quad \tau_0 = \langle \tau' \rangle = \int p(\tau')\tau' \, d\tau'.$$

4.3.1. Correlation approach to rotational isomerism

As was pointed out previously, certain relaxation features in polymers closely resemble the simple rotational isomeric transitions studied in small molecules. Many models have been proposed to deal with rotational isomerism. The simplest of these assumes that only two states need be considered and that the rate of exchange between states is controlled by a simple rate equation of the form discussed in § 4.2.4. This approach leads to a simple Debye relaxation. In polymers we have to consider the finite probability that the rotational isomeric process involves correlation of the motion of one unit with either that of its neighbours or with the motion of a side-group, and this may be observed as an increase in the breadth of the relaxation spectrum.

Rotational isomerism in polymers can be modelled more realistically in terms of multiple-site exchange, which we shall illustrate by considering a four-site system[31,32] (Fig. 4.7). Site 1 is of depth $(E_1 + E_2)$ and the other wells are of depth E_2. The barrier over which the chain segments must pass is equal to E_2 in either direction for the exchange 2 to 3, and 3 to 4, and in one

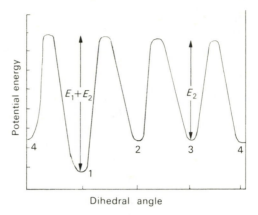

Fig. 4.7. Multiple-site relaxation model. Wells 2, 3, 4 are assumed to be essentially isoenergetic and Well 1 is assumed to be of lower energy.

direction only from 2 to 1 and 4 to 1. The barrier from 1 to 2 or 4 equals $E_1 + E_2$. This type of system is close to that found in real polymers. The transition probabilities are

$$P_{12} = P_{14} = P = A \exp(-(E_1 + E_2)/RT) \tag{4.30}$$

$$P_{41} = P_{21} = P_{23} = P_{32} = P_{34} = P' = A \exp -(E_2/RT). \tag{4.31}$$

If N_i is the population of site i then the rate equations (if only single jumps are allowed) are:

$$dN_1/t = -2PN_1 + P'N_2 + P'N_4$$

$$dN_2/t = PN_1 - 2P'N_2 + P'N_3$$

$$dN_3/t = P'N_2 - 2P'N_3 + P'N_4$$

$$dN_4/t = PN_1 + P'N_3 - 2P'N_4. \tag{4.32}$$

The eigenvalues of the operator (d/dt) in the characteristic determinant of eqs (4.32) are 0, $-2P'$, $-(2P' + P - \sqrt{Q})$ and $-(2P' + P + \sqrt{Q})$ where Q is $P'^2 + (P' - P)^2$. The relaxation times associated with this model are $\tau_2 = 1/2P'$ $\tau_3 = 1/(2P' + P - \sqrt{Q})$, $\tau_4 = 1/(2P' + P + \sqrt{Q})$. The model leads to multiple relaxation times which, at very high temperatures, approximate to a single relaxation model.

4.3.2. Correlation theory approach to the multiple-barrier problem

If the effects of inertia can be ignored, then the preceding formulism can be introduced into a generalized theory in terms of orientational time-correlation functions which are exponential, or are weighted sums of exponential functions.[33] In this approach the two-site model can be described in terms of a barrier separating the states (E) with a transfer probability of k. The occupational probabilities $p_1(t)$, $p_2(t)$ for sites 1 and 2 are given by

$$dp_1(t)/dt = -kp_1(t) + kp_2(t) : dp_2(t)/dt = kp_1(t) - kp_2(t). \quad (4.33)$$

The above may be written in matrix form, when the various $p_i(t)$ are represented by the matrix $\mathbf{p}(t)$, and

$$d\mathbf{p}(t)/dt = \mathbf{T} \cdot \mathbf{p}(t) \quad (4.34)$$

of which the general solution is[34]

$$\mathbf{p}(t) = [\exp \mathbf{T}t] \, \mathbf{p}(0) = \mathbf{S}[\exp \mathbf{D}t] \, \mathbf{S}^{-} \, \mathbf{p}(0) \quad (4.35)$$

where $\mathbf{S}$ is the column matrix of the $p_i(t)$.

$$\mathbf{T} = \begin{bmatrix} -k & +k \\ +k & -k \end{bmatrix} \mathbf{S}$$

is the matrix that performs the transformation $\mathbf{S}^{-1} \cdot \mathbf{T} \cdot \mathbf{S} = \mathbf{D}$ where $\mathbf{D}$ is a diagonal matrix. If $\mathbf{S}$ and $\mathbf{D}$ can be found, eqn (4.35) can be solved with the aid of the relation $\exp\{\text{diag } \lambda_m t\} = \{\text{diag}(\exp \lambda_m t)\}$. Solution of the problem can be aided for more complex site models if the matrix is simplified using a group theoretical approach. For example, in the two-site case, we have

$$\mathbf{S} = \frac{1}{\sqrt{2}} \begin{bmatrix} 1 & 1 \\ 1 & -1 \end{bmatrix}; \qquad \mathbf{D} = \begin{bmatrix} 0 & 0 \\ 0 & 2k \end{bmatrix} \quad (4.36)$$

$$\mathbf{p}(t) = \begin{bmatrix} p_1(t) \\ p_2(t) \end{bmatrix} = \frac{1}{2} \begin{bmatrix} 1 + \chi_2(t), & 1 - \chi_2(t) \\ 1 - \chi_2(t), & 1 + \chi_2(t) \end{bmatrix} \begin{bmatrix} p_1(0) \\ p_2(0) \end{bmatrix} \quad (4.37)$$

where $\chi_2(t) = \exp(-2kt)$. The application of this formulism to observed phenomena such as dipole orientation is discussed in § 4.5.

4.3.3. Correlation approach to normal-mode relaxation

In the preceding section the discrete exchange between two defined sites was modelled. However, correlation theory can also be used to model the relaxation described as continuous normal-mode motion. The bead and spring model is used, attention being focused on the coupling and interactions which occur between the solvent and polymer.

The position and conformation of a polymer chain is defined in terms of a distribution function $f(\mathbf{r}, t)$, at a time t, with coordinates of the $(N+1)$ beads being expressed by $\{\mathbf{r}_0, \mathbf{r}_1, \mathbf{r}_2 \cdots \mathbf{r}_N\}$. The symbol $\mathbf{r}$ denotes the column matrix whose elements are the cartesian coordinates of all the beads. The hydrodynamic interactions are introduced in terms of a column vector $\mathbf{v}_0$, which describes the fluid velocity of a bead with root mean square displacement length b, and friction coefficient, ξ. The bead is itself subjected to a potential energy constraint U_e due to the applied external field. The diffusion equation can then be constructed,[13] and has the form,

$$\frac{d}{dt}[f(\mathbf{r}, t)] = -\nabla f(\mathbf{r}, t) \cdot \mathbf{v}_0 + \nabla \left(\frac{kT}{\xi}\right) \mathbf{H}\left\{\nabla f(\mathbf{r}, t) + f(\mathbf{r}, t)\nabla \frac{U_e}{kT}\right.$$
$$\left. + 3b^{-2}f(\mathbf{r}, t)\mathbf{Ar}\right\} \tag{4.38}$$

where ∇ is the vector operator, grad, $\mathbf{H}$ is a hydrodynamic interaction matrix and $\mathbf{A}$ is a $N \times N$ matrix expressing the linear sequences of spring forces between bonds. The development and use of this equation really rests on an evaluation of the hydrodynamic interaction matrix. This can be done by introducing constraining forces opposing rotational interactions (internal viscosity). In such an example the hydrodynamic interaction is expressed as

$$\mathbf{H}_{jk} = \delta_{jk} + (1 - \delta_{jk}) \left(\frac{\pi\eta_0}{6}\right)\langle\mathbf{r}_{jk}^{-1}\rangle \tag{4.39}$$

$$= \delta_{jk} + (1 - \delta_{jk})(6\pi^3)^{\frac{1}{2}} b\eta_0|j - k|^{\frac{1}{2}} \tag{4.40}$$

where δ_{jk} is the Kronecker delta, $\mathbf{r}_{jk}$ is the bond vector, and η_0 is the solvent viscosity.

Quantities, such as solution viscosity, can now be written in terms of an appropriate correlation function

$$\eta = (1/6kT) \int_0^\infty \langle J^{xy}(t) \cdot J^{xy}(0)\rangle \exp(-i\omega t)\, dt \tag{4.41}$$

in which

$$J^{xy} = \sum_i m_i \dot{x}_i \dot{y}_i + \sum_i y_i F_i(x) \tag{4.42}$$

the summation being over all particles in the volume V. The ith particle has mass m_i, velocity components $\dot{x}_i$ and $\dot{y}_i$ and is acted on by $F_i(x)$, the x component of the force, F. The time-correlation function is over an equilibrium ensemble and the viscosity corresponds to that of an oscillating shear wave of circular frequency ω.

When considering the motion of the polymer, it is only necessary to compute the difference in J between the solution and the pure solvent. It should, however, be remembered that in certain situations the 'structure' of the solvent may be significantly altered by the presence of a polymer and whilst the above approximation is valid for consideration of shear relaxation, it is not adequate for treatment of the equilibrium thermodynamic properties of a polymer in solution.

It follows from the separation of the time-correlation functions for solvent and solution that

$$[\eta] = (N_A/M\eta_0 kT) \int_0^\infty \langle \Delta J^{xy}(t) \cdot J^{xy}(0) \rangle \exp(-i\omega t) \, dt \qquad (4.43)$$

where N_A is Avogadro's constant, M is the molecular weight of the polymer, and η_0 is again the solvent viscosity. Physically the integral is concerned with the mutual interactions among the elements of a single chain of the polymer. It is possible to define J^{xy} alternatively

$$J^{xy} = \sum_i y_i \{F_i'(x) + F_i''(x)\} \qquad (4.44)$$

where the sum is over all elements of a polymer chain and the two forces are called 'diffusion' and 'bond interaction' contributions respectively. The former is related directly to the viscosity and may be written as

$$F_i'(x) = -kT(d \ln f(\mathbf{r} \cdot t)/dx_i) \qquad (4.45)$$

while the second is

$$F_i''(x) = dU_{int}/dx_i \qquad (4.46)$$

in which U_{int} is the intramolecular potential energy. In the limit of field-free zero flow the 'diffusion' term vanishes and

$$\langle J^{xy}(t) \cdot J^{xy}(0) \rangle = (kT)^2 \sum_p \exp(-t/\tau_p) \qquad (4.47)$$

in which the relaxation times τ_p are given by

$$\tau_p = b^2 \xi / 6kT\lambda_p. \qquad (4.48)$$

Thus this treatment provides a method of bridging the gap between experimental observation and theoretical modelling of the long-range motions executed by the polymer.

4.4. Further models of chain motion

4.4.1. Internal motion as movement on a three-dimensional lattice

Attempts have been made to model the motions of a polymer more accurately by assuming that the backbone occupies a lattice structure, and that conformational changes correspond to movements on this lattice. If the bond angles are fixed at 90° we have a cubic lattice whereas if the angles are 109°24' we have a tetrahedral lattice. Three rotational isomeric states are possible—0°, +90°, and −90° for the cubic lattice and 0°, +120°, and −120° for the tetrahedral lattice (t, g, g'). The left- and right-hand rotational isomers (g and g' respectively) have the same statistical weighting. The matrix **B** then takes the following form[35-8]

$$\mathbf{B}_{kn} = \mathbf{B}_{|k-n|} = C_1 \Lambda_1^{|k-n|} + C_2 \Lambda_2^{|k-n|} \qquad (4.49)$$

$$\Lambda_{12} = \frac{\cos \beta (1 - \eta)}{2} \pm \sqrt{\left\{\frac{\cos^2 \beta (1 - \eta)^2 + 4\eta}{4}\right\}} \qquad (4.50)$$

$$C_{12} = \frac{\sqrt{\{(1 - \eta)^2 \cos^2 \beta + 4\eta\}} \pm (1 + \eta) \cos \beta}{2\sqrt{\{\cos^2 \beta (1 - \eta)^2 + 4\eta\}}} \qquad (4.51)$$

where η is the average change of the internal rotation angle in the chain and β is the bond angle. The rotational isomeric process is illustrated in Fig. 4.8.

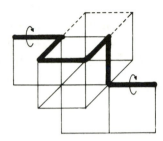

Cubic lattice

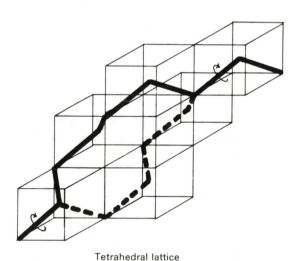

Tetrahedral lattice

Fig. 4.8. Relaxation using lattice dynamics—ideal model structures.

The mobility tensor $\mathbf{T}_{kl}$ can be obtained if the specific structure of the kinetic unit is known. This, of course, is where the molecular modelling is an approximation. Units of three and four bonds have been considered,[39,40] and often the minimum is considered as five monomers! However, the composite movement of large numbers of bonds can be resolved into equivalent three- or four-bond interchanges.

The general form of the mobility tensor is

$$\mathbf{T}_{kl} = \frac{b\bar{\omega}}{kT} \left\{ C_0 \delta_{k,l} + \sum_{m=1}^{d-l} C_m (\delta_{k,l-m} + \delta_{k,l+m}) \right\}$$ (4.52)

where b is a model constant, d is the number of monomer units in a kinetic unit, the coefficients C_m are related by the condition $C_0 + 2 \sum_{m=1}^{d-l} C_m = 0$ (which reflects the fact that the ends of the kinetic units are assumed to remain fixed during the transition), and δ_{kl} is the Kronecker delta. $\bar{\omega}$ is the mean frequency of the rotational isomeric transitions.

The mobility tensor for a cubic lattice has the form

$$\mathbf{T}_{k,l} = \frac{b_1 \bar{\omega}}{kT} \{ 2\delta_{k,l} (\delta_{k,l-3} + \delta_{k,l+3}) \}$$ (4.53)

and for the tetrahedral lattice

$$\mathbf{T}_{k,l} = \frac{b_2 \bar{\omega}}{kT} \{ 6\delta_{k,l} - 2(\delta_{k,l-2} + \delta_{k,l+2}) - (\delta_{k,l-4} + \delta_{k,l+4}) \}$$ (4.54)

where b_1, b_2 are constants characteristic of the model.

This lattice theory approach has provided experimentalists with a useful method for modelling observations such as dielectric, magnetic, and luminescence relaxation which are sensitive to the orientation of a relaxating vector in an applied field.[39,40]

4.4.2. Molecular dynamic modelling of polymer motions

Molecular dynamic modelling of molecular processes, although of necessity a rather idealistic approach, can bring some insight into the nature of the motions which give rise to the experimentally observed relaxation processes.[41-53]

Consider a system containing N particles, each having a pair interaction potential and enclosed in a limited volume. The coordinates and velocities of all the particles are defined at an initial time $t = 0$. The system is then allowed to change with time, the particles obeying the simple laws of mechanics,

$$d\mathbf{v}_i(t)/dt = \frac{1}{m_i} \sum_{j=i}^{N} F_{ij}(\mathbf{r}_i(t), \mathbf{r}_j(t))$$ (4.55)

where m_i, r_i and v_i are respectively the mass, coordinates, and velocity of the ith particle; $F_{ij}(\mathbf{r}_i(t), \mathbf{r}_j(t))$ is the time and separation-dependent force of interaction between the ith and jth particles. The variation of the coordinates and velocities of the particles during a time interval can provide information on both the equilibrium and non-equilibrium properties of the system. The coordinates and velocities at time t can be obtained by integrating the system of differential equations represented by (4.55). For instance, the diffusion coefficient, D, can be computed from the time dependence of the mean particle displacement

$$\langle \Delta \mathbf{r}^2(t) \rangle = \frac{1}{N} \sum_{i=1}^{N} \{ \mathbf{r}_i(t) - \mathbf{r}_i(0) \}^2. \tag{4.56}$$

In studies of polymers the problem is simplified by assuming that the particles are restricted to a lattice. A further restriction which can be placed on the calculation is that of volume exclusion (two elements of the same chain may not occupy the same junction points). Using this approach calculations of the time dependence of the mean square distance between chain ends have been made.[54-63] Such studies provide a means of predicting[63-70] the temperature coefficients of translational diffusion and chain dimensions.

The principal difficulty with the molecular dynamic approach is that whilst it appears to provide a means of calculating the conformation of a chain it is not readily extended to the study of dynamics. The problem can be summarized as follows: the interval of time over which the equations are integrated is of the order of 10^{-14} s whereas most relaxation processes have characteristic relaxation times of the order of 10^{-9} to 10^{-8} s and therefore would require the evaluation of approximately 10^7 integrals, each of which would require the computation of approximately 10^8 numerical integrals. Such calculations still require a prohibitive amount of time. To overcome this the polymer is approximated by a model similar to that outlined in § 4.2.3. The random conformational changes are assumed to be produced as a result of thermal fluctuations induced by the solvent. This type of modelling is still in its infancy and whilst it provides a visual representation of the changes which occur in a polymer it does not at present provide the possibility of understanding how the short-range and long-range motions arise from the same basic molecular processes.

4.4.3. Hydrodynamic screening and polymer–polymer interactions

The models outlined in this chapter assumed that only interactions between polymer 'beads' within the same chain need be considered.[71,72] This condition is realized in solutions only for concentrations, c, which obey the limit $[\eta]c \leqslant 1$, where $[\eta]$ is the intrinsic viscosity. At higher concentrations, the conformational changes of chain segments in one polymer cause solvent motions to be non-averaged before encountering segments of a second polymer molecule. These memory effects are referred to as hydrodynamic interactions and become severe when the concentration of polymer approximates to the spherical packing volume at $[\eta]c \simeq 4$. The nature and extent of the interchain interactions depend upon the molecular weight, concentration, and intrinsic flexibility. Above this limit the polymer chains undergo some degree of interpenetration. The precise form of the entangled network which is formed depends on the nature of the chemical interactions between the polymer chains, the flexibility of the chains, and the thermodynamic properties of the solvent. The network formed at concentrations in excess of $[\eta]c \simeq 10$ gives rise to viscoelastic properties.

Scaling theories have been used to characterize the changes which occcur with temperature, solvent, and concentration in the size of the polymer.[73-80]

In dilute solution the limiting case of zero hydrodynamic interaction, wherein the chain units do not perturb the local solvent velocity, has been extensively explored.[81-4] The Oseen tensor which arises from the hydrodynamic interaction between moving segments can be expressed in the relationship

$$\mathbf{v}_j' = \sum_i \mathbf{T}_{ji} \xi (\mathbf{v}_i - \mathbf{v}_i') + \mathbf{v}_j^0. \tag{4.57}$$

According to this approximation the velocity of the fluid at the position of the jth segment ($\mathbf{v}_j'$) is equal to the perturbed velocity ($\mathbf{v}_j^0$) plus contributions from all the other segments, i, proportional to their velocity relative to the solvent ($\mathbf{v}_i - \mathbf{v}_i'$). ξ is the hydrodynamic friction coefficient. The Oseen tensor is given as

$$\mathbf{T}_{ji} = \frac{1}{8\pi\eta_0} \left(\mathbf{I}_i + \frac{\mathbf{r}_{ij}\mathbf{r}_{ij}}{r_{ij}^2} \right) \tag{4.58}$$

where η_0 is the solvent viscosity, $\mathbf{r}_{ij}$ is the vector distance between the ith and jth segments and $\mathbf{I}_i$ is the unit matrix. It has been pointed out that the form of the Oseen tensor leads to singularities in the physical properties of the solution when the hydrodynamic interaction strength grows beyond a certain limit but these are not observed in practice. This topic has been discussed in detail and the errors introduced by the pre-averaging of the Oseen tensor considered.[82,84] For non-zero hydrodynamic interactions the lower frequency relaxation of a polymer can be described by the introduction of a single constant as the hydrodynamic interaction parameter.[85-96]

4.4.4. Excluded volume effects

As suggested first by Flory,[97] long chain polymers have dimensions which are not proportional to N, but may be described by

$$\langle r^2 \rangle \propto N^\delta \tag{4.59}$$

with $\delta > 1$. For theoretical treatments, the effects may be introduced into random-flight chain models composed of N effective bonds of length, a, in such a way that the effect of short-range interferences are absorbed into a, and the effective excluded volume between segments is expressed in terms of the binary cluster integral. Within the framework of the 'two-parameter theory' for long chains, the expansion factor is defined by the ratio of $\langle r^2 \rangle^{\frac{1}{2}}$ to its unperturbed value $\langle r_0^2 \rangle^{\frac{1}{2}}$ and depends only on the excluded volume parameter

$$z = (3/2\pi a^2)^{\frac{3}{2}} \beta N^{\frac{3}{2}} \tag{4.60}$$

where β is 'a parameter characteristic of the chain-type'.

The first statistical-mechanical calculation of the effect used perturbation theory. It is based on the Mayer cluster expansion, which allows one to write the expansion coefficient, α, as

$$\alpha^2 = 1 + \sum_{k=1}^{\infty} (-1)^{k+1} C_k z^k \tag{4.61}$$

where C_k is the kth coefficient of a virial expansion.

Higher-order coefficients in C have been proposed. However, all relationships have a common asymptotic form

$$\lim \alpha^\nu = Cz \qquad z \to \infty \tag{4.62}$$

with $1 \le \nu \le 5$. The lower limit, $\nu = 1$, corresponds to the upper limit $\delta = 2$ in eqn (4.59), and represents the behaviour of a rigid rod. Most theories[98,99] may be classified into two groups with $\nu = 3$ and 5 (corresponding to $\delta = 4/3$ and 6/5 respectively). There is evidence, although not decisive, to support the latter figures, first suggested by Flory.[97]

The most sophisticated approach to the problem of finding the asymptotic solution is the self-consistent field (SCF) theory first developed by Edwards[100-2] which basically also supports the conclusions of Flory.[97]

More detailed discussions of these topics are to be found in a number of specialized reviews.[81,83,84]

4.5. The connection between bulk physical observation and molecular correlation theory

In the preceding sections a number of correlation functions have been discussed in relation to various aspects of polymer molecular motion. However, information on the dynamic properties of macromolecules comes from observation of bulk properties. As we shall see below, interpretation of the observations in terms of molecular properties can be achieved only if the correct forms of the correlation functions are used. This is illustrated by the efforts made in recent years towards rationalizing the apparent discrepancies found in observations made using different techniques.

In many cases description of the 'movement' of a particle requires consideration of rotation, translation, and internal motions of the molecule.

Clearly the associated correlation function cannot be formulated in a simple fashion and it is the complexity of the many-body problem which leads to major uncertainty in the correct choice of model to be used. For instance, a diatomic molecule in the gas phase may be described by a combination of equations for the vibrational and whole-molecule rotational motion, when good agreement between experiment and theory is obtained. However, the introduction of a degree of internal freedom, even in a molecule such as ethane, leads to a much more complex problem, with no simple solution. This difficulty can be alleviated to some extent by the choice of a molecular probe to sense a selected feature of the motion. For instance, the orientation of a molecular dipole is vectorial in nature, and so the correlation function appropriate for macroscopic dielectric relaxation requires inclusion only of terms appropriate to the time dependence of that particular vector.

One approach to the problem of stochastic motional phenomena is through a generalized scattering function for a 'probe' of wave vector $\mathbf{k}$,

$$S(\mathbf{Q}, \omega) = (2l+1)j^2(\mathbf{Q_p}) S_l(\omega) \tag{4.63}$$

where $\mathbf{Q}$ is the momentum transfer vector with a particular value, $\mathbf{Q_p}$, for a positional orientation $\mathbf{p}$, $S_l(\omega)$ is the Fourier transform of the relaxation function $F_l(t)$ with the initial condition $F_l(t=0)=1$ for all l and j. For a spherical molecule macroscopic rotational diffusion has the description

$$D_r\Delta G_s(\omega - \omega_0; t) = G_s(\omega - \omega_0; t)\, t \tag{4.64}$$

in which the orientational self-correlation function, $G_s(\omega - \omega_0; t)$, is related to the relaxation function by

$$F_l(t) = \exp(-l(l+1)D_r t) \tag{4.65}$$

and D_r is the rotational diffusion constant. In the case of a polymer this constant contains terms which reflect both the over-all (normal-mode) motion and the local segmental processes. There is a tendency for the faster motions to dominate the observed correlated functions and in long-chain polymers these are often segmental in nature. In the following sections we shall consider the form of these various functions in more detail. However, at this point it is worth following the analysis presented in (4.63) a little further.

If a molecule experiences a strong interaction from some surrounding field, the motion of the particle can be considered as occurring in the presence of internal friction ($\xi(t)$) which modifies the correlation function

$$F_l(t) = \exp\{-l(l+1)D_r[t - \{1 - \exp(\xi(t)/\xi_0)\}]\}. \tag{4.66}$$

The form of (4.66) will appear familar to the reader, since it is the type of equation used to describe kinetic changes in many simple molecular systems.

In the transcription of these equations to describe particular experiments it is necessary to apply selection rules descriptive of the interaction between the applied probe or field and the molecular response. In the case of dielectric and nuclear magnetic relaxation l has values of 1 and 2 respectively. The form of (4.66) is attractive since it contains one adjustable parameter ξ. In practice, the nature of the many-body potential results in a change of the form of the correlation function with the strength of the interaction potential, and the choice of a universally acceptable form of $\xi(t)$ is a test of the validity of a particular molecular model.

The many-body problem can often be reduced by making a simplifying definition of the correlation function. Two types of function are commonly applied to changes in molecular coordinates, the auto-correlation function which describes the motion of a vector relative to its own initial value and the cross-correlation function which describes motion of a vector relative to the initial value of some other defined vector. Hence if $C(t)$ describes the auto-correlation function of the dynamic variable, A, molecular time auto-correlation functions would have a form $\langle A_i(0) . A(t)\rangle$, particular examples being

$$\langle \mathbf{v}_i(0) . \mathbf{v}_i(t)\rangle, \langle \mathbf{r}_i(0) . \mathbf{r}_i(t)\rangle, \langle \boldsymbol{\mu}_i(0) . \boldsymbol{\mu}_i(t)\rangle$$

where $\mathbf{v}_i$ is the velocity, $\mathbf{r}_i$ is the position and $\boldsymbol{\mu}_i$ is the dipole moment of the reference molecule i. Cross-correlation functions between molecules have the form $\langle \mathbf{v}_i(0) . \mathbf{v}_j(t)\rangle$, $i \neq j$ and are important in the analysis of the behaviour of many systems.

In many experiments observation is made of the reorientational motion of a vector which is chosen so as to have a particular initial orientation. The motions of the vector are completely described by a conditional orientational distribution function $\Xi[\boldsymbol{\Omega}(t)|\boldsymbol{\Omega}_0(0)]$, where $\Xi[\boldsymbol{\Omega}(t)|\boldsymbol{\Omega}_0(0)]\partial\Omega\partial\Omega_0$ is the probability that the unit vector is in the element of solid angle $\partial\Omega$ around $\boldsymbol{\Omega}$ at time t given that it was in $\partial\Omega_0$ around $\boldsymbol{\Omega}_0$ at $t = 0$. This function may be expanded in terms of spherical harmonics, $Y_{lm}(\boldsymbol{\Omega})$, and time correlation functions $\phi_{lm}(t)$ for angular motion so that

$$\Xi[\boldsymbol{\Omega}(t)|\boldsymbol{\Omega}_0(0)] = \sum_l \sum_m Y^*_{lm}(\boldsymbol{\Omega}) . Y_{lm}(\boldsymbol{\Omega}_0)\phi_{lm}(t) \tag{4.67}$$

$$\phi_{lm}(t) = \int_l \int_m Y_{lm*}(\boldsymbol{\Omega}_0) Y_{lm}(\boldsymbol{\Omega}) \Xi\Omega(t)[\boldsymbol{\Omega}_0(0)] \, \partial\Omega \, \partial\Omega_0$$

$$= \langle Y_{lm*}[\boldsymbol{\Omega}(0)] Y_{lm}[\boldsymbol{\Omega}(t)]\rangle. \tag{4.68}$$

For the special case where reorientation occurs with axial symmetry from an initial orientation $\boldsymbol{\Omega}_0$, $\Xi[\boldsymbol{\Omega}(t)|\boldsymbol{\Omega}_0(0)]$ becomes a function of the angle γ

between the orientations Ω and Ω_0. Eqns (4.67) and (4.68) become

$$\Xi[\Omega(t)|\Omega_0(0)] = \frac{1}{4\pi} \sum_{l=0}^{\infty} \{(2l+1)P_l \, (\cos \gamma) \, \psi_l(t)\} \qquad (4.69)$$

$$\psi_l(t) = \int \int P_l \, (\cos \gamma) \Xi[\Omega(t)|\Omega_0(0)] \partial\Omega \partial\Omega_0$$
$$= \langle P_l[\cos \gamma(t)] \rangle. \qquad (4.70)$$

The values P_l, the coefficients of the Legendre polynomial, depend on the experiment being considered and the constraints imposed on the motions which relax the vector quantity involved.

4.5.1. Dipole reorientational motion [142]

In order to illustrate the application of correlation to a particular experiment we start with dielectric relaxation since this provides the most direct connection between motion and the relaxation process.

Consider a system of N equivalent dipoles to which is applied a field $\mathbf{E}_z^N(t')$. The local field acting on the dipole $\mathbf{E}_z(t')$ is related to $\mathbf{E}_z^N(t')$ though $\mathbf{E}_z(t') = \mathbf{E}_z^N(t')((\varepsilon + 2)/3)$. The phase-space distribution function, $f(p, q)$, for N equivalent dipoles obeys the Liouville equation[103-5]

$$\frac{df}{dt} + \sum_i^N \left[\frac{dH}{dp_i} \frac{df}{dq_i} - \frac{dH}{dq_i} \frac{df}{dp_i} \right] = \frac{df}{dt} + \mathscr{L}f = 0 \qquad (4.71)$$

where H is the hamiltonian of the system, i refers to dipole i, and $\mathscr{L}$ is the Liouville operator: $\mathscr{L}f = \{f, H\} = -\{H, f\}$. In the presence of the applied field

$$H(p, q: t) = H_0(p, q) + \mathbf{M}_z(q) \, \mathbf{E}_z(t) \qquad (4.72)$$

where $-\mathbf{M}_z(q)\mathbf{E}_z(t) = -\sum_i \mathbf{m}_i(q) \, \mathbf{E}_z(t)$ is the energy of interaction between the dipole and the electric field. Expanding the Liouville operator, $\mathscr{L} = \mathscr{L}_0 + \mathscr{L}$ and noting that $\mathbf{M}_z(q)$ is independent of momenta p_i,

$$\mathscr{L}_0 = \sum_i^N \left\{ \frac{dH_0}{dp_i} \frac{d}{dq_i} - \frac{dH_0}{dq_i} \frac{d}{dp_i} \right\} \qquad (4.73)$$

$$\mathscr{L}_1 = \mathbf{E}_z(t) \sum_i^N \frac{d\mathbf{M}_z}{dq_i} \frac{d}{dp_i} \qquad (4.74)$$

which leads to

$$\frac{df_0}{dt} = -\mathscr{L}_0 f_0 \qquad \frac{df_1}{dt} = \{-\mathscr{L}_0 f_1 + \mathscr{L}_1 f_0\} \qquad (4.75)$$

and $\mathscr{L}_1 f_1$ is omitted from eqn (4.75) in order that $f_1 = O(\mathbf{E}_z)$. The solution to

(4.75) is the field-free Boltzmann relation

$$f_0 = A \exp -\beta H_0$$

where $\beta = (kT)^{-1}$ and A is a constant. This leads to

$$f_1(t) = \int_{-\infty}^{t} \{\exp -(t-t')\mathscr{L}_0\}\mathscr{L}_1 f_0 \, dt' \qquad (4.76)$$

The average moment in the field direction is

$$\begin{aligned}\langle \mathbf{M}_z(t) \rangle &= \iint \mathbf{M}_z(q) \cdot f(p, q, t) \\ &= -\iint dp \, dq \int_{-\infty}^{t} \mathbf{M}_z(q) \, (\exp -(t-t') \, \mathscr{L}_0)\mathscr{L}_1 \, dt' \qquad (4.77)\end{aligned}$$

which leads to

$$\langle \mathbf{M}_z(t) \rangle = \beta \int_{\infty}^{t} dt' \, \mathbf{E}_z(t) \iint \mathbf{M}_z(q) \, (\exp -(t-t') \, \mathscr{L}_0/f_0 \mathbf{M}_z \, dp \, dq. \qquad (4.78)$$

Since integrals of the form $\iint \mathscr{L}_0 f_0 \, dp \, dq$ vanish, this leads to

$$\langle \mathbf{M}_z(t) \rangle = \int_{-\infty}^{t} dt' \, \mathbf{E}(t') \iint f_0 \mathbf{M}_z \, (\exp(t-t') \, \mathscr{L}_0/\mathbf{M}_z(q) \, dp \, dq. \qquad (4.79)$$

The term $\exp(t-t') \, \mathscr{L}_0$ is the displacement operator describing the interaction of the solvent on the dipole which is undergoing reorientational relaxation. Then (4.79) can be rewritten as

$$\langle \mathbf{M}_z(t) \rangle = \beta \int_{-\infty}^{t} dt' \mathbf{E}(t') \iint f_0 \dot{\mathbf{M}}_z \mathbf{M}_z(t-t') \, dp \, dq. \qquad (4.80)$$

The field-free correlation function $\Phi_z(t)$, is defined as

$$\Phi_z(t) = \iint f_0 \mathbf{M}_z(0)\mathbf{M}_z(t) \, dp \, dq = \langle \mathbf{M}_z(0)\mathbf{M}_z(t) \rangle. \qquad (4.81)$$

Now

$$\dot{\Phi}_z(t) = \langle \mathbf{M}_z(0) \cdot \mathbf{M}_z(t) \rangle = -\langle \mathbf{M}_z(0)\mathbf{M}_z(t) \rangle. \qquad (4.82)$$

This leads to

$$\langle \mathbf{M}_z(t) \rangle = -\beta \int_{-\infty}^{t} \mathbf{E}_z(t')\dot{\Phi}_z(t-t') \, dt'. \qquad (4.83)$$

The total dipole moment correlation function $\Phi(t) = \langle \mathbf{M}(0) \cdot \mathbf{M}(t) \rangle / \langle \mathbf{M}(0) \cdot \mathbf{M}(0) \rangle$ where $\mathbf{M}(t) = \mathbf{M}_x(t) \, \mathbf{u}_x + \mathbf{M}_y(t)\mathbf{u}_y + \mathbf{M}_z(t)\mathbf{u}_z$ and $\mathbf{u}_x$, $\mathbf{u}_y$, and $\mathbf{u}_z$

are the unit vectors associated with the x, y, and z directions. This leads to

$$\mathbf{M}_z(t) = -\frac{\langle \mathbf{M}(0) \cdot \mathbf{M}(0) \rangle}{3kT} \int_{-\infty}^{t} \mathbf{E}_z(t') \Phi(t - t') \, dt' \tag{4.84}$$

There are three specific conditions of particular interest[142]:

(i) $\mathbf{E}_0$ is applied as a step at $t' = 0$

$$\mathbf{M}_z(t) = \frac{\langle \mathbf{M}(0) \cdot \mathbf{M}(0) \rangle}{3kT} \mathbf{E}_0 (1 - \Phi(t)); \tag{4.85}$$

(ii) $\mathbf{E}_0$ applied at $t = -\infty$ is removed as a step at $t' = 0$

$$\mathbf{M}_z(t) = \frac{\langle \mathbf{M}(0) \cdot \mathbf{M}(0) \rangle}{3kT} \mathbf{E}_0 \Phi(t) : t > 0; \tag{4.86}$$

(iii) Steady state: $\mathbf{E}(t') = \mathbf{E}_0 \exp(i\omega t')$: $-\infty < t < \infty$

$$\mathbf{M}_z(t) = \frac{\langle \mathbf{M}(0) \cdot \mathbf{M}(0) \rangle}{3kT} \mathbf{E}_0 \exp(i\omega t) \int_0^{\infty} \left[\frac{-\partial \Phi(t')}{\partial t} \right] \exp(= i\omega t) \, dt'$$

$$= \frac{\langle \mathbf{M}(0) \cdot \mathbf{M}(0) \rangle}{3kT} \mathbf{E}_0 \exp(i\omega t)(1 - i\omega)$$

$$\times \int_0^{\infty} \Phi(t') \exp(-i\omega t') \, dt'. \tag{4.87}$$

Since the electric polarization $\mathbf{P}_z(t) = N \langle \mathbf{M}_z(t) \rangle = (\varepsilon - \varepsilon_\infty) \mathbf{E}_z \kappa \varepsilon_v$, where ε_v is the permittivity of free space and κ is the internal field factor connecting the applied and local fields, in the frequency domain

$$\frac{K(\omega)}{K(0)} = \frac{(\varepsilon - \varepsilon_\infty)}{(\varepsilon_0 - \varepsilon_\infty)} = \mathscr{F}\left\{ -\frac{d}{dt} \Phi(t) \right\} \tag{4.88}$$

where $\mathscr{F}$ indicates the one-sided Fourier transform. For the simple case of a very dilute system of dipoles $\varepsilon_0 \approx \varepsilon_\infty$ and hence $K(\omega) \approx K(0)$, in which case (4.88) becomes

$$\frac{\varepsilon - \varepsilon_\infty}{\varepsilon_0 - \varepsilon_\infty} = \mathscr{F}\left\{ -\frac{d}{dt} \Phi(t) \right\} \approx 1. \qquad t \to 0 \tag{4.89}$$

If $\Phi(t) = \exp(t/\tau)$, the relaxation obeys the familiar expression for a single relaxation time. It can be shown, also, that the moment $\langle \mathbf{M}_z(t) \rangle$ is proportional to the applied field, the proportionality factor being the susceptibility which is determined by the equilibrium quantity $\langle \mathbf{M}(0) \cdot \mathbf{M}(0) \rangle / (3kT)$ and to a time-correlation function $\Phi(t)$ which in turn is defined by the motion of the solvent molecules in the system. The function $\Phi(t)$ contains the auto- and cross-correlation functions

$$\Phi(t) = \frac{\langle \sum_i^N \mu_i(0) \sum_i^N \mu_i(t) \rangle}{\sum_i^N \mu_i(0) \sum_i^N \mu_i(0)} + \frac{\langle \sum_i^N \sum_j^N \mu_i(0) \mu_j(t) \rangle}{\sum_i^N \sum_j^N \mu_i(0) \mu_j(0)}. \tag{4.90}$$

For the special case where cross-correlation terms, $\langle \mu_i(0) \mu_j(t) \rangle$, $i \neq j$, are zero, for identical dipoles $\Phi(t) = \langle \mu_i(0) \mu_i(t) \rangle / \langle \mu_i^2 \rangle = \langle P_1(\mu(t)) \rangle$. It will be apparent to the reader that, in principle, it is possible to obtain precise information on the motion of a specific atom or group within the perturbed system from a detailed analysis of relaxation data. For the dipole orientation relaxation, P_l corresponds to the first spherical harmonic, although other experiments probe higher harmonics.

The correlation function obtained using the above analysis must next be translated into a form which depicts the detailed molecular motion of the dipole being monitored. This can be done using a two-site model of dipole jump reorientation. The general time-correlation function, $\mu^2 \langle \mathbf{x}(0) . \mathbf{x}(t) \rangle$ may be expressed in terms of the decay functions for dipoles initially in sites 1 and 2 by $\xi_1(t)$ and $\xi_2(t)$. The non-normalized dipole vector time correlation function is given by

$$\langle \mu(0) . \mu(t) \rangle = \mu^2 \sum_{i=1}^{2} {}^0p_i \xi_i(t)$$

$$\xi_i(t) = \sum_{j=1}^{2} p_{ij}(t) \mu_i \mu_j \tag{4.91}$$

where ${}^0p_i$ is the equilibrium occupation probability of site i and the sum is taken over all sites. $p_{ij}(t)$ is the conditional probability that the dipole is in site j at t, given that it was in site i at $t = 0$. With $\mu_1 . \mu_1 = \mu_2 . \mu_2 = -\mu_1 . \mu_2$ the site interchange can be described by

$$\langle \mu(0) . \mu(t) \rangle = \mu^2 \exp(-2kt) \tag{4.92}$$

so that the correlation function is exponential in time with a relaxation time $(2k)^{-1}$. For more complicated interchange systems several relaxation times may arise. In addition $\langle \mu(0) . \mu(t) \rangle$ generally will not decay to zero for sites which are non-equivalent in energy.

Reorientational motion of a dipole may be constrained by the molecular structure to which it is attached. To describe such a situation it is usual to assume that the motion is associated with an appropriate lattice structure. This is often cubic or tetrahedral. The relaxation time of the longitudinal components of the individual dipole moment for any lattice model is:

$$\tau(\chi) = \frac{\tau^*(\chi)}{(1 - \cos \chi)} \frac{\langle M^2(\chi) \rangle}{N} \tag{4.93}$$

where $\tau^*(\chi)$ is a function of χ (specific for a model and type of kinetic unit) and $\mathbf{M}(\chi) = \sum_i \mathbf{e}_j \exp(\chi_{ij})$ where $\mathbf{e}_j$ is the vector of the jth element and N is the number of elements in the chain.

Application of this approach to a molecular system allows the discussion of the orientation of the relaxing vector relative to the direction of the molecular motion. Let us consider the dynamics of dipoles perpendicular to the chain backbone, i.e. motion associated with the transverse component of the individual dipole moments. These dipoles are directed along the vectors

$$\mathbf{f}_k = \frac{|\mathbf{e}_{k-1}, \mathbf{e}_k|}{|[\mathbf{e}_{k-1}, \mathbf{e}_k]|}, \tag{4.94}$$

where square brackets represent time averages.

The averaged projection of the transverse component of the dipole moment of the kth element on to the chosen direction of space $\mathbf{e}_0$ is expressed as a linear combination of the binary functions $f^{(k-1,k)}(\alpha_{k-1}, \alpha_k)$

$$m_k = \frac{\langle [\mathbf{e}_{k-1}, \mathbf{e}_k], \mathbf{e}_0 \rangle}{|[\mathbf{e}_{k-1}, \mathbf{e}_k]|} = \sum_{(\alpha_{k-1}, \alpha_k)} \frac{([\mathbf{e}_{k-1}, \mathbf{e}_k] \mathbf{e}_0)}{|[\mathbf{e}_{k-1}, \mathbf{e}_k]|} f^{k-1,k}(\alpha_{k-1}, \alpha_k). \tag{4.95}$$

The kinetic equations which come from these equations have the form:

$$\frac{\mathrm{d}}{\mathrm{d}t} \mathbf{m}_k = -\sum \boldsymbol{T}_{kl} C_{1s} \mathbf{m}_s. \tag{4.96}$$

The tensor $\boldsymbol{T}_{kl}$ is found as in the case of the longitudinal components by varying the transition frequencies in the exact kinetic equations: for the tetrahedral lattice

$$\boldsymbol{T}_{kl} = \frac{a\bar{\omega}}{kT} [12\delta_{k,l} - 4(\delta_{k,l-1} + \delta_{k,l+1}) - 4(\delta_{k,l-2} + \delta_{k,l+2})$$

$$+ 3(\delta_{k,l-3} + \delta_{k,l+3}) - 2(\delta_{k,l-4} + \delta_{k,l+4}) + (\delta_{k,l-5} + \delta_{k,l+5})] \tag{4.97}$$

with a reciprocal force constant tensor

$$\mathbf{B}_{s,l} \frac{\langle ([\mathbf{e}_{s-1}, \mathbf{e}_s], \mathbf{e}_0)([\mathbf{e}_{l-1}, \mathbf{e}_l] \mathbf{e}_0) \rangle}{[\mathbf{e}_{s-1}, \mathbf{e}_s][\mathbf{e}_{l+1}, \mathbf{e}_l]} = \frac{1}{3}(-\eta)^{|(l-s)|} \tag{4.98}$$

and a relaxation time for the transverse components

$$\tau(\chi) = \frac{kT\lambda_{\mathbf{B}}(\chi)}{\lambda_{\boldsymbol{T}}(\chi)} \tag{4.99}$$

The relaxation time of the longitudinal components tends to infinity in the case of in-phase excitation ($\chi = 0$), whatever the kinetic unit. The reason for this is that $\sum \mathbf{e}_k$ remains unchanged during a jump. For the transverse components $\sum \mathbf{e}_k$ changes during a jump. For example, in the transition (g',

g', g') → (g, g, g,) the eigenvalue of the matrix $\mathbf{T}$

$$\lambda_{\mathbf{T}} = \frac{a\bar{\omega}}{kT}\{6 + 5\cos\chi + 2\cos 2\chi + 3\cos 3\chi\} \qquad (4.100)$$

does not vanish at $\chi = 0$.

If different types of kinetic unit exist in a chain, and independent jumps occur singly,

$$\frac{\mathrm{d}m_k}{\mathrm{d}t} = \sum_p \sum_{l,s} \mathbf{T}_{kl} C_{ls} m_s \qquad (4.101)$$

where $\mathbf{T}_{kl}$ is the mobility tensor of the kinetic unit of type p. The corresponding relaxation time is

$$\tau(\chi) = \frac{k\mathbf{T}\lambda_B(\chi)}{\sum_p \lambda_{\mathbf{T}_p}(\chi)} \qquad (4.102)$$

and does not go to infinity at $\chi = 0$ if a non-vanishing $\lambda_T(0)$ exists for the kinetic unit.

The principles of this approach are general for a situation where a specific molecular vector can be identified and observed although the detailed theory is a little different for other vector reorientation such as nuclear magnetic spin, quadrupolar, and electron spin relaxation.

4.5.2. Hydrodynamic properties of a fluid

Molecular motions have a very significant effect on the hydrodynamic properties of a fluid. In Chapters 7 and 8 ultrasonic and viscoelastic relaxation phenomena will be discussed. Experimental techniques for studying these observe the simultaneous motions of collections of particles and thus interpretation presents a more complex problem than that associated with the 'direct' observation of a reference molecular dipole considered in the preceding section.

It is possible, using the Liouville or von Neumann equations, to describe the time-correlation functions for the transport coefficients of a molecular system. Then master equations can be developed from which the hydrodynamic equations, states of thermodynamic equilibrium, transport coefficients, and characteristic fluctuation and relaxation phenomena may be deduced.

In this approach,[106] the motion of a particle relative to the whole ensemble is described by the velocity auto-correlation function (Fig. 4.9),

$$\Gamma_v(t) = \frac{m}{3kT}\langle \mathbf{v}_i(0) \cdot \mathbf{v}_i(t)\rangle. \qquad (4.103)$$

The subscript, i, refers to the reference particle whose motion is studied, and

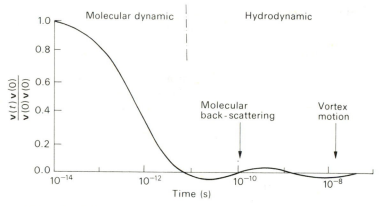

Fig. 4.9. Velocity auto-correlation function for a simple fluid.

the bracket implies an average over an equilibrium selection of initial times. When the time, t, is much larger than the mean collision time,[107] the hydrodynamic regime is reached and the description corresponds to a simple diffusion process. In this limit we can relate the diffusion coefficient to the position of the particle by

$$\langle \{\mathbf{r}_i(t) - \mathbf{r}_i(0)\}^2 \rangle = 6Dt \tag{4.104}$$

where D is the self-diffusion constant. It can be deduced that

$$D = \frac{kT}{m} \int_0^\infty \Gamma_v(t)\, \mathrm{d}t \tag{4.105}$$

which is the Einstein or Kubo formula for self-diffusion. The motion of a labelled particle, to a first approximation, can be described by a Langevin equation which has the form

$$\frac{\mathrm{d}\Gamma_v}{\mathrm{d}t} + \gamma\Gamma_v(t) = 0 \tag{4.106}$$

where γ is the friction coefficient. Taking the Laplace–Fourier transform

$$\Gamma_v(\omega) = \int_0^\infty \mathrm{e}^{-\mathrm{i}\omega t}\, \Gamma_v(t)\, \mathrm{d}t \tag{4.107}$$

leads to

$$\Gamma_v(\omega) = \frac{1}{\mathrm{i}\omega + \gamma}. \tag{4.108}$$

Inserting eqn (4.105) in order to satisfy the hydrodynamic limit, yields the

Stokes–Einstein relation

$$\gamma = \frac{kT}{mD}.$$ (4.109)

The velocity auto-correlation function corresponding to the Langevin equation has the form $e^{-\gamma t}$. This formulism is inconsistent with experiment, and rationalization is usually achieved by applying a frequency-dependent friction coefficient $\xi(\omega)$. This leads to a generalized form of the Langevin equation.[108]

$$\frac{d\Gamma_v}{dt} + \int_0^\infty \xi(t-\tau)\Gamma_v(\tau)\,d\tau = 0.$$ (4.110)

This equation includes description of the system at times prior to the time t and as a consequence $\xi(t)$ is often referred to as the memory function of the system. This artifact can be overcome if the form of $\Gamma_v(t)$ for small times is known and usually involves incorporation of a memory function of the form

$$\gamma(t) = \Omega_0^2 n(t)$$ (4.111)

where $n(0) = 1$, and Ω_0 is any operator.

If the self-motion of a particle does not induce appreciable collective motion, then the memory function extends only over a time interval of the order of the mean collision time. The assumption[109–11] that $n(t) = e^{-t/\tau}$, whilst leading to an analytical form for the function, fails to predict the long-time behaviour of the velocity auto-correlation function.

The effects of collective motions on the velocity auto-correlation function have been investigated by considering the motion of particles contained within a volume which is given an initial velocity relative to a static environment. A compression wave develops in front of this region and a rarefaction wave appears behind it. When the two waves have separated, the residual flow has the form of a double vortex in two dimensions or a vortex ring in three dimensions. The behaviour of the flow is described by a function of the form $1/t$ in two dimensions and $1/t^{\frac{3}{2}}$ in three dimensions. At long times the form of the correlation function must approach that of an incompressible fluid, and the flow velocity decays due to shear forces. Viscoelastic interactions in such a system are described by auto-correlation of the elements of the stress tensor.[112]

So the divergence between the acoustic properties predicted from simple theory and those observed in experiment arises because the principal decay time for the velocity auto-correlation function is of the order of 10^{-13} s, which is approximately the time for an acoustic wave to travel only a single molecular diameter. However the acoustic properties are intimately connected with the collective motions of the particles, and these motions are reflected only in the long-time tail of the auto-correlation function (Fig. 4.9). This point has been studied in detail.[113-18] Although some dispute exists as to the correct form of the long-time function, an expression of the form $At^{-\alpha}$, where $1\cdot34 \leqslant \alpha \leqslant 1\cdot65$. appears to provide a useful approximation.

In molecular terms the long-time behaviour corresponds to the reference particle attempting to push and drag its neighbours in its forward motion. The occurrence of a greater density of particles in front of the particle presents a barrier to forward motion, and can lead to its reversal. At a later stage the motion of the group of accompanying particles is itself reversed.

The spatial extent to which this correlated motion can occur is given by the Fourier transform of the self-current–current correlation function (whose $k = 0$ component is the velocity auto-correlation function). The current–current correlation function describes the motion of a small element of the total ensemble and hence reflects long time correlating effects of the total system. Examination of the total behaviour of a particle indicates that the two regimes considered above have quite different character. The short-time behaviour is simply described in terms of eqns (4.108)–(4.111) and the long-time hydrodynamic limit, when the motion becomes collective, is described by pseudo-continuum equations.

The spatial extent of the vortex motion[119] appears to be much larger than the volume of the group of particles originally involved in the collective motion. The long-time behaviour, described above and characterized by the slowly decaying tail in the correlation function, is reproduced rather well by a transient solution of the Navier–Stokes equation for times longer than ten mean collision times.[120,121]

4.5.3. Longitudinal wave propagation

A longitudinal wave may be considered in two parts, shear and compressional components.

Shear waves and viscosity. The shear viscosity can be calculated through the Kubo relation

$$\eta = \int_0^\infty \eta(t)\,\mathrm{d}t \tag{4.112}$$

with

$$\eta(t) = \frac{\rho}{kT} \langle \tau^{xz}(t)\tau^{xz}(0)\rangle \tag{4.113}$$

where $\tau^{xz}(t)$ is the xz component of the stress tensor at time t. Using the microscopic expressions for the stress tensor[122]

$$\tau^{xz} = \sum_{i=1}^{i=N}\left(mv_i^x v_i^z - \frac{1}{2}\sum_{j\neq i}\frac{x_{ij}z_{ij}}{r_{ij}}\frac{\mathrm{d}v(r)}{\mathrm{d}r_{ij}}\right), \tag{4.114}$$

one can calculate the function $\eta(t)$ from a knowledge of the short time scale motions of the particles. For a simple fluid the Maxwell viscoelastic theory can be assumed to apply and $\eta(t)$ may be approximated by a simple exponential

$$\eta(t) = G_\infty e^{-t/\tau} \tag{4.115}$$

where G_∞ is the infinite frequency shear modulus which corresponds to eqn (4.113) when τ is of the order of the mean collision time. From the previous discussion of the effects of collective wave motion $\eta(t)$ may be identified and expressed in terms of the transverse current correlation function and the wave vector $\mathbf{k}$,

$$C_t(\mathbf{k}, t) = \mathbf{k}^2 \left\langle \sum_{i=1}^{i=N} v_i^x(t) e^{-ikz_i(t)} \sum_{j=1}^{j=N} v_j^x(0) e^{-ikz_j(0)} \right\rangle. \tag{4.116}$$

The Fourier transform of (4.116) in the long-wavelength limit leads to

$$C_t(\mathbf{k}, \omega) = \frac{\mathbf{k}^2(kT/m)}{-i\omega + \eta(\omega)k^2/m}. \tag{4.117}$$

The usual hydrodynamic equations for damping of shear waves are then recovered in the low-frequency limit, and so

$$\eta(\omega) = G_\infty/(-i\omega + 1/\tau) \tag{4.118}$$

which at low $\mathbf{k}$ has the form of a diffusion equation, and at high $\mathbf{k}$ represents damped shear waves reminiscent of those present in a solid for all $\mathbf{k}$s. Thus, it can be shown that a large part of the total viscosity is contributed by the long-time tail of the cross current–current correlation function.

Longitudinal current correlation function. As indicated above, an acoustic wave will correspond to a longitudinal motion described in terms of appropriate current–current correlation function

$$C_1(\mathbf{k}, t) = \mathbf{k}^2 \left\langle \sum_{i=1}^{i=N} v_i^2(t) e^{-ikz_i(t)} \sum_{j=1}^{j=N} v_j^2(0) e^{ikz_j(0)} \right\rangle. \tag{4.119}$$

This has a Fourier transform which can be written as

$$C_1(k, \omega) = \frac{\mathbf{k}^2(kT/m)}{-i\omega + \left(\dfrac{\mathbf{k}^2(kT)}{-i\omega m S(\mathbf{k})}\right) + N_1(\mathbf{k}, \omega)}, \tag{4.120}$$

where $S(\mathbf{k})$ is the structure factor and $N_1(\mathbf{k}, \omega)$ is the appropriate memory function. The dynamic structure factor $S(\mathbf{k}, \omega)$ is available experimentally from coherent neutron scattering

$$S(\mathbf{k}, \omega) = \frac{1}{\pi} \mathrm{Re}\left[\frac{C_1(\mathbf{k}, \omega)}{\omega^2}\right] \tag{4.121}$$

In the hydrodynamic limit, the poles of $C_1(\mathbf{k}, \omega)$, for low $\mathbf{k}$ and ω values, lead to

$$N_1(\mathbf{k}, \omega) = \frac{\frac{4}{3}\eta_s + \eta_v}{m\rho} + \frac{\mathbf{k}^2(kT)}{mS(\mathbf{k})} \cdot \frac{\gamma - 1}{\mathrm{i}(\omega) + a\mathbf{k}^2} \qquad (4.122)$$

where η_s and η_v are the bulk shear and viscosities and a is the thermal conductivity, κ, divided by C_p. The hydrodynamic expression for $S(\mathbf{k}, \omega)$ does not show any peaks in this limit and can be given a simple form. Introduction of a generalized shear viscosity with a memory function involving two relaxation times provides an excellent fit to molecular dynamic data.[122]

Surprisingly the bulk viscosity correlation function is thought to have little or no tail, and so provides only a small contribution to the final function because $s/\eta \approx \frac{1}{4}$. As in the previous analysis γ can be replaced by $\gamma(\mathbf{k})$, which is found to tend to 1 as $\mathbf{k}$ is increased. This indicates that coupling with thermal modes disappears at long wave-vectors. Simplification of (4.122) in the limit of low $\mathbf{k}$ and ω yields again the classical result of the Navier–Stokes equation.

This discussion illustrates the complexity of the interactions connecting the acoustic observations with a specific model for molecular motion on a microscopic scale, but more importantly it points out that the long-time–low-frequency hydrodynamic limit may be described quite adequately by the usual classical continuum approach to the propagation of a sound wave in a fluid medium. It is important to note that the part of the velocity auto-correlation function which dominates spectroscopic observations, such as infrared, n.m.r., and Raman, will have much less importance in the acoustic and viscoelastic phenomena.

The time-correlation functions for a wide variety of experimental techniques are summarized in Table 4.2.

4.6. Relaxation distribution functions and models

4.6.1. Distribution functions

In a number of experiments it has been found that the observed relaxation spectrum is described not by simple exponential behaviour but rather by a distribution of relaxation times or processes.[123-5] The modelling of this distribution can be achieved using a phenomenological approach; often the form of the resulting equations reflect the background of the originator rather than a fundamental property of the molecular system.

In this treatment it will be assumed that some dynamic property is complex and has high- and low-frequency limiting values defined by the

Table 4.2

Time-correlation functions and experiments

Experimental observation	Dynamic quantity	Time-correlation function	P_1	Time/frequency range
Neutron scattering	$\mathbf{r}_l$, position of the lth nucleus in a fluid	$\dfrac{1}{N}\sum_{l=1}^{N} e^{i\mathbf{k}\cdot\mathbf{r}_l(0)}\cdot e^{i\mathbf{k}\cdot\mathbf{r}_e(\mathbf{k})}$	—	10^8–10^{12} Hz
Raman scattering	$\mathbf{u}$, unit vector along molecular transition vector	$\langle P_2(\mathbf{u}(0)\cdot\mathbf{u}(t))\rangle$	2	10^{13}–10^{14} Hz
Infrared absorption	$\mathbf{u}$, unit vector along molecular transition dipole	$\langle\mathbf{u}(0)\cdot\mathbf{u}(t)\rangle$	1	10^{10}–10^{14} Hz
Diffusion coefficient	$\mathbf{v}$, centre-of-mass velocity of tagged molecule	$\langle\mathbf{v}(0)\cdot\mathbf{v}(t)\rangle$	—	
Spin-rotation relaxation	J, angular momentum about molecular centre of mass	$\langle J(0)\cdot J(t)\rangle$	—	
NMR lineshape	M_x, x component of the magnetization of the system	$\langle\mathbf{M}_x(0)\cdot\mathbf{M}_x(t)\rangle$		10^3–10^8 Hz
Rotational diffusion coefficient	Ω angular velocity about molecular centre of mass	$\langle\Omega_\alpha(0)\cdot\Omega_\beta(t)\rangle$	—	
Fluorescence depolarization	r_{fi} the intensity of a line in fluorescence	$r_0\langle\tfrac{1}{2}(3\cos^2\alpha(t)-1)\rangle_{fi}$	2	10^{-6}–10^{-9} s
Sound absorption	$\zeta(\omega)$ frequency-dependent bulk viscosity	$\dfrac{1}{18VkT}\displaystyle\int_0^\infty dt\, e^{-i\omega t}\left\langle\sum_i \mathbf{T}_{ii}(0)-\bar{\mathbf{T}}_{ii}(t)\right\rangle\sum_j\langle\mathbf{T}_{ij}(0)-(\bar{\mathbf{T}}_{ij})\rangle$	—	10^5–10^9 Hz
Dielectric relaxation	$\gamma(t)$ the correlation function of a dipole at time t	$\dfrac{\langle\boldsymbol{\mu}(0)\cdot\boldsymbol{\mu}(t)\rangle}{\langle\boldsymbol{\mu}(0)\cdot\boldsymbol{\mu}(0)\rangle}$	1	10^{-4}–10^{-11} Hz
Tracer diffusion	Translational diffusion coefficient D_t	$(\tfrac{1}{3})\displaystyle\int_0^\infty\langle\mathbf{v}(0)\cdot\mathbf{v}(t)\rangle\, dt = \lim_{t\to\infty}\{\langle[\mathbf{r}(t)-\mathbf{r}(0)]^2\rangle\}/6t$	—	seconds to years
Kerr effect	Electric field-induced optical birefringence	$\langle P_1(\boldsymbol{\mu}(t))\rangle,\ \langle P_2(\boldsymbol{\mu}(t))\rangle$	1, 2	10^{-4}–10^7 Hz; 10^{-11}–10^{-12} s

Key to symbols: N, number of molecules per unit volume; $\mathbf{k}$, the wave vector; $r_{fi}=(I_\parallel^{fi}-I_\perp^{fi})/(I_\parallel^{fi}+2I_\perp^{fi})$ where $I_\parallel^{fi}$ is the intensity of a line in fluorescence when the emitted light has its electric field parallel to the electric field of the exciting light, and $I_\perp^{fi}$ is the corresponding emission intensity for perpendicular electric fields; Ω, the angular momentum vector; r_0, the polarization index for light emitted by an undisturbed molecule; $\mathbf{T}_{ij}$ is the momentum flux tensor; V, the volume; $\boldsymbol{\mu}(t)$ moment of the dipole at time t.

subscripts zero and infinity. The generalized function will be

$$\frac{K^*(\omega) - K_\infty}{K_0 - K_\infty} = \int_{-\infty}^{\infty} \frac{F(\tau/\tau_0)\, \mathrm{d} \ln(\tau/\tau_0)}{1 + i\omega\tau_0} = \{1 + (i\omega_0)^{(1-\alpha)}\}^{-\beta} \quad (4.123)$$

where

$$F(y) = \frac{1}{2i(K_0 - K_\infty)}\{K^*(1/y\ e^{i\pi}) - K^*(1/y\ e^{i\pi})\} \quad (4.124)$$

with $y = \tau/\tau_0$ solution of this integral equation is obtained via the normal coordinate transformation in which τ/τ_0 is replaced by $\exp(\pm i\omega)$ in the formula for $K^*(\omega)$. On substitution

$$K^*(1/y\ e^{\pm i\pi}) = (K_0 - K_\infty)/\{(1 + (1/y\ e^{\pm i\omega}))^{(1-\alpha)}\}^{-\beta} + K_\infty, \quad (4.125)$$

where K_∞ is the 'infinite' frequency limiting value.

The solution for $F(y)$ becomes

$$F(y) = (1/\pi)y^{(1-y)} \sin \beta\theta(y^{2(1-\alpha)} + 2y^{(1-\alpha)} \cos \pi(1-\alpha) + 1)^{-\beta/2} \quad (4.126)$$

where

$$y = \tau/\tau_0 \quad \text{and} \quad \theta = \arctan\{\sin \pi(1-\alpha)/y^{(1-\alpha)} + \cos(1-\alpha)\}. \quad (4.127)$$

Successive application of De Moivre's theorem to the above equations, followed by separation of real and imaginary parts leads to the following equations

$$K'(\omega) - K_\infty = r^{-\beta/2}(K_0 - K_\infty) \cos \beta\theta \quad (4.128)$$

$$K''(\omega) = r^{\beta/2}(K_0 - K_\infty) \sin \beta\theta \quad (4.129)$$

$$r = (1 + (\omega\tau_0)^{(1-\alpha)} \sin(\alpha\pi/2))^2 + ((\omega\tau_1)^{(1-\alpha)} \sin(\alpha\pi/2)) \quad (4.130)$$

$$\beta = \arctan \{(\omega\tau_0)^{(1-\alpha)} \cos(\alpha\pi/2)\}/\{1 + (\omega\tau_0)^{(1-\alpha)} \sin(\alpha\pi/2)\}. \quad (4.131)$$

If α is placed equal to zero, the equations describe a set of overlapping ideal relaxation processes, whereas if β is placed equal to unity the relaxation corresponds to a single process with a distribution of relaxation times. A number of variants of the above form are to be found in the literature. Certain of these have been developed from the treatment of particular experiments and have their origins in mechanical analogues. A selection of distribution functions and their associated diagrammatic representations are presented in Table 4.3. In practice the most popular representation has a form similar to that used in eqn (4.123). Other forms have been used,

Table 4.3.

Empirical distribution functions

Generalized functions

$$E_r(t) = \int_{-\infty}^{\infty} H(\tau)\exp(-t/\tau)\,d(\ln\tau); \qquad \eta(t) = \int_0^{\infty} H(\tau)\,d\tau$$

$$E'(\omega) = \int_{-\infty}^{\infty} H(\tau)\frac{\omega\tau}{1+\omega^2\tau^2}\,d(\ln\tau); \qquad E''(\omega) = \int_{-\infty}^{\infty} H(\tau)\frac{\omega\tau}{1+\omega^2\tau^2}\,d(\ln\tau)$$

Distribution	Limits	$E_r(t)$	$\eta(t)$	$E'(\omega)$	$E''(\omega)$
Box[†]	$H(\tau) = E_0$ $\tau_3 < \tau < \tau_m$ $H(\tau) = 0$ $\tau(\tau_3;\tau)\tau_m$	$E_0\left\{E_i\!\left(-\dfrac{t}{\tau_3}\right) - E_i\!\left(-\dfrac{t}{\tau_m}\right)\right\}$	$E_0(\tau_m - \tau_3)$ $\approx E_0\tau_m$	$E_0 \ln\dfrac{1+\omega^2\tau_m}{1+\omega^2\tau_3}$	$E_0(\arctan\omega\tau_m$ $-\arctan\omega\tau_3)$
Wedge–box[‡]	Wedge: $H(\tau) = M/\tau^{\frac12}$ $\tau_1 < \tau < \tau_2$ Box: $H(\tau) = E_0$ $\tau_3 < \tau < \tau_m$	$Mt^{-\frac12}\{\Gamma_{t/\tau_1}(\tfrac12) - \Gamma_{t/\tau_2}(\tfrac12)\}$	$2M(\tau_2^{\frac12} - \tau_1^{\frac12})$	$M\left(\dfrac{\omega}{2}\right)^{\frac12}\left\{\dfrac12\ln\dfrac{\omega\tau + (2\omega\tau)^{\frac12}+1}{\omega\tau - (2\omega\tau)^{\frac12}+1} \right.$ $\left. +\arctan\dfrac{(2\omega\tau)^{\frac12}}{\omega\tau - 1}\right\}_{\tau_1}^{\tau_2}$	$M\left(\dfrac{\omega}{2}\right)^{\frac12}\left[\arctan\dfrac{(2\omega\tau)^{\frac12}}{\omega\tau - 1} \right.$ $\left. -\dfrac12\ln\dfrac{\omega\tau + (2\omega\tau)^{\frac12}+1}{\omega\tau - (2\omega\tau)^{\frac12}+1}\right]_{\tau_1}^{\tau_2}$
Maxwell[§]	$H(\tau) = \tau E = \tau_p E$ $H(\tau) = 0,\ \tau \neq \tau_p$	$E\exp(-t/\tau_p)$	$\dfrac{E}{1+(\omega/\tau_p)^2}$	$\dfrac{E}{1+(1/\omega\tau_p)^2}$	$\dfrac{E(\omega\tau_p)}{1/\omega^2 + \tau_p^2}$

† Tobolsky, A. V., Dunell, B. A. and Andrews, R. D. *Textile Res. J.* **21**, 404 (1951).
‡ Tobolsky, A. V. *J. Amer. Chem. Soc.*, **74**, 3786 (1952); *J. appl. Phys.* **27**, 673 (1956).
§ Tobolsky, A. V. and Eyring, H. *J. Chem. Phys.*, **11**, 125 (1943).

although they do not lead to any greater insight into the nature of the relaxation process.

4.6.2. Local motions—free volume relaxation

Much of the discussion to this point has been concerned with relaxation of polymer molecules in solution, in which the availability of energy for activation is considered the controlling parameter defining the chain mobility. In solid polymers local motions dominate the dynamic spectrum and are sensitive to the volume available from motion.[126-31] At very low temperatures molecular motion will be almost completely frozen. On heating the lattice expands and sufficient energy also becomes available for short-range librational motions to become active. At slightly higher temperatures, rotational motions of spherically symmetric side-groups are observed. On further heating, motion of small elements of the chain or large side-groups becomes possible. For all the relaxation processes so far described there is little or no local volume fluctuation and these motions are observed in glassy solids. These motions are often characterized by broad relaxation features and a sensitivity to the perfection of the lattice structure. The occurrence of large-scale conformational changes is associated with the development of rubbery properties and is observed as the so-called glass transition.

The rotational isomeric processes attributed to small-scale conformational changes can often occur without any appreciable change in lattice dimensions. However, the main glass transition is usually associated with a significant change in the lattice dimensions, and has led to the concept of a 'free volume' limited relaxation process.[132-40] A definition of free volume used in polymer studies involves a relationship of the form

$$v_f = v - v_0 \qquad (4.132)$$

where v_f is the free volume per g and v the measured specific volume of the polymer at temperature T. v_0 is termed the 'occupied volume'. The free volume in a polymer system is a time-averaged quantity which can be determined from equilibrium properties—densiometric measurements. In a mobile system the local free volume is continually being redistributed throughout the medium, the redistribution occurring simultaneously with the random thermal motions of the polymer molecule. The free volume may be considered to be that part of the volume which can be redistributed without change in energy. If the interactions between polymer chains are defined by Lennard–Jones interactions it may be assumed that a critical volume will exist which must be achieved before redistribution of the volume can occur without a large contribution being required. These ideas give rise to a modified definition of the volume[141]

$$v = v_0 + \Delta v_c + v_f \qquad (4.133)$$

where Δv_c is that part of the excess volume, $v - v_0$, which requires an energy for redistribution. At temperatures below some critical temeperature, T_∞, where the 'cage' radius is small, the redistribution energy will be large and thus $v_f \approx 0$. At temperatures above T_∞ where the cage radius has increased to values corresponding to the linear region of the potential energy curve, most of the volume added by thermal expansion will be 'free' for redistribution. The precise nature of this limiting condition in a polymeric system has been widely discussed and the definition extended to include entropic and local interaction terms. In the simplest model

$$v_f = \alpha \bar{v}_m (T - T_\infty) \quad \text{for } T > T_\infty \qquad (4.134)$$

$$v_f = 0$$

where α is the average expansion coefficient of the medium and v_m the average of the 'molecular' volume, v_0, in the temperature range T to T_∞. The above equation can be rewritten to give

$$v_f = v_f(T_0) + v_g \Delta\alpha (T - T_0) \qquad (4.135)$$

where $\Delta\alpha$ is the difference in the thermal expansion above and below T_g and v_g is the specific volume of the polymer at T_g. Eqn (4.135) allows rationalization of the observed relationship between T_g and structural change in polymers. For instance, the T_g in isotactic poly(methylmethacrylate) has a value of 48 °C, whereas in the syndiotactic polymer it has a value of 160 °C. These observations have been rationalized on the basis of there being a greater steric interaction in the syndiotactic than in the isotactic polymer.

4.6.3. Defect diffusion models

In the discussion of the glass-transition process it was pointed out that relaxation of a particular segment of the polymer chain was achieved not just by that element of the polymer containing sufficient energy to overcome the intramolecular potential barrier opposing conformational change, but also that there must exist sufficient free volume for relaxation to occur. Diffusional models are based on the concept that above the glass transition temperature there is a rapid motion of the constituent chains and this motion constitutes a source of free volume which if concentrated at a point allows conformational change.[135,136] The defect diffusion model assumes that relaxation can occur at a site only when a defect arrives and that its motion is a random walk process. It has been demonstrated by a comparison of experiment with theory that the majority of relaxations either in glasses or melts are best described by a model which involves consideration of relaxation as a result of both nearest and next-nearest defect diffusion.[135] The hypothesis that a defect is a hole or a local increase in free volume can be tested for the model, and it has been shown that physically plausible values of relative free volume arise. This interpretation is very similar to other hypotheses adopted for relaxation in fluids.

References

1. Yamakawa, H. *Modern theory of polymer solutions*, p. 76. Harper and Row, New York (1971).
2. Bartenev, G. M. and Zelenev, Yu. Ya. *Relaxation phenomena in polymers*, p. 282. J. Wiley and Sons, New York and Toronto (1972).
3. Mizushima, S. *Structure of molecules and internal rotation.* Academic Press, New York (1954).
4. Sheppard, N. *Advan. Spectrosc.* **1**, 288 (1959).
5. Pethrick, R. A. and Wyn-Jones, E. *Quart. Rev.* **23**, 301 (1969).
6. Lowe, J. P. In *Progress in physical organic chemistry*, vol. 6 (ed. A. Streitweiser and R. W. Taft). Interscience, New York (1968).
7. Orville Thomas, W. J. *Internal rotation and molecular structure.* J. Wiley, New York (1975).
8. Pethrick, R. A. and Wyn-Jones, E. In *Topics in stereochemistry*, Vol. 5 (ed. E. L. Eliel and N. L. Allinger). Wiley Interscience, London and New York (1970).
9. North, A. M. *Chem. Soc. Rev.* **1**, 49 (1972).
10. Block, H. and North, A. M. *Advan. mol. rel. Processes* **1**, 309 (1970).
11. Pethrick, R. A., *Rev. macromol. Chem.* **C9**, 91 (1973).
12. Zwanzig, R. *Ann. Rev. phys. Chem.* **16**, 67–102 (1965).
13. Stockmayer, W. H., Gobush, W., Chikahisa, Y. and Carpenter, D. K. *Discuss. Faraday Soc.* **49**, 182–92 (1970).
14. Berne, B. J. and Harp, G. D. *Advan. chem. Phys.* **17**, 63–227 (1970).
15. Peterlin, A. *J. polym. Sci. A-2* **5**, 179–93 (1967).
16. Cerf, R. *J. chim. Phys.* **66**, 479–88 (1969).
17. Peterlin, A. *Polym. Lett.* **10**, 101–5 (1972).
18. Bazua, E. R. and Williams, M. C. *J. polym. Sci.* **12**, 825–48 (1974).
19. Brereton, M. G. and Davis, G. R. *Polymer* **18**, 764 (1977).
20. Hagasi, K. Dielectric relaxation and molecular structure. Research Inst. Appl. Electricity Hokkaido University Sapporo, Japan (1961).
21. Smyth, C. P. *Dielectric relaxation and structure.* Wiley Interscience, New York, Toronto, London (1955).
22. Kramers, A. *Physica* **7**, 284 (1940).
23. Chandrasekhar, S. *Stochastic problems in physics and astronomy.* Cambridge University Press (1947).
24. Gotlib, Yu. Ya. and Salikhov, K. H. *Fiz. Tvard Tela* **4**, 1166 (1962).
25. Birshtein, T. and Ptitsyn, O. *Conformation of macromolecules*, Wiley Interscience, New York (1966).
26. Thompson, D. S., Newark, R. A. and Sederholm, C. H. *J. chem. Phys.* **37**, 411 (1962).
27. Alger, T. D., Gutowsky, H. S. and Vold, R. L. *J. chem. Phys.* **47**, 3130 (1967).
28. Green, M. S. *J. chem. Phys.* **20**, 1281 (1952).
29. Kubo, R. *J. phys. Soc. Jpn* **12**, 570 (1957).
30. McQuarrie, D. A. *Statistical mechanics.* Harper and Row, New York (1976).
31. Hoffman, J. D. *Eighth Ampère Colloq.* **12**, 36 (1959).
32. Beyer, R. T. *J. acoust. Soc. Amer.* **29**, 243 (1957).
33. Hoffman, J. D. *J. chem. Phys.* **22**, 156 (1954).

34. Williams, G. and Cook, M. *Trans. Faraday Soc.* **67**, 990 (1971).
35. Verdier, P. H. and Stockmayer, W. H. *J. chem. Phys.* **36**, 227 (1962).
36. Flory, P. J. *Statistical mechanics of chain molecules.* Wiley Interscience, New York and London (1969).
37. Iwata, K. and Kurata, M. *J. chem. Phys.* **50**, 4008 (1969).
38. Schatzki, T. F. *J. polym. Sci.* **57**, 496 (1962).
39. Valeur, B. and Monnerie, L. *J. polym. Sci.*, **14**, 11 (1976).
40. Valeur, B. and Monnerie, L. *J. polym. Sci.*, **14**, 29 (1976).
41. Rahman, A. *Phys. Rev.*, **136A**, 405 (1964).
42. Verlet, L. *Phys. Rev.*, **154**, 98 (1967).
43. Einwohner, T. and Alder, B. J. *J. chem. Phys.*, **49**, 1458 (1968).
44. Verdier, P. H. *J. chem Phys.*, **45**, 2122 (1966).
45. Verdier, P. H. and Stockmayer, W. H., *J. chem. Phys.*, **36**, 227 (1962).
46. Bellemans, A. and Chervin, V. G. *Zhurn. fiz. Khim.*, **43**, 1069 (1969).
47. Clark, A. T. and Lal, M. *J. Phys. A.* **11**, L11 (1978).
48. Clark, A. T. and Lal, M. *Brit. polym. J.* **9**, 92 (1977).
49. Valleau, J. P. and Torrie, G. M. *Chem. Phys. Lett.* **28**, 578 (1964).
50. Wall, F. T. and Hioe, F. T. *J. phys. Chem.* **74**, 4410, 4416 (1970).
51. McCracken, F. L., Mazur, J. and Gutman, C. M. *Macromolecules* **6**, 859 (1973).
52. Mazur, J. and McCracken, F. L. *J. chem. Phys.* **49**, 648 (1968).
53. Lal, M. *R. Inst. Chem. Soc. Rev.*, **4**, 97 (1971).
54. Yamakawa, H. and Fujii, M. *Macromolecules* **6**, 407 (1973).
55. Domb, C. *Advan. chem. Phys.* **15**, 229 (1969).
56. Windwer, S. In *Markov chains and Monte Carlo calculations in Polymer Science* (ed. G. G. Lowry), chapter 5, p. 125. Dekker, New York, (1970).
57. Alexandrowicz, Z. *J. chem. Phys.* **51**, 561 (1969).
58. Alexandrowicz, Z. and Accad, Y. *Macromolecules* **6**, 251 (1973).
59. Alexandrowicz, Z. and Accad. Y. *J. chem. Phys.* **54**, 5338, (1971).
60. Wall, F. T. and Erpenbeck, J. J. *J. chem. Phys.* **30**, 637 (1959).
61. Kron, A. K. and Ptitsyn, O. B., *Vysokomol. Oyed* **A9**, 759 (1967).
62. Domb, C., Barrett, A. J. and Lax, M. *J. Phys. A* **6**, L82 (1973).
63. Bruns, W. *Makromol. Chem.* **124**, 91 (1969).
64. Stellman, S. D. and Gans, P. J. *Macromolecules* **5**, 516 (1972).
65. Grishman, R. *J. chem. Phys.* **58**, 220 (1973).
66. McKenzie, D. S. *J. Phys. A* **6**, 338 (1973).
67. Domb C., Gillis, J. and Wilmers, G. *Proc. phys. Soc.* **85**, 625 (1965).
68. Chay, T. R. *J. chem. Phys.* **52**, 1025 (1970).
69. McKenzie, D. S. and Moore, M. A. *J. Phys. A.* **4**, L82 (1971).
70. Chay, T. R. *J. chem. Phys.* **57**, 910 (1972).
71. Frisch, H. L. and Simha, K. In *Rheology* (ed. F. R. Eirich), vol. 1, chapter 14. Academic Press, New York (1965).
72. Utracki, I. and Simha, R. *J. polym. Sci.* **A1**, 1089 (1963).
73. Edwards, S. F. and Freed, K. F. *J. chem. Phys.* **61**, 1189 (1974).
74. Chompff, A. J. and Duiser, J. A. *J. chem. Phys.* **45**, 1505 (1966).
75. Edwards, S. F. *Proc. Phys. Soc.* **83**, 265 (1966).
76. Edwards, S. F. *J. Phys. A* **8**, 1670 (1975).
77. des Cloizeaux, J. *J. Physique* **36**, 281 (1975).
78. De Gennes, P. G. *J. Physique Lett.* **36**, L55 (1975).
79. Moore, M. A. *J. Physique* **38**, 265 (1977).
80. Daoud, M. and Jannink, G. *J. Physique* **37**, 973 (1976).

81. Yamakawa, H. *Modern theory of polymer solutions*, p. 419. Harper and Row, New York (1971).
82. Fixman, M. and Stockmayer, W. H. *Ann. Rev. phys. Chem.* **21**, 407 (1970).
83. Yamakawa, H. *Ann. Rev. phys. Chem.* **25**, 179 (1974).
84. Bixon, M. *Ann. Rev. phys. Chem.* **27**, 65 (1976).
95. De Wames, R. E., Hall, W. F. and Shen, M. C. *J. chem. Phys.* **46**, 2782 (1967).
86. Zwanzig, R., Kiefer, J. and Weiss, G. H. *Proc. Nat. Acad. Sci.* **60**, 381 (1968).
87. Batchelor, G. K. and Green, J. T. *J. fluid Mech.* **56**, 375 (1972).
88. Yamakawa, H. *J. chem. Phys.* **48**, 3845 (1968).
89. Burgers, J. M. *Second report on viscosity and plasticity*. North Holland, Amsterdam (1938).
90. Ullman, R. *Macromolecules* **7**, 300 (1974).
91. Edwards, S. F. and Freed K. F. *J. chem. Phys.* **61**, 1202 (1974).
92. Szu, S. C. and Hermans J. J. *J. polym. Sci.* **11**, 1941 (1973).
93. Stockmayer, W. H., Gobush, W., Chikahiza, Y. and Carpenter, D. K. *Discuss. Faraday Soc.* **59**, 182 (1970).
94. Bixon, M. *J. chem. Phys.* **58**, 1459 (1973).
95. Felderhorf, B. U., Deutsch, J. M. and Titulaer, V. M. *J. chem. Phys.* **63**, 740 (1975).
96. Wilemski, G. *J. chem. Phys.* **63**, 2540 (1975).
97. Flory, P. J. *Principles of polymer chemistry*, p. 672, Cornell University Press, Ithaca, New York (1953).
98. Isihara, A. *Advan. polymer. Sci.* **7**, 449 (1971).
99. Curro, J. G., Blatz, P. J. and Pings, C. J. *J. chem. Phys.* **50**, 2199 (1969).
100. Edwards, S. F. *Proc. Phys. Soc.* **85**, 613 (1965).
101. Freed, K. F. *J. chem. Phys.* **55**, 3910 (1971).
102. Whittingham, S. G. and Runfield, L. G. *J. Phys.* **6**, 484 (1973).
103. Berne, B. J. In *Physical chemistry, an advanced treatise*, Vol. VIIIB ('The liquid state') (ed. H. Eyring, D. Henderson and W. Jost), pp. 540–713. Academic Press, New York (1971).
104. Cole, R. H. *Mechanique statistique des mouvements angulaires en phase liquides*, Faculté Sciences, Orsay (1969).
105. Glarum, S. H. *J. chem. Phys.* **33**, 1371 (1960).
106. Verlet, L. *Phys. Rev.* **154**, 98 (1967).
107. Berne, B. J. *J. chem. Phys.* **56**, 2164 (1972).
108. Kadanoff, L. P. and Martin, P. C. *Amer. Phys.* (*NY*) **24**, 419 (1969).
109. Mori, H. *Prog. Theor. Phys.* (*Kyoto*) **33**, 423 (1965).
110. Rahman, A. *Phys. Rev* **136**, A405 (1964).
111. Levesque, D. and Verlet, L. *Phys. Rev.* **A2**, 2514 (1970).
112. Berne, B. J., Boon, J. P. and Rice, S. A. *J. chem. Phys.* **45**, 1086 (1966).
113. Alder, B. J. and Wainwright, T. E. *Phys. Rev.* **A1**, 18 (1970).
114. Dorfman, J. R. and Cohen, E. G. D. *Phys. Rev. Lett.* **25**, 1257 (1970).
115. Ernst, M. H., Hauge, E. H. and van Leeuwen, J. M. J. *Phys. Rev. Lett.* **25**, 1254 (1970).
116. Pomeau, Y. *Phys. Rev.* **A5**, 2569 (1972).
117. Verlet, L. *Faraday Symposia* (6), 116 (1972).
118. Verlet, L. *Phys. Rev.* **A7**, 1690 (1973).
119. Levesque, D. and Verlet, L. *Phys. Rev.* **A2**, 2514 (1970).
120. Schofield, P. *Computer Phys. Commun.* **5**, 17 (1973).
121. Alder, B. J. and Wainwright, T. E. *Phys. Rev* **A1**, 18 (1970).

122. Herzfeld, K. F. and Litovitz, T. A. *Absorption and dispersion of ultrasonic waves*. Academic Press, New York (1959).
123. Levesque, D., Verlet, L., and Kukijarvi, J. *Phys. Rev.* **A7**, 1960 (1973).
124. Cole, R. H. and Cole, K. S. *J. phys. Chem.* **9**, 341 (1941).
125. Cole, R. H. and Davidson, D. W. *J. chem. Phys.* **19**, 1484 (1951).
126. Emery, J. and Gasse, S. *Advan. mol. Rel.* **12**, 47 (1978).
127. Bueche, F. *J. polym. Sci.* **22**, 113 (1956).
128. Bueche, F. *J. chem. Phys.* **30**, 748 (1959).
129. Cohen, M. H. and Turnbull, D. *J. chem. Phys.* **31**, 1164 (1959).
130. Doolittle, A. K. *J. appl. Phys.* **22**, 1471 (1951).
131. Doolittle, A. K. *J. appl. Phys.* **23**, 236 (1952).
132. Williams, M. L., Landel, R. F. and Ferry, J. D. *J. Amer. chem. Soc.* **77**, 3701 (1955).
133. Miller, A. A. *J. polym. Sci.* **A1**, 1857, 1865 (1963).
134. Tammann, G. and Hesse, W. *Z. anorg. allgem. Chem.* **156**, 245 (1926).
135. Kovacs, A. J. *Advan. polym. Sci.* **3**, 394 (1963).
136. Phillips, M. C., Barlow, J. A. and Lamb, J. *Proc. Roy. Soc.* **A329**, 193 (1977).
137. Glarum, S. H. *J. chem. Phys.* **33**, 639, 1371 (1960).
138. Fox, T. G. and Flory, P. J. *J. appl. Phys.* **21**, 581 (1950).
139. Ferry, J. D. *Viscoelastic properties of polymers*. Wiley, New York (1961).
140. Turnbull, D. and Cohen, M. H. *J. chem. Phys.* **34**, 120 (1961).
141. Saito, S. *Rep. Prog. polym. Phys.* (*Jpn*) **5**, 205 (1962).
142. Williams, G. *Chem. Soc. Rev.* **7**, 89 (1977).

PART B

PHENOMENOLOGICAL TREATMENT

DIELECTRIC PROPERTIES

5.1. Ideal and non-ideal relaxation

In Chapter 1 it was pointed out how dielectric relaxation arose as the result of hindered movement of charged species, either through the rotational orientation of dipolar molecules or by migration and trapping of more mobile charge carriers. After removal of an electrical field, the polarization would, in an ideal case, obey the decay function

$$P(t) = P_0 \exp(-t/\tau) \tag{5.1}$$

where P_0, $P(t)$ represent the polarization at times zero and t respectively and τ is the single time constant. The normalized frequency-dependent complex permittivity would be given by

$$\frac{\varepsilon^*(\omega)}{\varepsilon_0 - \varepsilon_\infty} = \frac{1}{1 + i\omega\tau} \tag{5.2}$$

where subscripts zero and infinity refer to the permittivity at frequencies much below and above τ^{-1}, and τ is now called the dielectric relaxation time.

However it is very often found in practice that the complex permittivity of polymer systems does not vary with time or frequency according to these simple ideal equations, but instead exhibits what is called non-ideal relaxation. There are a number of approaches to the problem of such relaxations, but in general they fall into two categories. On the one hand are those methods that fit experimental observations to an empirical equation (describing either the complex permittivity in the frequency domain or the polarization decay in the time domain) which contains one or more adjustable constants over and above the single time constant, τ, of the ideal case. On the other hand are theories which start with a molecular model, and add to the simple picture of rotation or migration some other process or interaction so that the polarization equation is no longer a simple first-order linear differential equation, and consequently no longer has a simple (single constant) exponential solution.

5.1.1. Empirical equations

Generally, in the frequency domain, the decrease in the real permittivity and the maximum in the imaginary component extend over a wider frequency range than predicted by the ideal Debye equations. Quite often, too, the observations are not symmetrical in the log (frequency) plane, the relevant graphs presenting a 'skew' dependence. For many years now it has been conventional to fit such observations to empirical functions similar in form to the ideal expressions, but containing a new parameter as a measure of the departure from ideal behaviour.

Two of the best known are the Cole–Cole[1] and Cole–Davidson[2] relationships, both of which have been widely applied to polymer systems:

$$\text{Cole–Cole:} \quad \frac{\varepsilon^* - \varepsilon_\infty}{\varepsilon_0 - \varepsilon_\infty} = \frac{1}{1 + (i\omega\tau)^{1-\alpha}} \qquad (5.3)$$

$$\text{Cole–Davidson:} \quad \frac{\varepsilon^* - \varepsilon_\infty}{\varepsilon_0 - \varepsilon_\infty} = \frac{1}{(1 + i\omega\tau)^\alpha} \qquad (5.4)$$

and their combination[3]

$$\text{Havriliak–Negami:} \quad \frac{\varepsilon^* - \varepsilon_\infty}{\varepsilon_0 - \varepsilon_\infty} = \frac{1}{(1 + (i\omega\tau)^{1-\alpha})^\beta}. \qquad (5.5)$$

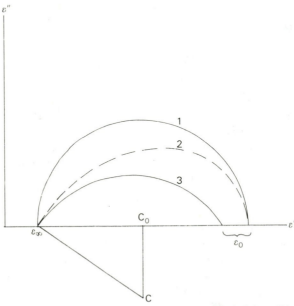

Fig. 5.1. Argand diagram of real and imaginary permittivity. (1) Ideal Debye relaxation; (2) Cole–Davidson skewed-arc relaxation; (3) Cole–Cole symmetric relaxation with semicircle centre depressed from C_0 to C.

The departures from ideal behaviour are often ascribed to the existence of a distribution of characteristic relaxation times, and so the adjustable parameter is often referred to in these terms. The symmetry (or lack of it) associated with these empirical functions is seen when an Argand diagram of ε'' against ε' is constructed. In this representation ideal Debye relaxation appears as a semi-circle, the symmetrical Cole–Cole distribution as the arc of a semi-circle of centre, C, depressed below the $\varepsilon'' = 0$ axis, and the Cole–Davidson distribution as a skewed arc (Fig. 5.1).

The representation of non-ideal relaxation can be carried back from the frequency domain to the underlying time-dependent decay function, by giving the latter a non-exponential form. One such has been proposed by Williams and Watts[4]

$$\phi(t) = \exp -(t/\tau)^{\beta}. \tag{5.6}$$

When the empirical 'distribution parameter', $\beta, = 0 \cdot 5$, the decay function generates behaviour in the frequency domain similar to the Cole–Davidson skewed arc. It must be stressed that introducing such a parameter in the decay function is every bit as empirical as in the complex permittivity but does aid the search for an explanation in molecular terms.

5.1.2. Two-process models

Turning to explanations of non-ideal behaviour other than those based upon a simple distribution of times for a single process, we find that many exhibit the common feature of a basis in two interrelated molecular processes of different time-dependence.

The most widely used of these is due to Glarum,[5] and is based upon the premise that the relaxation process of interest can occur only when some 'defect' diffuses to the neighbourhood of the molecule about to relax. In this way the over-all process is generated as a combination of diffusion (the time-dependence of which is naturally non-exponential) and relaxation (which may be ideal). The concept is certainly attractive when relaxation requires molecular translation or rotation and when a defect can be visualized as some vacancy or molecular free volume which is required for the appropriate movement. The type of relationship which can be derived from this model is

$$\phi(t) = \exp(-t/\tau)\{1 - P(t)\} \tag{5.7}$$

where $P(t)$ is the probability that arrival of a 'defect' by time t has caused relaxation.

If the diffusion of the defect is restricted in some way[6] and characterized by a time constant τ_D, then the case $\tau \ll \tau_D$ corresponds to ideal Debye relaxation, $\tau = \tau_D$ to Cole–Davidson 'skewed arc' relaxation with $\alpha = 0 \cdot 5$, and $\tau \gg \tau_D$ to the depressed circular arc with $\alpha = 0 \cdot 5$.

Another quite general explanation of non-ideal relaxation is based upon a consideration of the reaction field which surrounds any charged species in a dielectric environment. The starting point of this consideration is that the reaction field, due to the charged species, in its turn affects the movement of that species.

The situation when polarization is due to limited movement of charge carriers has been investigated by Jonscher.[7] The loss permittivity can be described as a power-law function of frequency and the loss tangent is a constant independent of frequency. Measurements on a wide variety of polymeric and semiconducting materials suggest that this approach may have considerable validity when dielectric behaviour is associated with carrier migration. The important equations are

$$\varepsilon'' = K(\omega/\omega_c)^{n-1} \qquad (0 < n < 1) \tag{5.8}$$

$$\frac{\varepsilon''(\omega)}{\varepsilon'(\omega)} = \tan\{(1-n)\pi/2\}. \tag{5.9}$$

where K is a constant and ω_c is the relaxation frequency.

An alternative approach is to consider polarization due to the rotation of molecular dipoles. Here, the reaction field due to a particular dipole exerts what is known as 'dielectric friction' on its rotational behaviour and this in turn leads to non-exponential decay functions. Considerable progress in this line has been made by Böttcher.[8] When a dipole, of electric moment $\boldsymbol{\mu}$ and inertial moment I in a spherical molecule experiences a torque T_r from the reaction field, $\mathbf{r}$, the lag in the reaction field enlarges the correlation $\langle \boldsymbol{\mu}(0) \cdot \boldsymbol{\mu}(t) \rangle$ and the relaxation time. The rotational rates are described by the time-correlation function

$$\frac{\langle \boldsymbol{\mu}(0) \cdot \mathbf{r}(t) \rangle}{\langle \boldsymbol{\mu}(0) \cdot \boldsymbol{\mu}(0) \rangle} = \frac{\langle \boldsymbol{\mu}(0) \cdot \mathbf{r}(t) \rangle}{\langle \boldsymbol{\mu}(0) \cdot \boldsymbol{\mu}(0) \rangle} - \frac{1}{I} \frac{\langle T_r^2 \rangle}{\langle \boldsymbol{\mu}(0) \cdot \boldsymbol{\mu}(0) \rangle}. \tag{5.10}$$

In these approaches, the same basic concept is employed, i.e. a differing time-dependent behaviour in an interacting variable.

The descriptions of non-ideal behaviour presented above illustrate the way in which the search for an explanation has moved from an empirical consideration of results in the frequency domain, through the responsible decay function in the time domain, to molecular explanations capable of generating some differential rate equation other than the ideal linear first-order equation.

The profitability of commencing any theoretical or interpretative study with the appropriate differential equation becomes very obvious in studies of phenomena occurring at very high frequencies. The equation of motion of any body will normally contain inertial (second-order differential), viscous (first-order differential), and elastic (linear) terms. At low frequencies, when dielectric relaxation has a molecular origin, inertial terms are insignificant and the equation of motion reduces to the familiar linear first-order form. Equally, at very high frequencies the inertial terms predominate and the familar second-order differential equations descriptive of resonance phenomena emerge. However, at intermediate frequencies both inertial and viscous terms may be important, when the resulting equation of motion will not have a solution in the form of a simple single exponential. Consequently, another type of non-ideal behaviour is observed at these frequencies where the macroscopic observation of relaxation takes on increasingly a resonance character. This particularly affects the real permittivity which passes through a maximum and minimum rather than exhibiting a simple decrease as frequency is increased.

5.2. Dipole-moment time-correlation functions for polymers

5.2.1. The basic function

The interrelating of molecular behaviour and macroscopic relaxation is further aided by the use of time-correlation functions. The application of the dipole-moment correlation function to dielectric phenomena has been set out in a most useful review by Williams.[9] In this case one is examining the time-dependent decay in the correlation of the angle made by a reference dipole with its orientation at zero time (the auto-correlation terms) or the angle made with some other dipole at zero time (cross-correlation terms).

It is usual to write the dipole correlation function in terms of the electric polarization $\mathbf{M}(t)$ of a macroscopic volume containing N molecular dipoles.

$$\Gamma(t) = \frac{\langle \mathbf{M}(0) . \mathbf{M}(t) \rangle}{\langle \mathbf{M}(0) . \mathbf{M}(0) \rangle} = \frac{\sum_{i'}^{N} \sum_{j}^{N} \langle \boldsymbol{\mu}_j(0) . \boldsymbol{\mu}_{j'}(t) \rangle}{\sum_{i'}^{N} \sum_{j}^{N} \langle \boldsymbol{\mu}_j(0) . \boldsymbol{\mu}_{j'}(0) \rangle} \tag{5.11}$$

where the vectors $\boldsymbol{\mu}_j$ represent the molecular dipole moments in the reference direction. In this expression the denominator terms $\langle \boldsymbol{\mu}_j(0) . \boldsymbol{\mu}_{j'}(0) \rangle$ describe the equilibrium orientation between dipoles. Separation of auto- and cross-correlation terms then gives

$$\Gamma(t) = \frac{\sum_{j=1}^{N} \langle \boldsymbol{\mu}_j(0) . \boldsymbol{\mu}_j(t) \rangle + \sum_{j=2}^{N} \sum_{j'=1}^{j-1} \langle \boldsymbol{\mu}_j(0) . \boldsymbol{\mu}_{j'}(t) \rangle}{\sum_{j=1}^{N} \boldsymbol{\mu}_j^2 + 2 \sum_{j=2}^{N} \sum_{j'=1}^{j-1} \langle \boldsymbol{\mu}_j(0) . \boldsymbol{\mu}_{j'}(0) \rangle} . \tag{5.12}$$

This correlation function is an expression of the decay of $\mu^2 \langle \cos \theta(t) \rangle$ where μ is the molecular dipole moment.

The application of the dipole moment correlation function requires evaluation of a connection between 'macroscopic' and 'microscopic' polarizations. The difficulty here is that the molecular dipole is surrounded by its own reaction field, and so experiences an 'internal field' which differs from the externally applied macroscopic field. When this is ignored, the correlation function $\Gamma(t)$ is simply the relaxation decay function $\phi(t)$, and

$$\frac{\varepsilon^* - \varepsilon_\infty}{\varepsilon_0 - \varepsilon_\infty} = \int_0^\infty \exp(-i\omega t)\left\{\frac{-d[\Gamma(t)]}{dt}\right\} dt \qquad (5.13)$$

in which we have written the Laplace transform in its integral representation. Inclusion of the internal field means that the macroscopic relaxation time τ_m, differs from the microscopic relaxation time τ_μ, characterizing the correlation function, although a single relaxation time on one scale implies that the same exists on the other. The two are related through an internal field factor,[10]

$$\tau_m = \frac{3\varepsilon_0}{(2\varepsilon_0 + \varepsilon_\infty)} \cdot \tau_\mu \qquad (5.14)$$

which lies between 1·0 and 1·5 and becomes less important when there is a distribution of relaxation times.

5.2.2. Application to polymer systems

The application of eqn (5.13) to polymer systems[11] can be simplified by considering only a single type of dipole, and examining only the dipoles on a single chain. Then

$$\Gamma(t) = \frac{\langle \boldsymbol{\mu}_j(0) \cdot \boldsymbol{\mu}_j(t)\rangle + \sum_{j'} \langle \boldsymbol{\mu}_j(0) \cdot \boldsymbol{\mu}_{j'}(t)\rangle}{\mu^2 + \sum_{j'} \langle \boldsymbol{\mu}_j(0) \cdot \boldsymbol{\mu}_{j'}(0)\rangle}. \qquad (5.15)$$

The correlation function can be obtained from the complex permittivity

$$\frac{\varepsilon^* - \varepsilon_\infty}{\varepsilon_0 - \varepsilon_\infty} \cdot p(\omega) = \mathscr{L} - \frac{d\Gamma(t)}{dt} \qquad (5.16)$$

where the internal field function is included and here given by[11,12]

$$p(\omega) = \frac{\varepsilon_0(2\varepsilon^* + \varepsilon_\infty)}{\varepsilon^*(2\varepsilon_0 + \varepsilon_\infty)}. \qquad (5.17)$$

In the study of polymers a number of observations under different conditions are often presented on a master plot of $\varepsilon''(\omega)/\varepsilon''_{max}$ against log (ω/ω_{max}) where $\omega_{max}\tau = 1$. Under these circumstances a reduced macroscopic decay function, $\phi(t/\tau)$ can be obtained,

$$\phi(t/\tau) = \frac{\int_{-\infty}^\infty (\varepsilon''/\varepsilon''_{max}) \cos\{(\omega/\omega_{max})(t/\tau)\}d\log(\omega/\omega_{max})}{\int_{-\infty}^\infty (\varepsilon''/\varepsilon''_{max}) d\log(\omega/\omega_{max})} \qquad (5.18)$$

which can be equated to the correlation functions in the absence of internal field effects.

An alternative picture of the dipole moment correlation function is obtained from the expression

$$\frac{\langle \mathbf{\mu}(0) \cdot \mathbf{\mu}(t) \rangle}{\mu^2} = \langle P_1[\cos \theta(t)] \rangle \tag{5.19}$$

where $P_1(x)$ is the first Legendre polynomial of x. It is possible[13,14] to estimate higher-order correlation functions, $P_l[\cos \theta(t)]$ $l = 2, 3, \ldots$ from the first Legendre polynomial, and so make a meaningful comparison of dielectric dipole orientation with rotational effects in other phenomena such as rotational contributions to spectroscopic resonance line shapes, or resonance emission depolarization.

5.3. Dielectric relaxation in polymer solutions

One of the objectives of any dielectric study is to evaluate the various parameters characterizing the motion of molecular dipoles. In the case of polymeric species with many unit dipoles this includes internal conformation change as well as overall molecular rotation. The internal motion can be charcterized to give a quantitative measure of dynamic flexibility. The ease of molecular motion is affected by intermolecular forces and interactions both of an internal (intrachain) and of an external (interchain) origin. In order to form the best assessment of the intrinsic flexibility of any chain, it is obviously desirable to study its conformational changes *in vacuo*, as can be done for certain small molecules. However this is impossible for species with such low vapour pressures, and the rather poor alternative is to study polymers in dilute solution, hoping to eliminate interchain effects and to deal sensibly with the solvation introduced instead. A study of solution properties should then yield information on intramolecular effects, and the comparison between solution and bulk behaviour should elucidate specifically interchain phenomena.

A large number of dielectric relaxation studies have been carried out in this way, and many of the significant results have been collected in three review articles.[15-17]

5.3.1. Dipole geometry and active modes

Dielectric polarization and relaxation arise from changes in the orientation of a molecular dipole, and in molecules composed of numerous groups of bond moments it is the resultant dipole vector which is of importance. Changes in the magnitude or direction of this resultant can be achieved both by rotation of the whole molecule and by conformational changes within the molecule. Which of these is of greater importance, or more specifically which modes of motion exhibit dielectric activity, will depend both on the chain flexibility and on the geometrical nature of the attachment of group dipoles in the chain.

The unit dipoles in a flexible polymer molecule may be classified into three major types according to the relative geometry of the dipole moment and the backbone contour. Thus, we may have:

(a) the unit dipoles attached rigidly parallel to the chain backbone;
(b) The unit dipoles attached rigidly perpendicular to the chain backbone;
(c) The unit dipoles attached in a side-group capable of movement independent of the chain backbone.

Of course any dipole moment fixed at an angle to the backbone contour may be resolved into parallel and perpendicular components, and so may be a combination of the above types.

When the dipole is parallel to the chain contour a further subdivision is possible where the dipoles may occur unidirectionally (consistently head-to-tail) along the chain, or where there is no correlation in dipole direction. The former situation is of interest because then the resultant dipole vector of the molecule always corresponds to its positional vector. Or, in simpler terms, the chain has a clearly defined beginning and end, and the 'electrical' length of the twisted chain always corresponds to the 'distance' length (Fig. 5.2).

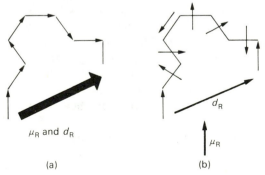

Fig. 5.2. Correlation of displacement and electric dipole vectors for unidirectional parallel dipole units, (a); but no correlation for random or perpendicular dipole units, (b).

These dipole geometries can be correlated with the three generally important modes of motion as follows. When the electric and positional vectors are identical, orientation of the resultant vector can be analysed using the normal mode equations of Chapter 3, and dielectric activity resides in the first normal mode.[18] When the unit dipoles are arranged perpendicular to the chain contour in a flexible coil, they can be considered as orienting almost independently of each other by the molecular weight-independent segmental modes of motion. Finally, of course, dipoles in side-groups capable of independent movement reflect the motion of those groups. It will be noted that the three situations represent decreasing correlation between the dipole and chain orientation (Table 5.1.).

Table 5.1

Relationship between dipole geometry and mode of motion observed

Dipole geometry	Mode of motion observed
In flexible side chain	Side chain movements
Rigid perpendicular or rigid randomly parallel	Independent segment rotation
Rigid unidirectional parallel	Normal modes of motion

Of course when a polymer chain is rigid, (either rod-like or coil-like) it will be the whole-molecule rotation that will be observed. It must be noted, too, that if a rod-like molecule has a resultant dipole across the rod as well as along it, both the end-over-end rotation and movement about the major axis will be observed. Also in the considerations which follow, it is assumed that polarization due to migration of charge carriers in an ion atmosphere, and other polyelectrolyte effects such as field-induced ionization, can be ignored.

5.3.2. Dielectric relaxation of rod-like molecules

Current interest in rod-like molecules has been stimulated by investigations into the structure and behaviour of naturally occurring molecules such as proteins and nucleic acids, and their synthetic analogues such as homopolymers of α-amino acids. In discussion of such molecules, it must be emphasized that, to a certain extent, the distinction between 'rod-like' and 'coil-like' is rather arbitrary because, for all long-chain polymers with finite energy barriers opposing segmental rotation, it must be possible to reduce the chain length to a size where the molecule is essentially rod-like. Conversely, any real rod must have a degree of flexibility, so that if the length can be increased sufficiently, a point must be reached when even slight curvature can build up to cause an over-all coil-like conformation. Consequently, the only difference between molecules conventionally described as 'coils' and those conventionally described as 'rods' is the chain length, and hence length-to-diameter ratio, at some value of which the observed behaviour changes from 'coil-like' to 'rod-like'. Some examples are given in Table 5.2.

Table 5.2

Form of certain polymers

Polymer (molecular weight $\sim 10^5$)	Flexibility
Poly(methylmethacrylate)	Flexible coil
Poly(vinylchloride)	Flexible coil
Cellulose esters	Stiff coil
Polysulphones	Stiff coil
Poly(n-butylisocyanate) (molecular weight $<10^4$)	Stiff rod
Poly(n-butyl isocyanate) (molecular weight $>10^5$)	Stiff coil
Poly(γ-benzyl L-glutamate) (α-helix)	Stiff rod
DNA (molecular weight $<10^5$)	Stiff rod
DNA (molecular weight $>10^6$)	Stiff coil

Since the rotational relaxation time of a rigid rod varies approximately as the cube of the rod length, and the length of a rod-like polymer molecule varies linearly as the number of monomer units, the observed dielectric relaxation time should vary approximately as the cube of the molecular weight. This is, of course, the end-over-end rotational time described in § 3.4.1. Such a dependence has been observed[19,20] for dilute solutions of fractionated samples of poly(n-butylisocyanate), of molecular weight below 10^5; however, if curvature gives rise to coil-like behaviour, the radius of gyration varies as the square root of the degree of polymerization, and the dielectric relaxation time for rotation of the resultant vector varies as the three-halves power of molecular weight. Such behaviour is observed for poly(n-butylisocyanate) when the molecular weight is above 10^5.

An intermediate state of affairs seems to exist[21] in solutions of the helical form of poly(γ-benzyl L-glutamate). In the low molecular-weight polymers studied the relaxation time varies approximately as the square of the molecular weight. This could be due to imperfections in the α-helix conferring a degree of curvature on the molecule (the effect is exaggerated by incorporation of the D-enantiomorph) or a ramification of the molecular-weight distribution in the polymers studied. Indeed for solutions of this polymer in chloroform the disparity between an ideal Debye relaxation and the observed relaxation has been used to estimate the molecular weight polydispersity. In addition, estimates of helix parameters were more in accord with a 3_{10} (repeat unit 3 turns containing 10 monomers) helix than the standard α-helix. The problem in making such an assignment is that curvature of a helical molecule will result in the molecule functioning as a prolate ellipsoid of revolution and so yield large values for the ellipsoid minor axis (rather than a true rod diameter). Consequently helix parameters estimated from these will always represent a lower pitch than is actually the case. The incorporation of γ-benzyl D-glutamate into the chain decreases the stability of the helix and so affects the relaxation behaviour. It has been suggested[21] that the helix dimensions are essentially unaltered up to a mole fraction of the D-enantiomorph of 0·1, but above this the helix is disrupted and smaller axial ratios result.

Much the same is found[20] in the case of poly(n-butylisocyanate). The monomer projection length along the rod is somewhat less than 0·1 nm and the rod diameter is between 0·5 and 1·0 nm, suggesting that the molecule exists as a loose imperfect helix and not as the tight helix described for the crystalline state. Increasing temperature decreases the apparent length to diameter ratio.

The curvature causing these helices to depart from a perfect rod-like shape can be estimated from the molecular dipole moments. In this case the summation of the unit vectors depends on the chain 'persistence length', i.e. a measure of the distance in one direction over which the vector property of interest extends essentially unchanged. In rough qualitative terms this is the projection of the distance along a chain before curvature reverses the direction of our property of interest. For poly(γ-benzyl L-glutamate) in chloroform this distance is about 20 nm and for poly(n-butylisocyanate) it is about 22 nm. These molecules have, therefore, rather similar curvature in solution. An interesting feature in the structure of these stiff chains is that the effective curvature seems to increase with molecular weight. The effect has been observed in both poly(n-butylisocyanate)[20] and in DNA.[22] Presumably this is due to the greater torques exerted on the longer chains by thermal fluctuations of solvent. It is not only polymers with a clearly defined helical structure that exhibit rod-like properties. In low molecular-weight poly(N-vinylcarbazole) the restrictions on internal rotation are sufficiently large for the molecule to be rod-like in behaviour. An analysis[23] of the dielectric relaxation times for polymers of molecular weight less than 5×10^3 suggests that the over-all profile of the molecule is better described as an arc than as either a rod or a coil, and that the persistence length is about 25 nm.

A number of early papers ascribed dielectric relaxation in solutions of biopolymers or polyelectrolytes to internal motion of the polymer chain. However it is now thought that the polarization occurs within the ion atmosphere surrounding the chain, and is better described by ion migration and surface polarization relationships.[16,24,25]

5.3.3. 'Stiff' and 'flexible' macromolecules

In the preceding section it was pointed out how increasing the molecular weight of a rod-like polymer brought it to a 'stiff-coil' geometry. As a result of this the rotational time for movement of the whole molecule changed from a cubic to a three-halves power of the molecular weight. Since for most polymers the end-to-end distances in solution for a given degree of polymerization or molecular-weight are rather similar (at least to an order of magnitude), the whole-molecule rotation times in a solvent of given viscosity should also be similar. However the dielectric polarization arises from an orientational change in the resultant vector moment of the molecule regardless of whether this is achieved by rotation of the molecule or by a conformational change within it. Thus the dielectric observation will 'see' whichever is the faster, and thus easier, process.

Consequently we are led to define 'flexibility' in a dielectric experiment by comparison of the rates of whole-molecule rotation and internal conformation change. This is really introducing a time- and molecular weight-dependent definition of molecular flexibility (or rigidity) in that we consider a molecule to be 'rigid' when the time required for changes in conformation by segmental rotation is greater than the time required for rotation of the whole molecule. Brownian whole-molecule rotation will be very rapid for small molecules (or short chains) and very slow for large molecules (or long chains). Consequently, for a given chemical structure there must be, in principle, a molecular weight at which the times required for segmental rearrangement and whole-molecule rotation are comparable. Below this value the molecule will be defined as 'rigid', and above this value the chain will be defined as 'flexible'. The significance of this can be gathered from Fig. 5.3. The broken lines enclose the frequency-molecular weight band in which should lie the rotational times of rigid coils in non-viscous solvents. Higher molecular weights so increase the rotational time that internal conformation

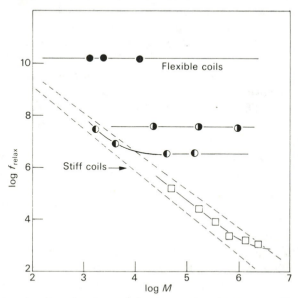

Fig. 5.3. Dielectric relaxation frequencies in solution, 25 °C, for: (●) poly(ethylene oxide); (◑) poly(methylmethacrylate);(◐) poly(N-vinylcarbazole); □ poly(γ-benzyl L-glutamate), α-helix.

change by segmental motion becomes the active process for dielectric relaxation.

It can be seen that the change to molecular weight independence occurs at the lowest molecular weights for the most flexible chains (e.g. below 10^2 for the very flexible poly(ethyleneoxide)[26] chain reorienting at about 10^{10} Hz, and above 10^6 for the very stiff poly(γ-benzyl L-glutamate)[21] helix reorienting below 10^4 Hz). The molecular weight dependences of the dielectric relaxation frequencies for some stiff-coil polymers are listed in Table 5.3. It can be seen that the relaxation frequency varies with molecular weight to a power $\frac{3}{2}$ or higher, showing that certain of the coils are more ellipsoidal than spherical in the molecular-weight range studied.

Table 5.3

Molecular-weight-dependence of relaxation frequencies for stiff coils at 25 °C

Polymer	Solvent	Molecular weight, M, or degree of polymerization, P, related to relaxation frequency, f_c. Log $f_c =$
Cellulose acetate[27]	Dioxan	$14 \cdot 9 - 2 \cdot 4 \log M$
Ethyl cellulose[27]	Dioxan	$5 \cdot 5 - 1 \cdot 9 \log P$
	Carbon tetrachloride	$4 \cdot 2 - 1 \cdot 2 \log P$
	Benzene	$4 \cdot 8 - 1 \cdot 4 \log P$
	Toluene	$5 \cdot 2 - 1 \cdot 6 \log P$
Poly(hexene 1-sulphone)[28]	Benzene or toluene	$12 \cdot 4 - 1 \cdot 5 \log M$
Poly(2-methylpentene 1- sulphone)[28]	Benzene	$14 \cdot 9 - 2 \cdot 0 \log M$

5.3.4. Dipole orientation in flexible coils

The 'stiff' coiled polymers mentioned in the preceding section evidenced a relatively low molecular weight-dependent relaxation frequency. This should be exhibited, too, by solutions of flexible-coil macromolecules with unidirectional parallel unit dipoles, although in this case the electric vector changes by way of the first normal hydrodynamic mode of motion.

No polymers with only unidirectional parallel components have been investigated, but a number of measurements have been made on molecules with both perpendicular and unidirectional parallel moments. Two such are poly(propyleneoxide)[18] and poly(p-substituted phenylacetylene)[29] (Fig. 5.4). In the former case the parallel component results from the unsymmetrical electron-donating powers of the $-CH_2-$ and $-CH(CH_3)-$ groups about the oxygen atom, whereas in the latter it results from the greater polarizability of the double bond than the single bond. In both cases the unidirectional character is preserved only as long as the chain is formed by consistent head-to-tail addition.

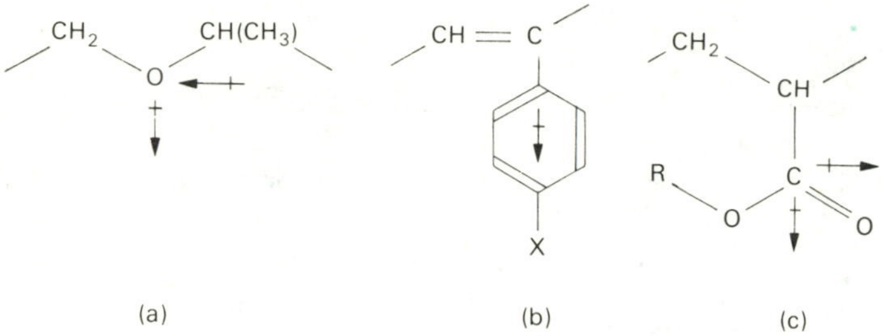

DIELECTRIC PROPERTIES

Fig. 5.4. Unidirectional parallel and perpendicular dipole components in: (a) poly(propyleneoxide); (b) poly(p-substituted phenylacetylene) with electron withdrawing substituents; (c) acrylic polymers.

In the case of poly(propyleneoxide) the measurements were made on undiluted low molecular-weight liquid polymers of low viscosity, and resulted in the very reasonable $5 \cdot 8 \times 10^{-31}$ and $3 \cdot 4 \times 10^{-30}$ C m ($0 \cdot 18$ and $1 \cdot 0$ Debye units) for the parallel and perpendicular moments respectively. In the case of the poly(p-substituted phenylacetylene)s, calculation of the dipole components could not be made because the polymers had broad molecular-weight distribution, and the low- and high-frequency processes (arising from the parallel and perpendicular components, respectively) were imperfectly resolved.

Moving from molecules with unidirectional parallel components where the normal mode exhibits dielectric activity to those with rigidly attached perpendicular components, we find that a very large number of different polymers have been investigated. Probably one of the most interesting aspects of the dielectric relaxation exhibited by such chains lies in a comparison of the relaxation frequencies with the structure of substituents on otherwise similar chains. Thus for vinyl (and related) polymers the relaxation frequency diminishes as the size and steric hindrance of the substituent is increased. Some representative examples of this phenomenon are listed in Table 5.4.

It can be seen that generally the frequency of relaxation diminishes as the steric interactions are increased. Interestingly the m-chlorostyrene exhibits a slightly higher relaxation frequency than the p-chloro analogue, possibly because in the former case rotation of the phenyl side-group is dielectrically active. Particularly surprising, in view of the formally conjugated backbone, is the high frequency exhibited by poly(p-chlorophenylacetylene). This is thought to arise by rotation about the backbone single bonds (which retain their identity with bond alternation) with the energy-angle profile for rotation being unusually flat. This latter phenomenon is thought to arise because steric constraints make it unlikely that lengths of the backbone chain and the phenyl side-groups are coplanar, and rotation of the backbone

Table 5.4

*Dielectric relaxation frequencies of some dissolved vinyl (and related)
polymers with rigid perpendicular unit dipoles at 25 °C*

Polymer	Solvent	f_{max}/Hz	$\Delta H\ddagger$/kJ mol^{-1}	Ref.
$-CH_2CH_2O-$	Benzene	$1 \cdot 5 \times 10^{10}$	$10 \cdot 3$	26
$-CH_2-CH-$ | Cl	Tetrahydrofuran	2×10^{8}	$9 \cdot 2$	30
$-CH_2-CH-$ | Br	Cyclohexane	3×10^{7}	—	31
$-CH_2-CH-$ (4-Cl-phenyl)	Toluene	3×10^{7}	$20 \cdot 2$	29
$-CH_2-CH-$ (3-Cl-phenyl)	Toluene	8×10^{7}	—	32
$-CH_2=CH-$ (4-Cl-phenyl)	Toluene	1×10^{10} (h.f. process)	6	29
$-CH_2-CH-$ (carbazolyl)	Toluene	9×10^{5}	42	33

to make or break the conjugation is associated with concomitant rotation of the side-group to break or make its conjugation to the backbone double bond.

The acrylic polymers represent a class in which the unit dipole, residing primarily on the ester carbonyl moiety, can be resolved into a perpendicular component rigidly attached to the backbone, and a parallel component ostensibly free to rotate with the side-group (Fig. 5.4). However, although these two components are ostensibly free to undergo separate motions with different relaxation times, all investigations of dilute solutions have detected only a single relaxation process[34]. It appears, therefore, that in solution the rotation of the backbone and first side-group bonds is a co-operative process. As a consequence, the dielectric relaxation time should, again, be a measure of the chain flexibility. The effect on this flexibility of increasing group steric interaction, both directly substituted onto the backbone and in the ester side chain is illustrated in Table 5.5. In this instance the increase in

Table 5.5

Dielectric relaxation frequencies of acrylic polymers in toluene at 27 °C

Polymer	f_{max} (Hz)	$\Delta H\ddagger$ (kJ mol^{-1})
$-CH_2-CH-$ $\quad\quad COOCH_3$	$1 \cdot 87 \times 10^9$	23
$\quad\quad CH_3$ $-CH_2-C-$ $\quad\quad COOCH_3$	$3 \cdot 9 \times 10^7$	27
$\quad\quad CH_3$ $-CH_2-C-$ $\quad\quad COOC_4H_9$	$3 \cdot 4 \times 10^7$	27
$\quad\quad CH_3$ $-CH_2-C-$ $\quad\quad COOC_9H_{10}$	$3 \cdot 0 \times 10^7$	27
$\quad\quad CH_3$ $-CH_2-C\quad\quad CH\quad\quad CH-$ $\quad\quad COOCH_3CO\quad CO$ $\quad\quad\quad\quad\quad O$	$1 \cdot 4 \times 10^7$	32

(2 per cent citraconic anhydride)

steric hindrance on the backbone is effective in reducing the segmental reorientation frequency, but increasing the size of the ester group has very little effect. Presumably this is because the relatively small reorientation of the carbonyl dipole in the electric field can be accommodated by partial rotation of the neighbouring bonds, and does not require movement of groups more than two or three atoms distant on the side chain.

The dielectric relaxation in solutions of polyvinyl acetate (see ref. 16 for details of the many observations) and polyvinyl isobutyl ether[35] shows the same characteristics as discussed for the acrylic polymers, and so a co-operative segmental side chain is presumably again involved. The comparable relaxation frequencies in these cases are 2×10^9 and 3×10^8 Hz respectively.

One of the problems in analysing the observations on polymer solutions arises because the test of theories incorporating specific models usually requires an analysis of the way in which the observed relaxation departs from ideal form. In this respect the experimental results are usually inadequate in that measurements made on the most dilute solutions have generally been at the sensitivity limits of the apparatus used. On the other hand more concentrated solutions complicate the relaxation behaviour through entanglement and related viscosity effects. However there is an indication[17] that in the most dilute solutions, at reasonable temperatures, the relaxation is almost ideal, becoming less so as the concentration is raised or the temperature lowered. This implies that the decay of segment dipole angular orientation is characterized by an exponential time-correlation or decay function, in contrast to the situation in solid polymers.

5.4. Dielectric relaxation in solid polymers

An enormous amount of work has been carried out on the study of dielectric relaxation in solid polymers. The principal impetus for this has derived from the use of such materials as insulators and dielectrics. The major objective of such studies has been to characterize the various loss and dispersion processes in the time (or frequency) and temperature domains, and to provide an explanation in terms of various polarization (usually dipole orientation) processes.

5.4.1. Nomenclature

Considerable confusion exists in the literature with regard to the naming of the different relaxation or loss processes observed in polymers. This has arisen because of the custom of naming the process observed at higher temperatures or lowest frequencies as the 'α-process' and then naming subsequent lower temperature (higher frequency) processes β, γ, δ, etc. Of course this has depended, in a somewhat arbitrary way, on the initial temperature or frequency selected and on the dielectric activity of the processes observed. The authors prefer the system which designates the process associated with the main glass transition in amorphous polymer as the α-relaxation, retaining a clear β-relaxation at appropriate temperatures and frequencies as detailed in Chapter 1, and naming subsequent relaxations of amorphous polymer γ, δ, etc. The possible relaxations in crystalline regions can be designated by appropriate subscript, viz. α_c, and higher temperature relaxations in rubbery or molten material by appropriate superscripts viz α^+.

5.4.2. Origin of dielectric activity

The dielectric activity of polymers is usually discussed in terms of partial or complete dipole orientation polarization. Thus movement of non-polar segments between orientations of different energy may constitute a relaxation observable by dynamic mechanical or acoustic techniques, and yet not be observable as dielectric relaxation. Conversely, reorientation of dipolar segments through iso-energetic conformations may create dielectric polarization relaxation with no analogous mechanical relaxation.

The extent of reorientational freedom of the resultant dipole vector of the polymer chains determines the degree of polarization. The various component-group dipole moments, which together constitute the total resultant moment, may have differing degrees of freedom, and consequently achieve reorientational freedom at different temperatures so giving rise to the different transition or relaxation processes. It is important to note that this includes not only different chemical groupings but different degrees of co-operation in the movement of neighbouring identical groups, or different amplitudes or degrees of orientational freedom of individual groups. Indeed the application of time-correlation functions to dielectric relaxation in polymers has yielded, through the cross-correlation terms, insight into the extent of interaction between neighbouring group dipoles.

5.4.3. Relaxations in the temperature or frequency domains

A catalogue of the many relaxations reported for an enormous variety of polymers is obviously beyond the scope of this volume. The subject has been extensively reviewed.[36-9] Consequently, we present here a selection of results for some common types of polymers, pointing out how these illustrate some basic principles or effects.

In many respects the concepts discussed in § 5.3 apply to bulk polymers in much the same way as to dissolved polymers. The principal difference is that in solids the molecular motion is controlled not only by the intrinsic intrachain, intersegment interactions, but also by interchain interactions. These latter intermolecular forces influence chain dynamics in a variety of ways, one particularly important aspect arising in the morphology of the solid material.

Examples of the relaxation parameters of a selection of characteristic polymers are given in Table 5.6, and the following general concepts can be identified.

Polarity of relaxing group. Generally the greater the polarity of the relaxing group, the stronger will be the intergroup interactions on both the intra- and interchain levels. Consequently in polymers where other steric or geometric factors are similar, the temperature and activation energy of a relaxation will increase (and at any temperature the relaxation frequency

Table 5.6

Dielectric relaxation in some common polymers

Polymer	Relaxation	Temperature (K)	Frequency	Ref.
Polyethylene	α-motion in the crystalline phase	350	1 kHz	43–7
	β-primary motions of the amorphous phase, probably associated with chain branches	260	1 kHz	43–7
	γ-combination of processes associated with defect migration and reorientation motion of the amorphous phase	160	1 kHz	43–7
Polypropylene	α-motion in the crystalline phase	390	1 kHz	48–50
	β-relaxation in the amorphous region	333	1 kHz	48–50
Poly(vinylchloride)	α-segmental motion of the main backbone chain	373	1 kHz	51–4
	β-local motions of short elements of the chain backbone	273	1 kHz	51–4
Polytetrafluero-ethylene	α-decreases in magnitude with increasing crystallinity-motion in the amorphous phase	400	1 Hz	55–7
	β-torsional motions of chains in the crystalline phase about the crystalline axis.	320	1 Hz	55–7
	γ-associated with local motion of elements of the backbone in the amorphous phase	180	1 Hz	55–7
Poly(methyl-acrylate)	α-glass transition. Co-operative motions of segments of the polymer. Sensitive to the tacticity of the polymer.	383	20 Hz	58–62
	β-local motions of elements of the chain backbone.	308	20 Hz	58–62
Poly(vinylacetate)	α-glass transition temperature. Co-operative motions of segments of the polymer.	338	1 kHz	53, 63–5
	β-motion of the acetate side-group independent of the backbone motion.	233	1 kHz	53, 63–5
Polystyrene	α-glass transition segmental motion about the backbone.	388	20 Hz	66–9
	β-local motions of the backbone.	333	20 Hz	66–9
Poly(acry-methylene)	Relaxation of the amorphous region.	210	100 Hz	70–2
	Motion of end groups in the crystalline phase.	190	100 Hz	70–2

will decrease) with increasing polarity. An obvious comparison is between relaxations in amorphous poly(vinylchloride) and polypropylene, where the chloro and methyl groups exert similar steric effects, but are of quite different polarity.

Although the dipole–dipole interactions involved in such forces are considered to be 'long-range' compared with other intermolecular forces, nevertheless their strength is still quite sensitive to intermolecular separation. Consequently any chemical or isomeric modification which prevents close approach of neighbouring dipoles will weaken the intersegment interactions, and facilitate relaxation characterized by lower 'transition' temperatures and activation energies, or higher relaxation frequencies. This is illustrated by the poly (alkylmethacrylates), where the larger alkyl ester residues control the intergroup separation of the carbonyl dipoles. It must be pointed out, however, that the inclusion of flexible alkyl moieties facilitates intramolecular motion by other effects as well as that of a simple spacer.

Steric interactions of relaxing group. Again, as expected, the polymer chains with the greatest steric constraints to rotational isomeric change of the backbone geometry exhibit the highest relaxation 'transition' temperatures. Although it is difficult to separate steric (or repulsive interaction) effects from intermolecular attractive effects arising from changes in group polarizability, the series polystyrene, poly (vinylnaphthalene), poly (vinylanthracene) illustrates how increasing size of a rigid backbone substituent hinders the necessary molecular motion and so raises the relaxation transition temperature.

Energetics of conformational change. One of the difficulties inherent in the simple qualitative approaches outlined in the preceding paragraphs, is that consideration of forces opposing group rotation leads naturally to the concept of activation energy barriers and a kinetic approach to relaxation. It is not quite so easy to construct 'rule of thumb' principles which lead to an estimate of relaxation parameters via the concept of 'flex energy' involved in the statistical mechanical approach. The importance of this can be seen immediately by comparing isotactic and syndiotactic poly(methylmethacrylate) or poly(acrylonitrile), where the tactic form of higher density (and so ostensibly closer dipole–dipole separation) has the lower relaxation temperature. For such comparisons the relaxation behaviour can only be completely understood when considered against possible pathways on a multi-dimensional energy–angle surface, which detail does not yet exist for the majority of polymer systems.

Morphology and block copolymers. When chain segments of an inherently flexible chain are attached to either less flexible chain blocks, or to domains of inherently greater rigidity (such as crystallites or a second phase of different chemical nature) the mobility of units close to the attachment is considerably restricted. The observed dielectric relaxation is averaged over all active dipolar units, so that the effect appears to increase as the number of units between attachment points decreases. This is particularly clear in the orientation polarization of ether segments in a series of polyether-poly-urethanes, or in polymers of different degrees of crystallinity.

5.4.4. Form of the relaxation function

One of the most immediately apparent features of the dielectric relaxation of solid polymers is that the relaxations very seldom (in the case of the principal α- and β- transitions, never) conform to the ideal Debye model. Observations of such behaviour have been reported and interpreted on a variety of levels.

An enormous volume of literature exists in which an observed relaxation is fitted to one of the empirical functions discussed in § 5.1, and the empirical parameter (often called the 'distribution parameter') reported without further elucidation. Often the width at half height of the α-loss peak is about two decades of frequency, which is almost double that of an ideal relaxation. In many cases the relaxation is asymmetric in the frequency plane, and can be fitted by the Cole–Davidson 'skewed arc' empirical function (with the distribution parameter, α, of eqn (5.4) about 0·3 to 0·5), or in the time domain by the Williams–Watts empirical function (eqn (5.6)) with β between 0·4 and 0·6. The β-relaxation shows even more extreme departures from ideality, often extending over more than five decades of frequency, and being characterized by distribution parameters ($1-\alpha$, α, and β in eqns (5.4)–(5.6), respectively) much less than unity.

The next stage of interpretation is the analysis of the relaxation function in such a way as to yield a specific distribution function. The general relation-ships involved have been extensively discussed in connection with the viscoelastic and dynamic mechanical properties of polymers, and have been less widely used in the interpretation of dielectric data. Indeed the hope, widely held in the two decades 1950–70, that evaluation of the phenomenological distribution function would lead to the revelation of some basic molecular characteristics has not been realized. Nevertheless, the distribution of relaxation times can be derived from the time-dependent electric modulus (inverse permittivity), $M(t)$, as

$$M(t) = M(0) + \int_0^\infty f(\tau) \exp\left(\frac{-t}{\tau}\right) d\tau \qquad (5.20)$$

where $f(\tau)$ is the 'relaxation-time spectrum'. More usefully a logarithmic scale is employed such that

$$M(t) = M(0) + \int_{-\infty}^{\infty} M(\tau) \exp\left(\frac{-t}{\tau}\right) d(\ln \tau) \qquad (5.21)$$

where $M(\tau) d(\ln \tau)$ gives the contribution to the relaxation associated with relaxation times between $\ln \tau$ and $(\ln \tau + d(\ln \tau))$. In the more familiar permittivity presentation the spectrum is analogous to the dynamic mechanical retardation-time spectrum, $L(\tau)$,

$$\varepsilon(t) = \varepsilon_{\infty} + \int_{-\infty}^{\infty} L(\tau) \left\{ 1 - \exp\left(\frac{-t}{\tau}\right) \right\} d(\ln \tau). \qquad (5.22)$$

The spectra $M(\tau)$ and $L(\tau)$ can be computed from the time-dependent dielectric parameters by the usual Fourier and Laplace transformations.

Early attempts to compare distribution functions so observed with those relating to molecular-weight distribution (or other related polymeric properties) failed because for high molecular-weight solids the α- and β-processes are relatively insensitive to molecular weight (or more specifically to a long wave-length mode of motion of the whole chain). Similarly little progress derived from attempts to fit simplified distribution functions ('box' or 'wedge' functions) to the relaxation observations. Consequently the reader is referred to Chapter 4 for further discussion of this approach as applied to the viscoelastic functions.

An alternative approach is one in which the empiricism of the distribution function is replaced by selection of an arbitrary molecular model for the relaxation process. The defect diffusion model[6] discussed earlier (eqn (5.7)) and the many-body interaction model[7] are two such in which the over-all relaxation is modelled by two interacting processes.[40,41] However, here again, the application of such models has not permitted evaluation of a specific molecular quantity, but only evidence that such models can (presumably with others) provide a self-consistent explanation of, or fit to, the experimental observations.

Even the use of correlation functions, by which it was hoped to express the non-ideal behaviour in terms of auto- and cross-correlations has not yielded unambiguous correlation parameters which provide greater insight than the Kirkwood 'g'-factor[42] used to differentiate dipole orientation in condensed and gaseous phases.

5.5. Dielectric relaxation in two-phase systems

5.5.1. Interfacial polarization in polymers

We move now to examine processes some of which exhibit extremely low relaxation frequencies (below 1 Hz in many cases). These are the dielectric polarizations brought about when mobile charge carriers, migrating under the influence of the applied field, become trapped or localized at an interface arising from some two-phase structure in the polymer (Fig. 5.5). The time

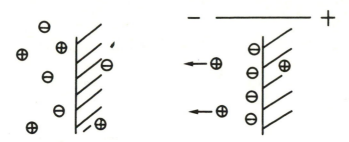

Fig. 5.5. Polarization by trapping of migrating charge carriers.

constant for the dielectric process is related to the charge carrier mobility, and so will be high in systems of very low conductivity.

This interfacial polarization has been known for many years, and its study is usually associated with the names of Maxwell[73], Wagner[74], and Sillars.[75] Maxwell considered the case where carrier localization occurred close to the electrodes, and indeed almost all 'non-conducting' polymers seem to exhibit polarization phenomena due to inadequate carrier discharge at 'blocking' or 'non-ohmic' electrodes. However, for many systems of interest (other than thin films or relatively highly conducting materials) these processes require long times or equivalent frequencies below 10^{-3} Hz. Of greater relevance in polymer systems are the interfacial polarizations which occur internally due to an intrinsic two-phase structure in the sample under investigation. Very large permittivities and losses can occur, which are often characterized by frequencies between 10^{-3} and 10^{+3} Hz (or times between 10^{+3} and 10^{-3} s), and so may critically affect the electrical properties of supposedly insulating materials in technical use.

5.5.2. Theoretical treatment of interfacial polarization

A variety of theories have been proposed to describe this type of phenomenon.[76–94] In practice, apart from a very few rather specialized and somewhat artificial situations,[74,75] quantitative comparison of theory and experiment has not been very successful, particularly for polymers.[90] One reason for this failure has been an inability to describe accurately the complex morphologies found in most polymers.

The permittivity increment. The simplest models of an heterogeneous dielectric consider the solid as an ensemble of isotropic ellipsoids of permittivity ε_2 and volume fraction v_2 embedded in a matrix of permittivity ε_1, both phases having zero conductivity and $\varepsilon_2 > \varepsilon_1$. The orientation of the ellipsoid axes (a, b, c) relative to the applied electric field can be defined by the angles γ_i $(i = a, b, c)$. The permittivity, ε_s, of this simple solid is described by the Maxwell–Wagner–Sillars (MWS) model

$$\varepsilon_s = \varepsilon_1 \left\{ 1 + v_2(\varepsilon_2 - \varepsilon_1) \sum_i \frac{\cos^2 \gamma_i}{\varepsilon_1 + A_i(1 - v_2)(\varepsilon_2 - \varepsilon_1)} \right\} \tag{5.23}$$

where A_i is the depolarization ratio of the constituent ellipsoids along the i-axis and is defined by

$$A_a = \frac{abc}{2} \int_0^\infty \frac{\delta s}{(s + a^2)[(s + a^2)(s + b^2)(s + c^2)]^{\frac{1}{2}}} \tag{5.24}$$

where $s = (a^2 - c^2)/(z^2 - a^2)$ for A_a and $s = (b^2 z^2 - c^2)/(1 - z^2)$ for A_c, z being the projection onto the a axis, which is always chosen to lie parallel to the electric field direction. The coefficient A_a has a value of unity for lamellae, is one-third for spheres, and tends to zero for long thin rods. Orientations which lead to simple analytical forms of eqn (5.23) are:

(i) The a-axis orientated parallel to the electric field:

$$\cos^2 \gamma_a = 1 : \cos^2 \gamma_b = \cos^2 \gamma_c = 0;$$

(ii) The a-axis orientated perpendicular to the electric field:

$$\cos^2 \gamma_a = \cos^2 \gamma_b = 0 : \cos^2 \gamma_c = 1;$$

(iii) A random orientation of ellipsoids:

$$\cos^2 \gamma_a = \cos^2 \gamma_b = \cos^2 \gamma_c = \tfrac{1}{3}.$$

In general, with the a-axis parallel to the electric field, an ensemble of rods gives a larger permittivity than a system of spheres or random ellipsoids, which in turn are higher than perpendicular ellipsoids or a perpendicular lamellar structure. Systems polydisperse in the size of the occluded domains can be described by a modification of eqn (5.23) provided $v_2 < 0.2$. In fact eqn (5.23) with $A_a = \tfrac{1}{3}$ has been shown to be valid only for samples in which there is a size gradation of the disperse phase.

At higher concentrations of the dispersed phase, electrostatic interactions between domains become significant and modified expressions are required, viz:

(a) Bruggeman[78] equation:

$$\frac{\varepsilon_2 - \varepsilon}{\varepsilon_2 - \varepsilon_1} \left(\frac{\varepsilon_1}{\varepsilon} \right)^{\frac{1}{3}} = 1 - v_2; \tag{5.25}$$

(b) Looyenga[79] equation:

$$\frac{\varepsilon^{\frac{1}{3}} - \varepsilon_1^{\frac{1}{3}}}{\varepsilon_2^{\frac{1}{3}} - \varepsilon_1^{\frac{1}{3}}} = v_2 . \tag{5.26}$$

Both relationships are applicable to concentrated dispersions of spheres, but the distinction between phases and validity of these equations is lost when contact between the dispersed domains creates a semi-continuous path for charge transport.

The dielectric dispersion. The position in the frequency or time domain of the dielectric dispersion, and its shape, may be obtained by substituting the complex admittances ($\Delta_i = \sigma_i + i\omega\varepsilon_i(i\omega)$) for the ε_i in the permittivity equations appropriate to the morphology being studied. For dilute dispersions in which the conductivity of the occluded phase, σ_2, is greater than that of the matrix phase, σ_1, the relaxation equations obtained from (5.23) are:

$$\varepsilon_0 = \varepsilon_1 \left\{ 1 + v_2(\sigma_2 - \sigma_1) \sum_1 \frac{\cos^2 \gamma_i}{\sigma_1 + A_i(1 - v_2)(\sigma_2 - \sigma_1)} \right.$$
$$\left. + v_2\sigma_1 \sum_i \frac{\cos^2 \gamma_i(\sigma_1\varepsilon_2 - \sigma_2\varepsilon_1)}{\sigma_1 + A_i(1 - v_2)(\sigma_2 - \sigma_1)^2} \right\} \tag{5.27}$$

$$\tau_i = \varepsilon \left\{ \frac{\varepsilon_1 + A_i(1 - v_2)(\varepsilon_2 - \varepsilon_1)}{\sigma_1 + A_i(1 - v_2)(\sigma_2 - \sigma_1)} \right\} . \tag{5.28}$$

For an axially isotropic dispersion of phases the Debye approximation of a single relaxation time, τ, may be used

$$\frac{\varepsilon_0' - \varepsilon_\infty}{\varepsilon_0 - \varepsilon_\infty} = \frac{1}{1 + \omega^2\tau^2}$$

$$\frac{\varepsilon''}{\varepsilon_0 - \varepsilon_\infty} = \frac{\omega\tau}{1 + \omega^2\tau^2} . \tag{5.29}$$

Randomly orientated dispersed phases may be represented by an equal combination of three Debye processes.[83] The relaxation amplitude and time are both increased by ellipsoid elongation in the field direction, and this can lead to increases of several orders of magnitude.

Reversal of the relative magnitudes of the conductivities of the continuous and discontinuous phases ($\sigma_1 > \sigma_2$) leads to a reduction in the magnitude of the dielectric dispersion in all cases except the lamellar morphology (when both phases are continuous over relatively large distances). These simple models for low-volume fractions of second phase have been discussed extensively by Sillars, and the various relationships are summarized in Table 5.7.

Table 5.7

(a) MWS relations for low-volume fraction dispersions: $v_2 < 0.2$: $Xl < 1$

Morphology	Author	Relationships in eqns (5.27) and (5.28)
spheres	Wagner	$A_a = \frac{1}{3}$; $\quad \cos^2 \gamma_a = 1$; $\quad \cos^2 \gamma_b = \cos^2 \gamma_c = 0$
ellipsoids *a*-axis parallel	Sillars	all A_a: $\quad \cos^2 \gamma_a = 1$; $\quad \cos^2 \gamma_b = \cos^2 \gamma_c = 0$
ellipsoids *a*-axis perpendicular	Sillars	$0 \leqslant A_b \leqslant 0.5$: $\quad \cos^2 \gamma_b = \frac{1}{2}$; $\quad \cos^2 \gamma_a = \cos^2 \gamma_c = 0$
ellipsoids random	Fricke	all A_i: $\quad \cos^2 \gamma_i = \frac{1}{3}$

(b) MWS relations for concentrated dispersions; all v_2: $Xl < 1$

Morphology	Author	
spheres	Bruggemaan Hanai	$\varepsilon_0 \left\{ \dfrac{3}{\sigma - \sigma_2} - \dfrac{1}{\sigma} \right\} = 3 \left\{ \dfrac{\varepsilon_2 - \varepsilon_1}{\sigma_2 - \sigma_2} + \dfrac{\varepsilon_2}{\sigma - \sigma_2} \right\} - \dfrac{\varepsilon_2}{\sigma_1}$

(c) MWS relations in the presence of space charge layer: $v_2 < 0.05$: all Xl: $\sigma_1 \ll \sigma_2$

Morphology	Author	ε_∞	ε_0	τ
spheres (radius $\frac{1}{2}\alpha$)	Trukhan	$\varepsilon_\infty = \varepsilon_1\left[1 + 3v_2\dfrac{\varepsilon_2 - \varepsilon_1}{2\varepsilon_1 + \varepsilon_2}\right]$	$\varepsilon_0 = \left\{1 + 3v_2 \times \dfrac{(Xl)^2\tanh\left(\frac{Xl}{2}\right) + \frac{2(2\varepsilon_2+\varepsilon_1)}{\varepsilon_2}\left[2\tanh\left(\frac{Xl}{2}\right) - Xl\right]}{(Xl)^2\tanh\left(\frac{Xl}{2}\right) + \frac{4(\varepsilon_2-\varepsilon_1)}{\varepsilon_2}\left\{2\tanh\left(\frac{Xl}{2}\right) - Xl\right\}}\right\}$	$\tau = \dfrac{\varepsilon(2\varepsilon_1+\varepsilon_2)}{(2\sigma_1+\sigma_2)} \times \dfrac{(Xl)^2\tanh\left(\frac{Xl}{2}\right) + 6\left(2\tanh\left(\frac{Xl}{2}\right) - Xl\right)}{(Xl)^2\tanh\left(\frac{Xl}{2}\right) - \frac{4(\varepsilon_1-\varepsilon_2)}{\varepsilon_2}\left\{2\tanh\left(\frac{Xl}{2}\right) - Xl\right\}}$

(d) Relations for systems with surface conductivities

Morphology	Author	ε_∞	ε_0	τ
spheres (radius a)	O'Konski	In the MWS relations of Table 5.7 (a) σ_2 is replaced by $\sigma_2 + \lambda_s^2/a$		
ellipsoids	O'Konski	In the MWS relations of Table 5.7 (a) σ_2 is replaced by $\sigma_2 + Ki\lambda_s$, where $Ka = \sigma^2/b$ and $Kb = Kc = \sigma^4/\pi b$.		
spheres (radius a)	Schwarz	eqn (5.23) $Aa = \frac{1}{3}$ $\cos^2\gamma_a = 1$ $\cos^2\gamma_b$ $= \cos^2\gamma_c = 0$	$\varepsilon_\infty + \dfrac{9e^2\sigma' a v_1}{4\varepsilon kT(1+v_1/2)^2}$	$\tau = a^2/(2\eta^s kT)$[†]
ellipsoids (a-axis parallel)	Takashima	eqn (5.23) all A_i, $\cos^2\gamma_a = 1$ $\cos^2\gamma_b$ $= \cos^2\gamma_c = 0$	$\varepsilon_\infty + \dfrac{9e^2\sigma' a^2 v_1}{8\varepsilon kTb(1+v_1)^2}$	$\tau = \dfrac{a^2 + b^2}{2\eta^s kT}$[†]
ellipsoids (random)	Takashima	eqn (5.23) all A_i $\cos^2\gamma_a = \frac{1}{3}$	$\varepsilon_\infty + \dfrac{e^2\sigma' v_1}{3\varepsilon kT(1+v_1)^2}\left\{\dfrac{9}{8}\dfrac{a^2}{b} + 2b\right\}$	$\tau_a = \dfrac{a^2 + b^2}{2\eta^s kT}$[†] τ_b not given

† η^s is the charge mobility in the surface layer.

In polymer solids with a high-volume fraction of dispersed phase, $\Delta\varepsilon/v_2$ is larger than for dilute suspensions, although the relaxation time is unaffected.[85-8]

Effect of space charge. In the simplest models (MWS) the interfacial polarization is assumed to arise from charges located on an infinitely thin layer surrounding the dispersed phase. In practice, the layer may have a finite thickness (l) of the order of the Debye shielding radius χ^{-1},[89]

$$\chi^{-1} = \left(\frac{\varepsilon\varepsilon_2 kT}{2e^2 n}\right)^{\frac{1}{2}} \tag{5.30}$$

where ε is the permittivity of free space, n is the charge carrier density, e is the charge on an electron, and ε_2, k, and T have their usual meaning. This diffusion layer leads to a decrease (with increasing χ^{-1}) in both the magnitude of the dispersion and its relaxation time.

Surface conductivity. An obvious inadequacy of the simple theories is the assumption of sharp boundaries between the constituent phases. In reality, the occluded domain will be surrounded by a diffuse region with dielectric properties (ε_s, σ_s) different from either of the pure phases. Analysis[91] of this predicts that two distinct dispersion regions should be observed. In the limit, when the boundary becomes very thin and is of extremely high conductivity, ($\sigma_s \gg \sigma_1$, σ_2), one dispersion region predominates and occurs at higher frequency with a larger amplitude than would be expected for a system without the conducting layer.

Two refinements of this model are worth mentioning. Both involve the surface conductivity, which in the first[92] is independent of frequency and in the second[93] exhibits a frequency dependence. In the first case with an infinitely thin interfacial region the surface conductivity is given by

$$\sigma_s = \sum_i n_i \eta_i^s q_i. \tag{5.31}$$

n_i is the charge carrier density at the surface and η_i^s is the mobility of the charge q_i. The relaxation behaviour is obtained by replacing σ_2 in the simple model by $(\sigma_2 + k\sigma_s)$ where k is dependent on both the shape and size of the dispersed phase. For spheres of radius a, $k = 2/a$, while for long thin rods ($a \gg b = c$), $k_a = 2/b$ and $k_b = k_c = 4/\pi b$.

It has been pointed out that the conductivity may be frequency-dependent due to localization of charge by strong electrostatic interaction with the surface of the dispersed phase. This theory predicts that the equilibrium polarization is governed by surface diffusion, and the relaxation time is directly related to the time required for the charge carrier to diffuse over a distance of the order of the particle diameter. A further limitation of this approach is that it ignores the possibility of exchange between bound charge carriers and the bulk of the matrix phase. Additionally, as a result of the charge carrier motion, electric fluxes are induced normal to the surface, and these give rise to an oscillating diffuse volume close to the interface reducing the relaxation time and dielectric increment by $(1 + 2^{x n_s}/n)$, where n_s is the charge carrier density in the interface and n is the bulk charge carrier density. Such analyses are included in Table 5.7.

5.5.3. Comparison of theory and experimental observation

Almost all polymer systems studied exhibit interfacial polarization with the following characteristics.[95-8] The loss is often of high magnitude ($\varepsilon'' \gg$ 100) and has a temperature coefficient (activation energy) which mirrors that for d.c. conduction. While the conductivity and relaxation frequency are increased by adsorbed moisture, the relaxation magnitude may be increased or decreased.[99] This seems to indicate that in these polymers, generally non-polar and non-conducting, the interfacial polarization involves the same charge carriers (perhaps arising from spurious ionic or ionizable impurities) as does d.c. conduction. There is a general lack of data relating the dielectric increment to the volume fraction of the dispersed phase. This reflects the lack of reproducibility of morphology in many samples, even when obtained from the same polymeric material.

A study of the relaxation magnitude in a series of nylons indicated an increase with increasing crystallinity. However the observed dielectric increment was much higher than predicted by simply MWS theory.[95,96] Better agreement would have been obtained if the dispersed phase had been assumed to be ellipsoidal, which is consistent with current theories of the morphology in these systems.

Measurements made on surfactant-doped polyethylene[100] in which the detergent forms a highly conducting layer at the crystallite interface have given reasonable agreement with theory. Similar close correspondence between theory and experiment has been observed in poly(vinyl-idenefluoride)[101] where a diffuse double layer may exist at crystallite surfaces.

One of the difficulties in such work has been in obtaining samples with a simple, clearly defined, two-phase geometry.

The styrene-diene-styrene three-block copolymers form materials in which glassy polystyrene domains are dispersed in a polydiene matrix in a very regular geometrical array of lamellae, cylinders, or spheres. The morphology of such systems has been extensively studied and is easily characterized by electron microscopy. An additional attraction is that the geometry can be altered or distorted by application of an appropriate strain, and this can be done during the course of dielectric examination.

The interfacial polarization process in a film-cast copolymer with 30 weight per cent styrene was found[102] to be in reasonable agreement with

MWS predictions for a morphology of parallel cylinders oriented at random, whereas that for a 50 weight per cent styrene copolymer was compatible with predictions for a random arrangement of partially parallel lamellae. These structures correspond to morphologies observed by electron microscopic observation. No evidence was found for a dielectric effect specifically requiring a diffuse conducting interfacial zone between the polystyrene and polybutadiene domains. However in polar two-phase polyester, polyether systems the experimental results cannot be fitted to theory without postulation of such a zone.[103]

A striking example of the technical significance of this phenomenon is illustrated by the phenol-formaldehyde resin-cellulose laminate used as baseboard for the construction of electronic circuitry.[104] Losses due to localized conduction in the 'wet' regions of the cellulose–resin boundary, are the cause of failure when used under adverse (temperature and humidity) operating conditions. Modification of the resin formulation so as to reduce moisture absorption gives correspondingly superior electrical performance.

Of course the existence of Maxwell–Wagner–Sillars interfacial polarization is not always undesirable. When a conducting phase is dispersed in a non-conducting matrix, then polarization of the dispersed phase can occur up to a frequency associated with the carrier mobility in the dispersed phase. For dispersed metals, for example, this means optical frequencies, so high permittivities or refractive indices are exhibited throughout the radio and microwave spectral regions. This provides a useful trick for the manufacture of highly refracting microwave lenses!

In a rather similar way high permittivities can be achieved up to megahertz frequencies by the dispersion of graphite in a polymer matrix. Such systems are interesting because the use of oriented carbon fibres permits achievement of anisotropic polarization effects.

5.6. Thermally stimulated depolarization

5.6.1. Introduction

Closely related to dipole orientation and interfacial polarization is the phenomenon of thermally stimulated depolarization. In this case a sample is polarized by an electric field at a temperature permitting the rotational or translational movement of polar entities. The polarizing voltage is maintained while the sample is cooled to temperatures when the polarizing, or depolarizing, molecular movements are impossible. When the field is removed and the sample warmed, a depolarization current is observed at those temperatures corresponding to the onset of mobility in the polar entities. In this way a depolarization thermogram is obtained which can yield information on orientation transition phenomena, or energy trap depths for charge carriers.

This technique of thermally stimulated depolarization was pioneered by Bucci and Fieschi[105] for the study of trapping and liberation of charge carriers in ionic crystals. Early studies of polymers,[106,107] too, concentrated on the evaluation of trap depths, but over recent years the technique has been extended to investigation[108,109] of the various transitions associated with polymer molecular motions.

5.6.2. Theory for ideal relaxation

We consider a number of molecular polar species, N, polarized by an electric field applied at temperature, T_p, for an effectively infinite time. This initial polarization, P_0 is retained on cooling in the field to a measurement temperature $T(t)$, which then increases in the absence of a field. The instantaneous variation of N with time is given by the ideal Debye expression,

$$-\mathrm{d}N/N = \mathrm{d}t/\tau(T)$$

where $\tau(T)$ is the depolarization relaxation time at temperature, T. Consequently, the polarization remaining at temperature $T(t)$ is

$$P(t) = P_0 \exp\left[-\int_0^t \{\mathrm{d}t'/\tau(T(t'))\}\right]. \tag{5.32}$$

The instantaneous depolarization current at time, t, is

$$j(t) = -\mathrm{d}P(t)/\mathrm{d}t = P(t)/\tau(T(t)). \tag{5.33}$$

The residual polarization can also be written in terms of a temperature increase to infinity,

$$P(T(t)) = \int_T^\infty j(t)\,\mathrm{d}t \tag{5.34}$$

so that the relaxation time at a given temperature is

$$\tau(T) = \int_T^\infty j(t)\,\mathrm{d}t/j(T(t)). \tag{5.35}$$

The object of such an experiment is often to characterize the relaxation time (and amplitude) as a function of temperature. In order to achieve this a relationship between temperature and time must be established. The simplest case is a linear temperature change, $\mathrm{d}T/\mathrm{d}t = C$. The depolarization thermocurrent passes through a maximum at a temperature determined by the heating rate, the relaxation time, and its temperature coefficient. Insertion of a linear temperature variation in (5.35), followed by differentiation yields the current–temperature profile as

$$\frac{\mathrm{d}\ln j(T)}{\mathrm{d}T} = \frac{1}{\tau(T)}\left(\frac{1}{C} + \frac{\mathrm{d}\tau}{\mathrm{d}T}\right) \tag{5.36}$$

and the conditions at the temperature of maximum current, $T_{\max}$, as

$$\left(\frac{d\tau}{dT}\right)_{T_{\max}} = \frac{-dt}{dT} = \frac{-1}{C}. \tag{5.37}$$

The calculation can be extended, in the same way, to non-linear temperature programmes and depolarization processes which obey more complex relaxation-decay functions.

5.6.3. Resolution of multiple processes

One interesting aspect of the technique is found in systems exhibiting more than one relaxation process. Here there is scope for examination (and often complete resolution) of the individual processes by alteration of the temperature of initial polarization. As an example we consider the case of two superposed thermocurrent peaks, $T_{m1} < T_{m2}$ being the temperatures of current maxima. The important point is that for appropriately activated rate processes the relaxation time is a sensitive function of temperature, so that

$$\tau_1(T_{m2}) \ll \tau_1(T_{m1}) \sim \tau_2(T_{m2}) \ll \tau_2(T_{m1}).$$

For simple, ideal processes the polarization and depolarization times are the same. Then polarizing at T_{m1} for the time $\tau_1(T_{m1})$ will result in extensive (*ca.* 63 per cent) polarization of the easier 'low-temperature' process while leaving the second process virtually unpolarized. Cooling and observing the thermogram in the usual way then exhibits the low-temperature peak at slightly reduced intensity. The opposite effect is achieved by carrying out the usual polarization–depolarization cycle until T_{m2} is reached. The temperature is maintained for the short period $\tau_1(T_{m2})$, when the first process is depolarized in a time permitting little change by the second polarization mechanism. The sample is then recooled, when the ensuing thermogram shows the relatively undistorted second relaxation.

The fundamental requirement that the relaxation time should be very sensitive to temperature in the depolarization region is met in the segmental motions associated with the glass-to-rubber transition. However this is not always the case with facile low-temperature transitions, when the technique must be applied with caution.

Once individual depolarization peaks have been resolved, their profiles can be compared with the 'ideal' predictions of (5.36). Because the thermogram peaks have their origin in the temperature dependence of the relaxation time, they allow immediate observation of this quantity without the necessity of carrying out repetition measurements at a variety of observation frequencies or times (as in dielectric and mechanical relaxation). Individual peaks can then be used in a discussion[110] of free volume and enthalpic orientation parameters for the molecular movement involved.

Again it must be stressed that erroneous conclusions on the activation process can be reached by incautious assumption that the relaxation follows an ideal Debye decay law at each temperature.

This resolving power of the thermally stimulated depolarization technique makes it particularly suited for examination of those broad transitions lying below the main glass transition. The depolarization current peaks over a range of relaxation times corresponding to a very wide frequency range in isothermal dielectric relaxation studies. An interesting application of this is found in an examination of the low-temperature transition (30 °C at 10^4 Hz, -100 °C at 2.5×10^{-3} Hz) in poly(ethyleneisophthalimide).[6] Although dielectric measurements at room temperature illustrate this as a broad featureless process, analysis of the corresponding low temperature thermogram indicates a structure of contributing processes, each with a discrete relaxation time exhibiting an Arrhenius exponential temperature dependence.

While the thermally stimulated current technique often provides a convenient method for studying those transitions observed in dielectric and dynamic mechanical experiments, its sensitivity to conduction effects means that the observed peaks may contain additional information. This may take the form of peaks not seen in other experiments. These would have their origin in the energization, by a mechanism quite independent of that causing the familiar transitions, of charge carriers out of immobilizing traps. However it is often the case that changes in charge carrier mobility are closely associated with the molecular motions responsible for the normal relaxations. Under these circumstances the depolarization thermograms reflect both processes, and analysis of their amplitude and shape is much more difficult.

5.7. Dielectric relaxation at very high frequencies

5.7.1. Aspects of very high-frequency observations

As polymeric materials become more widely used as dielectric waveguides and as fillers in high-frequency communication wave guides, so an understanding of the appropriate dielectric properties assumes a technological importance. In this context we are concerned with frequencies between 10^{10} and 3×10^{12} Hz (100 cm^{-1}). There are a number of factors which complicate the observation and interpretation of dielectric phenomena in this frequency range. In the first place experimental techniques (submillimetre microwave and Fourier transform spectroscopy) may be difficult, and indeed have only recently become available. In the second place the equations of motion describing any molecular movement responsible for dielectric absorption contain both inertial (second derivative) and viscous (first derivative) terms.

Often it is not possible to neglect one of these, as is the case in ideal relaxation (when inertial terms are ignored) and resonance (when viscous terms are ignored) phenomena.

A further difficulty in the comparison of data in this spectral region arises in the method of presentation. The familiar relaxation procedure (ε' and ε'' against $\log f$) gives a decrease in the real, and maximum in the imaginary, parts of the complex permittivity. However, if the loss is portrayed as a spectroscopic absorption coefficient against linear frequency (α against ν) then an ideal relaxation exhibits a monotonously increasing value, approaching a limiting asymptote, α_∞. In this presentation a maximum in α and an associated maximum and minimum in refractive index, are indicative of a resonance process, or at least of a significant contribution from inertial terms in a quasi-relaxation process. Very often, of course, the relative importance of the inertial and 'viscous' terms changes in the relaxation/resonance region, so that a particular polarization process may have the characteristics of relaxation at low frequencies, but those of resonance at the higher frequencies.

In addition to these difficulties, more than one molecular process may be responsible for the observed relaxation, and it is often necessary to resolve experimental observations into their component parts. It is now generally accepted that at least part of the orientation polarization of a permanent molecular dipole may occur at these high frequencies. This occurs in such a way that the ideal Debye relaxation, even allowing for inertial terms[111,112] is not an adequate description of the process.

The residual dielectric absorption exhibited by dipolar molecules after subtraction of the Debye rotational process (including inertial terms) is usually designated the Poley absorption.[113] It is thought[114-8] to arise by a restricted librational movement of the dipole within minima in the non-spherically symmetric force field established by neighbouring molecules. Under these circumstances it contributes to the full orientational change of the reference dipole (Fig. 5.6(a)). Because of this, the integrated intensity of absorption under both the Debye orientation and Poley libration losses should be comparable with the integrated intensity of the rotational spectrum observed in gases,[118] as is indeed found to be approximately the case.

Since the full rotational intensity extends into the submillimetre wavelength region, the 'infinite frequency' permittivity is more correctly replaced by the square of the refractive index at infrared frequencies.[119] Then the rotational auto-correlation function is written as

$$I_A(t) = \frac{\langle \boldsymbol{\mu}(0) \cdot \boldsymbol{\mu}(t) \rangle}{\mu^2} = \langle \cos \beta(t) \rangle \qquad (5.38)$$

where β is the angle made by the dipole at time t with respect to its

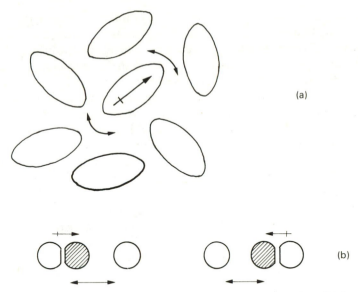

Fig. 5.6. Diagrammatic representation of: (a) Poley dipolar libration; (b) fluctuating collisonal polarization.

orientation at $t = 0$, and the frequency-dependent spectroscopic absorption coefficient is

$$\alpha(\omega) = \frac{\varepsilon_0 - n_{i.r.}^2}{nc}\left[-\dot{\Gamma}_A(0)\frac{+\ddot{\Gamma}(0)}{2}\frac{\Gamma(0)}{4} + \text{higher derivatives}\right] \quad (5.39)$$

and depends on the odd derivatives of the orientation decay function

$$\dot{\Gamma}(t) = \dot{\beta}(t)\sin\beta(t)$$
$$\ddot{\Gamma}(t) = \dot{\beta}^2(t)\cos\beta(t) - \ddot{\beta}(t)\sin\beta(t), \quad \text{etc.} \quad (5.40)$$

n is the refractive index at frequency, ω, and c is the speed of the electromagnetic wave in the medium under investigation.

However the full dielectric absorption of condensed phases is not adequately explained by the combined Debye-Poley processes. This is not too apparent in polar media, but in non-polar materials the dipole orientation processes are sufficiently small for a further dielectric loss process to become apparent. This appears to be intimately associated with collisional phenomena, and has been observed in gases[120,121] as well as in a variety of condensed phases.

Three processes have been postulated to explain its origin. These are energy transfer to molecular translation, energy transfer to molecular rotation, and electromagnetic interaction with a fluctuating 'collision-induced' dipole. While the assignment to any of these three processes is by no means unambiguous, the way in which both the intensities and frequencies vary in liquid mixtures[122-4] leads these authors to prefer the third explanation. The picture is that, as two molecules approach to within a separation closer than the equilibrium value, the pair-interaction potential rises and the multipole on one molecule induces a dipole on the other. The magnitude and direction of this induced polarity fluctuate as the molecules undergo thermal motion and 'collide' with their nearest neighbours, and the fluctuating dipoles then interact with electromagnetic radiation (Fig. 5.6(b)). The strongest evidence for this model is that the intensity of absorption is related to the concentration of compounds in a mixture in a way which is at least bimolecular and is strongly dependent on the polarizability of the molecules and any electron transfer processes which may occur on collision,[124] while the frequency of the process correlates well with that expected from considerations of collision dynamics governed by the slope of the interaction pair potential.

At frequencies above those expected for librational and collisional polarizations, absorption processes are still possible. However these are now far-infrared resonance modes as discussed in Chapters 2 and 11.

5.7.2. Observations on non-polar polymers

The first extensive measurements of very far-infrared and submillimetre microwave absorption in non-polar polymers have been reviewed by Chantry and Chamberlain.[125] When polyethylene, polypropylene and poly(4-methylpent-1-ene) are examined below 50 cm^{-1} it can be seen that the amplitude of the absorption depends on the size (and so the polarizability) of the substituents on the chain backbone (Fig. 5.7).

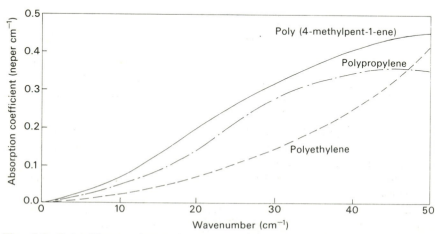

Fig. 5.7. Submillimetre losses (as absorption coefficients) in three non-polar polymers.[125]

The importance of side-group polarizability can be seen in polystyrene[126] which has a higher absorption than the aliphatic polymers. Interestingly the process in polystyrene is extremely similar to that in benzene, indicating that in times shorter than those required for diffusive separation of molecules the distinction between 'liquids' and 'amorphous solids' is virtually non-existent. Values of the frequency of maximum dielectric loss and the calculated induced dipole moments for benzene and polystyrene are listed in Table 5.8.

Table 5.8

Collisional polarization 'relaxation' frequencies and induced dipole moments in benzene and polystyrene

Molecule	Temperature (K)	f_{max} (cm^{-1})	μ_{eff} 10^{31} C m	(D)
Benzene	298	50	5·6	(0·17)
Polystyrene	323	49	6·9	(0·21)
	298	49	6·6	(0·20)
	196	42	5·3	(0·16)
	77	36	3·6	(0·11)

One of the interesting features is that the effective dipole moment varies as the square root of absolute temperature, as would be expected for a perturbation proportional to displacement in a quadratic potential well. This is observed also in poly(fluorinated styrenes)[127] where the intensities and frequencies of the process lend further support to the collisional polarization mechanism.

5.8. Conduction relaxation

5.8.1. Introduction

Although it is usual[128] to describe the electrical conduction of polymeric materials using either 'band' or 'hopping' models (or some combination of these), very little information on the time dependence of discontinuous charge carrier movement is available. If the migration of charge carriers has the characteristics of 'hopping' between regions of low and high mobility (or between 'traps') then the time dependence of the phenomenon should be observable in the electric field, or conduction relaxation within the sample.

5.8.2. The electric modulus (inverse permittivity) presentation

The problem of observing charge-carrier conduction relaxation is formally similar to the study of a relaxing shear modulus in a viscoelastic liquid undergoing continuous or alternating shear flow. In this case the procedure adopted is to separate the elastic and flow components assuming some reasonable mathematical model combination of the two, such as a linear combination of shear strains (Maxwell) or shear stresses (Voigt).

In the electrical case the observed complex permittivity is defined by

$$\varepsilon^*(\omega) = \varepsilon'(\omega) - i\varepsilon''(\omega) = \varepsilon'(\omega) - i(\sigma(\omega)/\omega\varepsilon) \qquad (5.41)$$

where ε^*, ε', ε'' are the complex, real, and imaginary permittivities, $\sigma(\omega)$ is the conductivity measured at frequency ω, and ε is the permittivity of free space $(8\cdot854 \times 10^{-12}\ \mathrm{F\ m^{-1}})$. For materials of relatively low conduction it is usual to resolve the imaginary permittivity into a linear sum of relaxation and conduction components, and to assume that the real permittivity is virtually unaffected

$$\varepsilon^*(\omega) = \varepsilon'(\omega) - i(\varepsilon''_{relax} + \sigma_0/\omega\varepsilon) \qquad (5.42)$$

where σ_0 is the d.c. conductance.

However observation of relaxation, in either the time or the frequency domain, is complicated by the frequency dependence of observed quantities (such as capacitance and conductance) arising from the conductivity and electrode polarization. In materials where conduction is important, and analysis of a conductivity relaxation is required, it is more realistic to consider a series addition of electrical impedances. For this purpose we define the complex inverse permittivity or electrical modulus

$$M^*(\omega) = 1/\varepsilon^*(\omega) = M'(\omega) + iM''(\omega). \qquad (5.43)$$

Then

$$M'(\omega) = \frac{\varepsilon'(\omega)}{[\varepsilon'(\omega)]^2 + [\varepsilon''(\omega)]^2} = \frac{\varepsilon'(\omega)}{[\varepsilon'(\omega)]^2 + [\sigma(\omega)/\omega\varepsilon]^2}$$

$$M''(\omega) = \frac{\varepsilon''(\omega)}{[\varepsilon'(\omega)]^2 + [\varepsilon''(\omega)]^2} = \frac{\sigma(\omega)/\omega\varepsilon}{[\varepsilon'(\omega)]^2 + [\sigma(\omega)/\omega\varepsilon]^2}. \qquad (5.44)$$

For the ideal (Maxwell) RC circuit, in which ε' is again independent of frequency and the Maxwell relaxation time, τ_M, is defined by

$$\tau_M = \frac{\varepsilon_0\varepsilon}{\sigma_0} \qquad (5.45)$$

where ε_0, σ_0 are the static permittivity and d.c. conductivity respectively, then

$$M^*(\omega) = M_\infty\left[\frac{i\omega\tau_M}{1 + i\omega\tau_M}\right] = M_\infty\left[\frac{(\omega\tau_M)^2 + i\omega\tau_M}{1 + (\omega\tau_M)^2}\right] \qquad (5.46)$$

where $M_\infty = 1/\varepsilon_\infty$ is the electric modulus at frequencies higher than those of the relaxation region.

Just as in viscoelastic relaxation, a variety of empirical methods can be used to accomodate a non-ideal frequency dependence in ε' and σ. The method adopted by Moynihan and his co-workers,[129,130] for the study of alkali silicate glasses, is to assume a linear superposition of independent contributions to the modulus. For example overlapping bulk conductivity processes are accommodated in a distribution of Maxwell relaxation times

$$M^* = M_s \int_0^\infty g(\tau_M) \, d\tau_M \left[\frac{(\omega\tau_M)^2 + i\omega\tau_M}{1 + (\omega\tau_M)^2} \right] \qquad (5.47)$$

where $g(\tau_M) d\tau_M$ is the distribution function. Electrode polarization, too, can be included as an additive term in the modulus by considering appropriate combinations of electrode and bulk impedances.

One very powerful advantage of using the electric modulus to interpret bulk relaxation properties is that variations in the very large, but low frequency, permittivity, and conductivity are minimized. In this way the familiar difficulties of electrode nature and contact, space-charge injection phenomena, and absorbed impurity conduction effects, which appear to swamp relaxation in the permittivity presentation, can be resolved, or even ignored.

Just as the usual permittivity–frequency presentation represents the Laplace transformation of a current/voltage/charge time-dependent decay, so too does the frequency-dependent complex modulus reflect an electric field decay. This can be written[131] as an empirical relaxation function

$$M^*(\omega) = M_\infty [1 - N^*(\omega)]$$
$$N^*(\omega) = \mathscr{L}[-d\phi(t)/dt] \qquad (5.48)$$
$$\phi(t) = E(t)/E_0 = \exp[-(t/\tau_M)^\beta]$$

where β is 1 in the ideal (Maxwell) case. This particular decay function is familiar in the discussion of non-ideal dipole orientation processes.

Although consideration of electric field relaxation portrayed as a frequency (time)-dependent complex modulus is common in the study of inorganic semiconductors and glasses, the technique has not yet been widely applied to partially conducting organic polymers. A study of conjugated arylene–vinylene polymers has confirmed the usefulness of the technique. The clearly defined relaxations have been used to establish the importance of electronic charge carrier hopping, and to characterize some of the 'hopping' parameters.[132]

References

1. Cole, K. S. and Cole, R. H. *J. chem. Phys.* **9**, 341 (1941).
2. Davidson, D. W. and Cole, R. H. *J. chem. Phys.* **19**, 1484 (1951).
3. Havriliak, S. and Negami, S. *Polymer*, **8**, 161 (1967).
4. Williams, G. and Watts, D. C. *Trans. Faraday Soc.* **66**, 80 (1970).

5. Glarum, S. H. *J. chem. Phys.* **33**, 1371 (1960).
6. Jonscher, A. K. *1972 Annual Report on Conference on Electrical Insulation and Dielectric Phenomena*, p. 418. Nat. Acad. Sci., Washington, DC (1973).
7. Bordewijk, P. *Chem. Phys. Lett.* **32**, 592 (1975).
8. Tjia, T. H., Bordewijk, P. and Böttcher, C. J. F. *Advan. mol. relaxation Processes* **6**, 19 (1974).
9. Williams, G. *chem. Rev.* **72**, 55 (1972).
10. Powles, J. G. *J. chem. Phys.* **21**, 633 (1953).
11. Cook, M., Watts, D. C. and Williams, G. *Trans. Faraday Soc.* **66**, 2503 (1970).
12. Fatuzzo, E. and Mason, P. R. *Proc. phys. Soc.* **90**, 741 (1967).
13. Berne, B. J., Pechukas, P. and Harp, G. D. *J. chem. Phys.* **49**, 3125 (1968).
14. Berne, B. J. and Harp, G. D. *Advan. chem. Phys.* **17**, 63 (1970).
15. De Brouckère, L. and Mandel, M. *Advan. chem. Phys.* **1**, 77 (1958).
16. Block, H. and North, A. M. *Advan. mol. relaxation Processes* **1**, 309 (1970).
17. North, A. M. *Chem. Soc. Rev. (Lond.)* **1**, 51 (1972).
18. Stockmayer, W. H., *Pure appl. Chem.* **15**, 539 (1967).
19. Bur, A. J. and Roberts, D. E. *J. chem. Phys.* **51**, 406 (1969).
20. Lochhead, R. Y. and North, A. M. *J. chem. Soc. Faraday Trans. I.* **68**, 1089 (1972).
21. Block, H., Hayes, E. F. and North, A. M. *Trans. Faraday Soc.* **66**, 1095 (1970).
22. Eisenberg, H. *Discuss. Faraday Soc.* **49**, 286, 1970); *Biopolymers* **8**, 545, (1969).
23. Dev, S. B., Lochhead, R. Y. and North, A. M. *Discuss. Faraday Soc.* **49**, 244 (1970).
24. O'Konski, C. T. *J. phys. Chem.* **64**, 605 (1960).
25. Kirkwood, J. G. and Schumaker, J. B. *Proc. Nat. Acad. Sci. US* **38**, 855 (1952).
26. Davies, M., Williams, G. and Loveluck, G. D. *Z. Elektrochem.* **64**, 575 (1970).
27. Scherer, P. C., Levi, D. W. and Hawkins, M. C. *J. polym. Sci.* **31**, 105 (1958).
28. Bates, T. W., Ivin, K. J. and Williams, G. *Trans. Faraday Soc.* **63**, 1964 (1967).
29. North, A. M. and Phillips, P. J. *Trans. Faraday Soc.* **64**, 3235 (1968).
30. De Brouckère, L. and van Nechel, R. *Bull Soc. Chim. Belges* **61**, 262 and 452 (1952).
31. Kryszewski, M. and Marchal, J. *J. polym. Sci.* **29**, 103 (1958).
32. Phillips, P. J. Ph.D. Thesis. University of Liverpool (1968).
33. North, A. M. and Phillips, P. J. *Chem. Commun.* 1340 (1968).
34. North, A. M. and Phillips, P. J. *Trans. Faraday Soc.* **63**, 1537 (1967).
35. Pohl, M. A. and Zabusky, M. M. *J. phys. Chem.* **66**, 1390 (1962).
36. McCrum, N. G., Read, B. E. and Williams, G. *Anelastic and dielectric effects in polymeric solids.* Wiley, London and New York (1967).
37. Hill, N., Vaughan, W. E., Price, A. H. and Davies, M. *Dielectric properties and molecular behaviour*, p. 411 et seq. Van Nostrand, London (1969).
38. Parker, T. G. In *Polymer science* (ed. A. D. Jenkins, Chapter 19. North-Holland, Amsterdam (1972).
39. Baird, M. *Electrical properties of polymeric materials.* Plastics Institute, London (1973).
40. Bordewijk, P. *Chem. Phys. Lett.* **32**, 592 (1975).
41. Jonscher, A. K. *Nature* **267**, 673 (1977).
42. Kirkwood, J. G. *Trans. Faraday Soc.* **42A**, 7 (1946).
43. Kakizaki, M. and Hideshima, T. *J. Macromol. Sci. Phys.* **B8**, 367 (1973).
44. Matsuoka, S., Roe, R. J. and Cole, H. F. *Dielectric properties of polymers* (ed. F. E. Karusz), p. 255. Plenum, New York, (1972).

45. Phillips, P. J., Wilkes G. L., Delf, B. W. and Stein, R. S. *J. Polymer Sci.* **A2**, 499 (1971).
46. Kakizaki, M., Saito M. and Hideshima, T. *Abstracts 25th Annual Meeting of the Japan Polymer Society, Tokyo* **25** (2), 388 (1976).
47. Stoll, B., Peckhold, W. and Blasenberg, S. *Kolloid Z Polymer* **250**, 1111 (1972).
48. Kishi, N. and Uchinda, N. *Rept. Prog. Polym. Phys. (Japan)* **VI**, 233 (1963).
49. Kramer, H. and Helf, K. E. *Kolloid Z.* **180**, 114 (1962).
50. Sazhin, B. I., Skurikhina, V. A. and Illin, Y. I. *Vysokomolecul. Svedin* **1**, 1383 (1959).
51. Reddish, W. *Am. Chem. Soc. Polymer Preprints* **6**, (2) 571 Sept (1965).
52. Kastner, S. E., Schlosser, E. and Pohl, G. *Kolloid Z.* **192**, 21 (1963).
53. Saito, S. *Kolloid Z.* **189**, 116 (1963).
54. Koppleman, J. *Kolloid Z.* **189**, 1 (1963).
55. Bruens, O. and Muller, F. H. *Kolloid, Z.* **140**, 1121; **141**, 20 (1955).
56. Kabin, S. P. *Zh. Tekhn. Fiz.* **26**, 2628 (1956); *Sov. Phys. Tech. Phys.* **1**, 2542.
57. Takayanagi, M. *Rept. Prog. Polymer Phys. (Japan)* **6**, 121 (1963).
58. Mikhailov, G. P., Borisova, T. I. and Dmitrochenko, D. A. *Sov. Phys. Tech. Phys.* **26**, 1924 (1956).
59. Heijboer, J. *Kolloid Z.* **134**, 149 (1956).
60. Deutsch, K., Hoff, E. A. W. and Reddish, W. *J. Polymer Sci.* **13**, 565 (1954).
61. Mikhailov, G. P. and Borisova, T. I. *Sov. Phys. Tech. Phys.* **28**, 137 (1958).
62. Mikhailov, G. P. and Borisova, T. I. *Polymer Sci. USSR* **2**, 387 (1961).
63. Mead, D. J. and Fuoss, R. M. *J. Am. Chem. Soc.* **63**, 2832 (1941).
64. Ishida, Y., Amano, O. and Takayanagi, M. *Kolloid Z.* **176**, 62 (1962).
65. Hikichi, K. and Furnichi, J. *Rept. Prog. Polymer Phys. (Japan)* **4**, 69 (1961).
66. Mikhailov, G. P. and Burshtein, L. L. *Vysokomolekul. Soedin* **4**, 270 (1962).
67. Baker, E. B., Auty, R. P. and Ritenour, G. J. *J. Chem. Phys.* **21**, 159 (1953).
68. Saito, S. and Nakajima, T. *J. Appl. Polymer Sci.* **2**, 93 (1959).
69. Tanaka, A. and Ishioa, Y. *J. Polymer Sci.* A2 **10**, 1029 (1972).
70. Reneker, D. H., Martin, G. M. and Broadhurst, M. G. *J. Appl. Phys.* **45**, 4172 (1974).
71. Porter, C. H. and Boyd, R. H. *Macromolecules* **4**, 589 (1971).
72. Miller, S., Tomozawa, M. and MacCrone, R. K. *Amorphous Materials* (ed. R. W. Douglas and B. Ellis) **89**. J. Wiley, New York (1972).
73. Maxwell J. C. *Electricity and magnetism*, Vol. 1, p. 452. Clarendon Press, Oxford (1892).
74. Wagner, K. W. *Arch. Electrotech.* **2**, 378 (1914).
75. Sillars, R. W. *J. Inst. Elect. Eng.* **80**, 378 (1937).
76. Van Beek, L. K. H. *Progr. Dielectrics* **7**, 69 (1967).
77. Davies, W. E. A. *J. Phys. D, appl. Phys.* **7**, 120 (1974).
78. Bruggeman, D. A. G. *Ann. Phys. Lpz.* **24**, 636 (1935).
79. Looyenga, H. *Physica* **31**, 401 (1975).
80. Davies, W. E. A. *J. Phys. D, appl. Phys.* **7** 1016 (1974).
81. MacDonald, J. R. *J. chem. Phys.* **54**, 2026 (1971).
82. Adamec, V. *J. polym. Sci.* A2 **6**, 1241 (1968).
83. Wylie, G. In *Dielectric and related molecular processes* (ed. M. Davies). Specialist Reports, Chemical Society, London (1972).
84. Fricke, H. *J. phys. Chem.* **57**, 934, (1953).
85. Hanai, T. *Kolloid Z.* **171**, 23 (1960).

86. Hanai, T. In *Emulsion science* (ed. P. Sherman). Academic Press, London (1968).
87. Van Beek, L. K. H., Boog, J. and Looyenga, H. *Appl. Sci. Res. B.* **12**, 57 (1965).
88. Mandel, M. *Physica* **27**, 827 (1961).
89. Trukhan, E. M. *Sov. Phys.—Solid State* **4**, 2560 (1963).
90. Dukhin, S. S. and Shilov, V. N. *Dielectric phenomena and the double layer in disperse systems and poly-electrolytes.* John Wiley and Sons, New York (1974).
91. Pauly, H. and Schwan, H. P. *Z. Naturforsch.* **14b**, 125 (1959).
92. O'Konski, C. T. *J. phys. Chem.* **64**, 605 (1960).
93. Schwarz, G. *J. phys. Chem.* **66**, 2636 (1962).
94. Takashima, S. *Advan. Chem. Ser.* **63**, 232 (1967).
95. Hirota, S., Saito, S. and Nakajima, T. *Kolloid Z. u Z. Polymere* **213**, 109 (1966).
96. Hirota, S., Saito, S. and Nakajima, T. *Rep. Progr. polym. Phys. Jpn* **10**, 425 (1967).
97. Baird, M., Goldsworthy, G. T. and Creasy, C. J. *Polymer* **12**, 159 (1971).
98. North, A. M. and Reid, J. C. *Eur. polym. J.* **8**, 1129 (1972).
99. Michel, R., Seytre, G. and Maitrot, M. *J. polym. Sci., polym. Phys. Ed.* **13**, 1333 (1975).
100. Kosaki, M. and Ieda, M. *J. phys. Soc. Jpn* **27**, 1604 (1969).
101. Yano, S., Tadano, K., Aoki, K. and Koizumi, N. *J. polym. Sci., polym. Phys. Ed.* **12**, 1875 (1974).
102. North, A. M., Pethrick, R. A. and Wilson, A. D. *Polymer*, **19**, 913 (1978).
103. North, A. M., Pethrick, R. A. and Wilson, A. D. *Polymer*, **19**, 923 (1978).
104. Jux, J. T., North, A. M. and Kay, R. M. *Polymer* **15**, 799 (1974).
105. Bucci, C. and Fieschi, F. *Phys. Rev. Lett.* **12**, 16 (1974).
106. Creswell, R. A. and Perlman, M. M. *J. appl. Phys.* **41**, 2365 (1970).
107. Kryszewski, M., Kasika, J., Patora, J. and Piotrowski, J. *J. Polym. Sci. C (Polymer Symposia)* **30**, 243 (1970).
108. Reichle, M., Nedetzka, T., Mayer, A. and Vogel, H. *J. phys. Chem.* **74**, 2659, (1970).
109. Lacabanne, C. and Chatain, D. *J. polym. Sci., Polymer Phys. Edn.* **11**, 2315 (1973).
110. Lacabanne, C., Chatain, D., Guillet, J., Seytre, G. and May, J. F. *J. polym. Sci., polymer Phys. Edn.* **13**, 445 (1975).
111. Powles, J. G. *Trans. Faraday Soc.* **44**, 802 (1948).
112. Steele, W. *J. chem. Phys.* **38**, 2404 (1963).
113. Poley, J. P. *Appl. Sci. Res.* **4B**, 337 (1955).
114. Hill, N. E. *Proc. phys. Soc.* **82**, 723 (1963).
115. Hill, N. E. *Chem. Phys. Lett.* **2**, 5 (1968).
116. Davies, M., Pardoe, G. W. F., Chamberlain, J. E. and Gebbie, H. A. *Trans. Faraday Soc.* **64**, 847 (1968).
117. Davies, M., Pardoe, G. W. F., Chamberlain, J. E. and Gebbie, H. A. *Trans. Faraday Soc.* **66**, 273 (1970).
118. Kroon, S. G. and van der Elsken, J. *Chem. phys. Lett.* **1**, 285 (1967).
119. Davies, M. *Ann. Rep. Chem. Soc.* **67A**, 65 (1970).
120. Whiffen, D. H. *Trans. Faraday Soc.* **46**, 124 (1950).
121. Gusewell, D. *Z. angew. Phys.* **22**, 461 (1967).
122. Pardoe, G. W. F. *Trans. Faraday Soc.* **66**, 2699 (1970).
123. Rosenberg, A. and Birnbaum, G. *J. chem. Phys.* **52**, 683 (1970).

124. North, A. M. and Parker, T. G. *Trans. Faraday Soc.* **67**, 2234 (1971).
125. Chantry, G. W. and Chamberlain, J. Far infra-red spectra of polymers. In *Polymer science* (ed. A. D. Jenkins), Vol. 2, p. 1330. North-Holland, Amsterdam and London (1972).
126. North, A. M. and Parker, T. G. *J. chem. Soc., Faraday Trans. II* **68**, 1094 (1972).
127. North, A. M., Parker, T. G., Pethrick, R. A. and Towland, M. *J. chem. Soc., Faraday Trans. II* **71**, 1473 (1975).
128. See, for example H. A. Pohl, *J. polym. Sci. C* **17** 13 (1967).
129. Macedo, P. B., Moynihan, C. T. and Bose, R. *Phys. Chem. Glasses* **13**, 171, (1972).
130. Moynihan, C. T., Boesch, L. P. and Laberge, N. L. *Phys. Chem. Glasses* **14**, 122 (1973).
131. Williams, G. and Watts, D. C. *Trans. Faraday Soc.* **66**, 80 (1970).
132. Abu-Bakr, A., North, A. M. and Kossmehl, G. *Eur. polym. J.* **13**, 799 (1977).

MOLECULAR MOTION AND PHOTOLUMINESCENCE

6.1. General characteristics of fluorescence and phosphorescence

Many of the significant features observed in the photophysics and photo-chemistry of polymer systems have their origin in the way the behaviour of electronically excited states is affected by molecular motion. In the present context we shall be concerned with the photophysical processes which occur after absorption of a photon, but before any resulting chemical reactions (some of which are treated in Chapter 13). Most of the interesting phenomena involve a competition between a physical change and emission of the excitation energy, and so are most easily observed in the characteristics of, or modifications to, fluorescence and phosphorescence.

The three important properties of luminescence radiation are frequency, intensity, and polarization. The frequency depends on the nature of the emitting state, whether this is a singlet or triplet state of the primary absorbing chromophore, whether it is an excited state of some other chromophore achieved by energy transfer, or whether it is some 'complex' state obtained by a bimolecular interaction. The intensity depends on spectroscopic selection rules, and also upon the efficiency of competing non-radiative, quenching, or energy transfer processes. The polarization depends on the polarization of the exciting radiation, on the geometrical changes in electron distribution which occur on absorption and emission, and on the mobility of either the excited molecules, or the energy quanta, during the excited state lifetime.

Some of the various unimolecular processes which may follow the absorption of light by a molecule with a singlet ground state are presented in the Jablonski diagram (Fig. 6.1). The initial absorption raises the molecule to some excited singlet state, but internal conversion between excited singlet states is usually very efficient, so that the molecule relaxes to the ground vibrational level of the first excited singlet, (S_1) in times of the order of a picosecond. The energy may then be emitted as fluorescence, or intersystem crossing may take place to the triplet state (IC_1). These competitive proces-

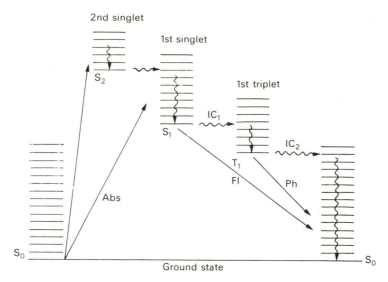

Fig. 6.1. Energy level diagram of important unimolecular photophysical processes. Straight lines represent radiative, wavy lines non-radiative processes. Abs, absorption; Fl, fluorescence; Ph, phosphorescence; IC, intersystem crossing.

ses occur in times between about one and ten nanoseconds. The triplet state may then lose its energy and return to the ground state either by emitting radiation (phosphorescence) or by a second radiationless intersystem crossing (IC_2). These processes, involving changes in multiplicity, are spectroscopically 'forbidden' and require times ranging from fractions of milliseconds to seconds. Consequently the net result is that the three routes, fluorescence, phosphorescence, and radiationless transition, represent competitive relaxation processes.

The situation becomes rather more complex when bimolecular interactions are included (Fig. 6.2). Once again we consider excitation of the ground singlet state of a molecule A, which is able to interact with a molecule B of the same, or different kind. Then, as well as the various radiative and non-radiative processes exhibited by A, there are further possibilities involving B. The first is that energy transfer from A to B may occur, so that fluorescence, phosphorescence, and non-radiative relaxation of B are possible. In addition, it may be that the excited state of one molecule can interact with the ground state of the other, to form an excited complex. If this involves singlet states, the complex can undergo fluorescence, or non-radiative relaxation (both associated with dissociation). This complex, which exists only in the excited state, is called an 'excimer' when A and B are the same type of molecule, and an 'exciplex' when they differ.

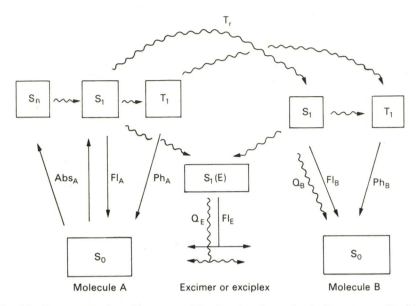

Fig. 6.2. Representation of important bimolecular photophysical processes. Straight lines represent radiative, wavy lines non-radiative processes. Molecule A is the primary absorbing species. Abs_A, absorption by A; $Fl_{A,B,E}$ fluorescence from A, B, or excimer/exciplex; $Ph_{A,B}$, phosphorescence from A or B; T_r non-radiative energy transfer; Q_B non-radiative quenching processes which may be undergone by B; Q_E non-radiative relaxation of excimer/exciplex.

The competition between two processes can be controlled by either thermodynamic or kinetic parameters, and the latter are of prime importance in a consideration of molecular motion and luminescence. It will be recalled that first excited singlet states generally have lifetimes of the order of a few nanoseconds, while triplet-state lifetimes can vary from a fraction of a millisecond to seconds. As significant competition to radiation can come only from processes capable of occurring in the appropriate time scales, changes in fluorescence are likely to be caused primarily by the fastest molecular motions, whereas phosphorescence can be significantly affected by a very wide variety of relatively slower processes.

A more detailed discussion of the photophysics of organic molecules has been presented by Birks,[1] and basic aspects of the photoluminescence of synthetic polymers have been reviewed by Fox[2] and by Somersall and Guillet.[3] Consequently we have selected for further discussion only those phenomena intimately affected by, or revealing of, polymer molecular motion.

6.2. Excimer formation in polymers

6.2.1. Inter- and intramolecular excimers

The fluorescent emission of small-molecule aromatic hydrocarbons is markedly concentration-dependent. In the gas phase, or in dilute solution, the emission is essentially a structured mirror image of absorption, as would be expected for relaxation from the excited singlet states of the isolated molecules. However in concentrated solution or in certain crystals, a featureless broad excimer emission at larger wavelengths is evidenced. In general this excimer formation can occur whenever the aromatic chromophores can come into a face-to-face coplanar arrangement with a separation of 0·3 to 0·35 nm. In certain systems these geometrical constraints for excited-state dimer (excimer) formation are difficult to achieve, so that the quantum yield for normal, or 'monomer' emission is greater than that for excimer emission.

When a molecule contains two or more aromatic moieties, connected by covalent bonds, then the relative positions of the chromophores may be fixed, or at least much affected by, the molecular framework. The result may be an intramolecular geometry holding two ground-state chromophores in a position which closely resembles that in the excimer, which is then formed readily, or alternatively a geometry which makes intramolecular excimer formation impossible. In the former situation high quantum yields for excimer emission are observed, even in dilute solution. In this way,[4-7] for rather simple aromatic chromophores, a particularly favourable face-to-face conformation can be achieved in 1,3-diarylpropanes and 2,2' or 4,4'-paracyclophanes (Fig. 6.3).

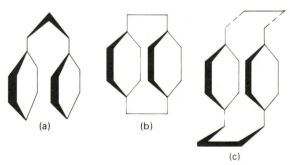

Fig. 6.3. Excimer-forming conformations of: (a) 1,3-diarylpropane; (b) 2,2'-paracyclophane; (c) 4,4'-paracyclophane.

Vinyl polymers represent a continuing sequence of 1,3-disubstituted propane moieties, and so excimer formation involving neighbouring groups is a possibility in all poly(vinylarene)s. In addition, when geometrical constraints inhibit near-neighbour excimer formation, coiled conformations of the polymer chain may still facilitate the formation of intra-chain excimers from non-nearest neighbours. Thus in polymer systems there are the three possibilities for excimer formation, near-neighbour and non-near-neighbour intramolecular interaction, and intermolecular interaction.

6.2.2. Segmental motion and intramolecular excimer formation

Excimer emission has now been detected from a number of aromatic vinyl polymers.[3,6,8] The aromatic chromophores have included phenyl, toluene, naphthalene, biphenyl, pyrene, carbazole, and acridine. A most important feature of the intramolecular excimer emission from dilute solutions of these polymers is that it occurs in fluid solvents, but not when the polymers are dissolved in a glass which solidified at low temperatures.[8-11] Obviously, therefore, the conformation of the chain backbone required to bring neighbouring aromatic chromophores into a face-to-face geometry is one of unfavourably high energy. It is consequently almost entirely absent from those systems where a low-temperature distribution of chain conformations becomes trapped in a rigid medium, but it can have a transient existence when segmental motion at higher temperatures rapidly rotates chain backbone units through a variety of conformational states.

According to this model the polymer chains dissolved in fluid media will have a Boltzmann distribution of suitable excimer-forming sites, the concentration of which will be one factor determining the ratio of excimer to 'monomer' emission intensities. The emission characteristics of solid solutions will then depend on the conformational equilibrium existing at the temperature of solidification.

An elegant demonstration of this phenomenon in solid solutions of poly(2-vinylnaphthalene)[10] and poly(4-vinylbiphenyl)[11] has been made by Frank and Harrah. Since emission measurements in the solid matrix reflect a static equilibrium distribution, the ratios of excimer to 'monomer' intensities at a given temperature can be related to the energy difference between the 'normal' and 'excimer-forming' conformations, ΔE_E at the temperature of solidification, T_S, through

$$I_E/T_M \propto \exp(-\Delta E_E/RT_S). \tag{6.1}$$

In this way the excimer site energy is found to lie 10 kJ mol^{-1} above that of the normal conformation in poly(2-vinylnaphthalene) and 3·4 kJ mol^{-1} in poly(vinylbiphenyl).

The fraction of excimer-forming sites, f_E is given by

$$f_E = \frac{g_E \exp(-\Delta E_E/RT}{g_G + g_E \exp(-\Delta E_E/RT)} \tag{6.2}$$

where g_E and g_G are the degeneracies of the excimer and 'ground' conformational states. This fraction appears to be of order 10^{-1} in poly(4-vinylbiphenyl) and 10^{-2} in the rather more hindered poly(2-vinyl-naphthalene). Energy migration studies[12] of poly(N-vinyl carbazole) suggest that the site fraction is about 10^{-3} for this relatively highly sterically hindered chain.

The analysis of excimer and monomer emission in fluid systems is much more complex, because here the excimer fluorescence functions as a dynamic sampling mechanism. It is usual[6,10,13-16] to discuss the various photophysical processes using a quasichemical kinetic scheme in which each step is described by a first-order rate constant. Of particular significance in this model is a term descriptive of energy migration along a sequence of chromophores as will be discussed in § 6.3. The picture is then one of energy movement, either to a preferred excimer site, or to a point where segmental motion creates the excimer geometry during the 'residence' of the electronic excitation.

A general kinetic scheme for excimer formation (inter- and intramolecular) can be presented in terms of a singlet ground and excited state 'monomeric' chromophores, ^1M, $^1M^*$, triplet 'monomeric' chromophores, $^3M^*$, singlet dimeric excimers, $^1D^*$, and quencher, Q.

Energy absorption	^1M	$\xrightarrow{h\nu_A} {}^1M^*$	I_a
'Monomer' fluorescence	$^1M^*$	$\rightarrow {}^1M + h\nu_{FM}$	k_{FM}
Radiationless relaxation	$^1M^*$	$\rightarrow {}^1M$	k_{IM}
Intersystem crossing	$^1M^*$	$\rightarrow {}^3M^*$	k_{IM}
'Monomer' phosphorescence	$^3M^*$	$\rightarrow {}^1M + h\nu_{PM}$	k_{PM}
Radiationless relaxation	$^3M^*$	$\rightarrow {}^1M$	k_{IT}
Singlet migration	$^1M_a^* + {}^1M_b$	$\rightarrow {}^1M_a + {}^1M_b^*$	k_E
Excimer formation (dynamic)	$^1M^* + {}^1M$	$\rightarrow {}^1D^*$	k_{DM}
(preformed sites)	$^1M^* + {}^1M . M$	$\rightarrow {}^1D^* + {}^1M$	
Excimer dissociation	$^1D^*$	$\rightarrow {}^1M^* + {}^1M$	k_{MD}
Excimer fluorescence	$^1D^*$	$\rightarrow 2{}^1M + h\nu_{FD}$	k_{FD}
Intersystem crossing	$^1D^*$	$\rightarrow {}^3D^*$	k_{ID}
Radiationless relaxation	$^1D^*$	$\rightarrow 2{}^1M$	
'Monomer' quenching	$^1M^* + Q$	$\rightarrow {}^1M + Q^*$	k_{QM}
Excimer quenching	$^1D^* + Q$	$\rightarrow 2{}^1M + Q^*$	k_{QD}

(with grouped rate constants k_M comprising k_{FM}, k_{IM}, k_{IM}; k_T comprising k_{PM}, k_{IT}; and k_D comprising k_{MD}, k_{FD}, k_{ID}.)

The intensities of 'monomer' and 'excimer' emission I_{FM} and I_{FD}, are then given by

$$I_{FM} = k_{FM}[^1M^*], \qquad I_{FD} = k_{FD}[^1D^*].$$

Insertion of a stationary-state concentration of excited states, $^1M^*$ and $^1D^*$, and solution of the rate equations in the usual way, yields[16] for the ratio of 'monomer' to excimer emission intensities,

$$\frac{I_{FM}}{I_{FD}} = \left\{\frac{k_{FM}}{k_{FD}}\right\} \frac{(k_D + k_{MD} + k_{QD}[Q])}{k_{DM}[^1M]} \tag{6.4}$$

and the effect of quencher,

$$\frac{I_{FM}^0}{I_{FM}} = \left\{\frac{k_D + k_{MD}}{k_D(k_M + k_{DM}[^1M])}\right\}$$

$$\times \left\{(k_M + k_{DM}[^1M] + k_{QM}[Q]) - \frac{k_{MD}k_{DM}[^1M]}{(k_D + k_{MD} + k_{QD}[Q])}\right\} \tag{6.5}$$

$$\frac{I_{FD}^0}{I_{FD}} = 1 + \left\{\frac{k_{QM}(k_D + k_{MD})[Q] + k_{QD}k_{QM}[Q]^2}{k_D(k_M + k_{DM}[^1M]) + k_M k_{MD}}\right\} \tag{6.6}$$

where superscript, 0, represents intensities in absence of an energy transfer acceptor.

Application of this scheme to fluorescence observed under a variety of conditions leads to a number of interesting conclusions. For example, a non-linear dependence of I_{FD}^0/I_{FD} on $[Q]$ shows[9,16] that excimer energy, as well as 'monomer' excitation, can be transferred to suitable energy acceptors by resonance non-radiative transfer. For transfer of this type the quenching coefficients k_{QM} and k_{QD} will be independent of time and reactant concentration (as assumed for kinetic rate 'constants') only when energy or molecular diffusion maintains a quasi-statistical mixing of donor and acceptor (see Chapter 13 for further discussion of diffusion-controlled processes).

When intramolecular excimers are formed the concentration of 'monomeric' chromophore is that existing in the molecule under consideration, so that the formation is better described by a first-order process,

Intramolecular excimer formation: $^1M^* \to ^1D^*$ k_{DM}^1.

This description applies both to the dynamic formation of an excimer site by conformational change at the excited chromophore, and the intramolecular energy migration to preformed sites.

A serious problem in the use of eqns (6.4)–(6.6) for evaluation of excimer-formation and dissociation rate coefficients is that the ratio of 'monomer' to excimer emission must be known very precisely. However such an analysis has shown[15] that in the model compounds 1,3-diphenyl propane and 1,3-bis-carbazolyl propane the intramolecular formation constants (10^9 and $3 \times 10^8 \, s^{-1}$ respectively) are much greater than the dissociation constant, which is estimated as approximately $10^7 \, s^{-1}$. These values are reasonably in line with rates of hindered rotation in such species.

Such studies have been extended[6,14,15] to vinyl polymers. For polystyrene the formation constant in dilute solution at room temperature is about

$2 \times 10^9 \, \text{s}^{-1}$. This is rather more rapid than the rate of segmental rotation measured by ultrasonic and dielectric techniques, and indicates the significance of energy migration to sites of particularly favourable geometry.

The temperature dependence of these emission characteristics can be used to obtain the activation energy barriers to excimer formation and dissociation. These are not the same, of course, as the energy difference between the 'normal' and excimer-forming geometries of the electronic ground state. In poly(2-vinylnaphthalene)[17] and poly(4-vinylbiphenyl),[11] excimer formation is much faster than excimer dissociation so that over an appropriate range the temperature dependence of the emission ratio derives principally from that of excimer formation. Comparison of high- and low-temperature (when formation is quenched) intensity ratios then yields the energy barrier to excimer formation. This is $15 \, \text{kJ} \, \text{mol}^{-1}$ in poly(2-vinylnaphthalene) and $8 \, \text{kJ} \, \text{mol}^{-1}$ in the less sterically hindered poly(4-vinylbiphenyl). These compare with the 'equilibrium' conformational energy differences of 10 and $3 \cdot 3 \, \text{kJ} \, \text{mol}^{-1}$, respectively.

6.2.3. Vibrational modes and excimer emission

The large-scale segmental rotations discussed in the last section are not the only modes of motion which can influence the relative orientation and separation of chain chromophores. Vibrational modes such as the longitudinal 'accordion' vibration can also modulate[10,11] energy migration characteristics and thus the relative amounts of absorbed energy which arrive at, and are radiated from, excimer sites.

The effect of the vibration is to increase the rate of Förster resonance energy migration by the sixth power (since energy transfer is a transition dipole-dipole interaction, as set out in eqn (6.10) below) of the diminution in interchromophore separation, r,

$$r = r_0 + \Delta r(t, T). \tag{6.7}$$

Insertion of the harmonic vibration, integration over the period of modulation, and ignoring high-order terms, gives[10] the factor f_{MOD} increasing the ratio of excimer to monomer emission as

$$f_{\text{MOD}} = 1 + 21 \left(\frac{\Delta r}{r_0}\right)^2. \tag{6.8}$$

Comparison of data at 4 K, when vibrational modulation is assumed to be negligible, with those at higher temperatures allows evaluation of the $(\Delta r^2/r_0)$ term, and thence an estimate of the 'accordion' mode frequency. This appears to be about $157 \, \text{cm}^{-1}$ for polystyrene, $135 \, \text{cm}^{-1}$ for poly(4-vinylbiphenyl) and $84 \, \text{cm}^{-1}$ for poly(2-vinylnaphthalene). These frequencies correspond to longitudinal vibrations of about 20 monomer units, in agreement with resonances observed in the low frequency Raman spectra of polymers.

6.3. Energy migration in polymers

6.3.1. Modes of exciton migration

When two identical chromophores are brought into proximity in a suitable relative orientation, there is the possibility that excitation energy may move as an exciton from one to the other by some non-radiative process. In a polymer molecule containing spectroscopically active repeat units, a large number of chromophores may be so placed, and thus participate in movement of the site of excitation. Consequently even dilute solutions of polymers may exhibit intramolecular or 'down chain' energy migration resembling that observed in only concentrated systems of analogous low molecular weight species. Of course in concentrated or bulk polymer systems both inter- and intramolecular migration may be possible. When movement of excitation is being considered, it is conventional to describe as 'migration' diffusion or delocalization of energy over identical chromophores, and to use the word 'transfer' when the energy moves between different chemical species.

As an example, some competing radiative, migration, and transfer processes which can be undergone by singlet excitation of a polymer chain are illustrated in Fig. 6.4.

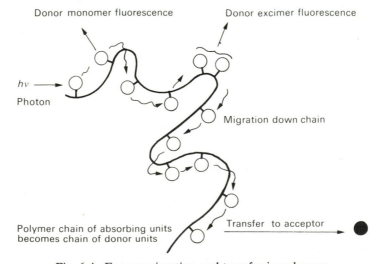

Fig. 6.4. Energy migration and transfer in polymers.

Migration of energy along polymer chains can occur by two mechanisms. When the electronic interactions between adjacent chromophores are weak, and the orientational correlations are low, the energy undergoes a random walk sequence of jumps from one chromophore to another. Under these circumstances there is a finite time between jumps, and there is no phase coherence between the different localizations of the exciton. Each individual jump has the characteristics of non-radiative resonance transfer as described by Förster[18] or Dexter.[19] The incoherent, or Förster–Dexter, 'hopping' exciton can be characterized by an effective energy diffusion distance, $\bar{r}_i$,

$$\bar{r}_i = (2[D + \Lambda]\gamma)^{\frac{1}{2}} \tag{6.9}$$

where D is a diffusion coefficient descriptive of molecular movement, Λ is the coefficient (in the same units) describing exciton migration, and γ is the excited state lifetime.

When there are strong interactions between neighbouring chromophores, and orientational correlations are appropriate, the energy may be delocalized as a wave over a number of chromophore units. For such a coherent, or Frenkel,[20] exciton, it is inappropriate to discuss the time required for energy to move down the delocalization length, but it is still possible to discuss, and observe, an effective interaction distance, $\bar{r}_c$, which is achieved almost instantaneously upon photon absorption.

Much of the evidence for energy migration down polymer chains comes from observations of excimer emission. In many polymers the intensity of excimer emission is greater than can be explained by absorption at pre-formed excimer sites (as would be necessary in rigid systems) or by conversion of an absorption site to suitable geometry by conformation change (as can happen in mobile systems). The explanation is that absorption occurs at normal 'monomer' sites, and that the energy then migrates down a sequence of such chromophores until it reaches an excimer-forming site, which then functions as an exciton trap. In polystyrene,[9,10] poly(1-vinyl-naphthalene),[9] poly(2-vinylnaphthalene),[9,10] poly(4-vinylbiphenyl),[11] the traps seem to be of such a depth that, although excimer dissociation is considerably less rapid than excimer formation, they do not form a major barrier to the transfer of energy to much deeper traps such as co-monomer units in the chain or accepter molecules. However there has been a suggestion[21] that the excimer sites are significant 'deep traps' in poly(N-vinyl-carbazole). This polymer appears to be unique in showing[22] two separate singlet excimer states. The less stable excimer occurs at pre-existing sites, and the lower energy state is formed by segmental rearrangement during the excited state lifetime.

Although excimer emission occurs as fluorescence rather than phosphorescence, migration of 'monomer' triplet energy allows triplet–triplet annihilation, creating[23,24] delayed fluorescence with either 'monomer' or excimer characteristics.

6.3.2. Energy transfer enhanced by exciton migration

Perhaps the most convincing quantitative evaluations of down-chain energy migration come from the efficiency of polymer luminescence quenching by acceptor species present in concentrations which would be most inefficient if the absorbed energy were localized. The enhanced transfer arises from an approach of the excitation and the energy acceptor during the excited state lifetime. This may occur by translational diffusion of the molecular species, by a segmental rearrangement of the chain which also moves an activated chromophore through space, or by exciton migration or delocalization.

In order to make a quantitative appraisal of this phenomenon it is necessary to solve a diffusion equation, with boundary conditions determined by the distance dependence of the energy transfer step. This is considerably more difficult than solution with nearest-neighbour collisional boundary conditions as is discussed in Chapter 13. The applicability of various solutions has been reviewed,[25] and it has been pointed out that those involving a step-function approximation to the probability of donor–acceptor energy transfer can introduce a serious overestimate of the transfer efficiency. A number of solutions are now available in which the transfer probability for singlet energy is given an inverse sixth-power dependence on donor–acceptor separation, as required for Förster interaction between dipolar transition moments. In this way the mobility coefficients can be evaluated from the observed increase in energy transfer over that expected for a Förster process between completely static donor and acceptor moieties.

The migration coefficients for random walk of incoherent singlet excitons down a number of polymer chains have been measured[25,26] using a numerical solution to the diffusion equation suggested by Yokota and Tanimoto.[27] A selection, observed in dilute solution at room temperature, is presented in Table 6.1. These illustrate a number of points. First, in poly(vinyl aromatics) the exciton mobility in the polymer chain is slightly greater than the diffusivity of a small molecule (analogous to the chromophore) in a mobile solvent. Secondly, the presence of excimers does not destroy quite distinct energy migration, even in poly(N-vinylcarbazole). Thirdly comonomer units which are spectroscopically inactive at the energies involved present observable barriers to the exciton random walk. Finally the migration is facilitated when the chain can undergo appreciable segmental rotation and conformation change during the excited state lifetime (2-vinylnaphthalene copolymers in fluid solution). On the other hand the fluidity of the medium is

Table 6.1

Some incoherent singlet exciton migration coefficients[25,26]

Polymer	Solvent	Migration coefficient $\times 10^9$ $(m^2 s^{-1})$
Polystyrene	Tetrahydrofuran (fluid)	3
Poly(1-vinylnaphthalene)	Tetrahydrofuran (fluid)	28
Poly(2-vinylnaphthalene)	Tetrahydrofuran (fluid)	13
Styrene-2-vinylnaphthalene copolymer, 6:94	Tetrahydrofuran (fluid)	2·5
Styrene-2-vinylnaphthalene copolymer, 6:94	Poly(methylmethacrylate) (rigid glass)	0·2
Poly(N-vinylcarbazole)	Tetrahydrofuran (fluid)	9
Poly(N-vinylcarbazole)	Poly(methylmethacrylate) (rigid glass)	7
N-vinyl carbazole methylacrylate 70:30 copolymer	Tetrahydrofuran (fluid)	4
N-vinylcarbazole methylacrylate 70:30 copolymer	Poly(methylmethacrylate) (rigid glass)	4

relatively unimportant when the chain stiffness inhibits rapid conformational change (N-vinylcarbazole polymers).

Similar observations of quenching efficiencies have permitted the evaluation[28,29] of effective interaction distances for excitons which may very well have a coherent character. These have been made on polymers where conjugation in the chain backbone provides a strong electronic interaction between aromatic chain substituents. Some representative values are listed in Table 6.2. In these cases the effective migration distances in fluid media

Table 6.2

Some interaction distances of singlet excitons which could have coherent character[28,29]

Polymer	Molecular weight	Solvent	Interaction distance (nm)
Poly(phenylacetylene)	$\bar{M}_n \, 5 \times 10^2$	Tetrahydrofuran (fluid)	very small
Poly(phenylacetylene)	$\bar{M}_n \, 7\cdot5 \times 10^3$ $\bar{M}_w \, 2\cdot5 \times 10^4$	Tetrahydrofuran (fluid)	31
Poly(p-methoxy-phenylacetylene	$\bar{M}_n \, 2 \times 10^3$ $\bar{M}_z \, 2 \times 10^5$	Tetrahydrofuran (fluid)	70
Poly(p-methoxy-phenylacetylene	$\bar{M}_n \, 2 \times 10^3$ $\bar{M}_z \, 2 \times 10^5$	Poly(methylmethacrylate) (rigid glass)	7

are comparable with the polymer end-to-end distances in an extended conformation, and the importance of segmental orientational freedom is most pronounced. The role of segmental motion in facilitating energy migration is thought to be rather different in the polymers with saturated and conjugated backbones. In the former case the rotation brings neighbouring side-group chromophores from a separation and relative orientation unfavourable for an exciton jump into a closer, more nearly parallel, conformation, so raising the individual jump probabilities and number of jumps in the excited-state lifetime. In conjugated polymers, entropic and other effects dictate that the number of alternating double and single bonds in a sequence permitting electron delocalization is usually smaller than about fifteen.[30] Consequently an ensemble of such long-chain molecules exists as a distribution of conjugated sequence lengths. Segmental reorientation ensures that this distribution is in a state of dynamic flux, shorter sequence lengths being converted to larger sequences, which in turn are truncated by continuing rotation. In this way energy localized over a short sequence length can be converted to greater delocalization distances, with a resulting increase in the effective interaction distance.

Precise measurements of quenching efficiencies provide evidence, too, of triplet energy migration. Thus the delayed fluorescence in poly(1-vinyl-naphthalene)[23] and poly(naphthylmethacrylate)[24] ascribed to triplet–triplet annihilation is reduced by quenchers which interfere with triplet energy migration. More directly, situations in which the quenching of phosphorescence is more efficient than predicted by the Terenin–Ermolaev[31] model for static donor–acceptor pairs, have been illustrated in a variety of polymers containing keto-groups. Thus there is now direct evidence of triplet energy migration in poly(vinylbenzophenone)[32,33] and poly(phenylvinylketone)[34,35] although not in poly(methylvinylketone)[35,36] and poly(methylisopropenylketone).[35] In a series of styrenevinylbenzophenone copolymers the quenching efficiencies have been used to estimate effective interaction distances for the triplet excitons. Although solution of the diffusion equation was carried out using boundary conditions which tend to underestimate the migration distances, these ranged from 11 to 29 nm as the vinylbenzophenone content increased from 9 to 77 mol per cent. Thus triplet excitons, primarily because of their relatively long lifetimes, may migrate over larger distances than comparable incoherent singlet excitons.

Because these observations of phosphorescence intensities were carried out in rigid glasses at 77 K, they yield no information on the role of segmental motion in facilitating (or otherwise) migration of triplet excitons. However in a related experiment triplet quenchers have been used to retard polymer chain degradation consequent upon formation of excited triplet states.[37] In dilute solutions of poly(phenylvinylketone) at 298 K the high efficiency of this quenching can be ascribed to triplet energy migration.

However insufficient data are available for meaningful comparison of migration characteristics in the high-temperature fluid and low-temperature rigid environments.

6.3.3. Theoretical calculations of exciton migration

The migration of both coherent and incoherent molecular excitons in lattices resembling polymer chains has been considered in a number of sophisticated theoretical expositions. Many of these have been listed by Pearlstein,[38] who has made significant contributions to the theory of quenched random walks on linear chains,[39] as well as to the quenching of coherent[40] and partially coherent[41] excitons. However most of these treatments are restricted to static lattices, and so have not yet been applied to the phenomena of interest in this text, namely polymer chain motions.

On the other hand it is possible to interpret many of the reported experimental observations using a more simplistic approach to a sequence of resonance transfer steps. The starting point is the equation for a single resonance transfer such as that developed by Förster.[18]

$$k_{AD} = \frac{CK^2}{r^6} \int_0^\infty f_D(\nu)\varepsilon_A(\nu)\frac{d\nu}{\nu^4} \tag{6.10}$$

where k_{AD} is a rate constant for donor-to-acceptor transfer, C is a constant embodying the quantum efficiency and lifetime of donor fluorescence and the refractive index of the surrounding medium, and the overlap integral covers the normalized donor fluorescence and acceptor absorbance intensities. In the context of molecular movement, however, the important terms in this expression are the interchromophore separation, r, and relative orientation factor K^2, defined by

$$K = \cos\theta_{AD} - 3\cos\theta_A\cos\theta_D \tag{6.11}$$

where θ_{AD} is the angle between the donor and acceptor transition moments and θ_A and θ_D are the angles between the transition moment vectors and the line of centres. When the chromophores are part of the same molecule, the effect of conformation changes can be investigated as variation or modulation of the orientation-distance factor K^2/r^6.

For random walk along a one-dimensional chain of chromophores the migration coefficient, Λ is

$$\Lambda = \tfrac{1}{2}k_{AD}r^2 \tag{6.12}$$

and the effective migration distance, $\bar{r}_i$ is

$$\bar{r}_i = (2\Lambda\tau)^{\frac{1}{2}}. \tag{6.13}$$

where τ is the excited state lifetime. Combination of eqns (6.10)–(6.13) allows prediction of the migration characteristics Λ and $\bar{r}_i$ from the spectroscopic and geometric parameters.

When the polymer chain has a well defined static conformation the calculation is straightforward. A typical example is the calculation of chromophore–chromophore transfer rates in the right-handed α-helix of poly(4-tyrosine).[42] In this helix three aromatic chromophores are attached to rather less than a single turn, and the tyrosyl–tyrosyl transfer rates (expressed as the number of jumps per lifetime) from unit zero to units numbered in sequence along the helix reach a maximum for transfer between third nearest neighbours (Table 6.3). Since jumps in both positive

Table 6.3

Tyrosyl–tyrosyl energy transfer rates in poly(L-tyrosine)[42]

Residue number (+ve direction only)	Separation r (nm)	Orientation factor K	Exciton jumps per lifetime
1	0·90	+1·03	0·32
2	1·18	−0·86	0·04
3	0·73	+1·90	3·70
4	0·72	+1·57	2·85
5	1·32	−0·53	0·01
6	1·35	−0·08	0·00
7	1·07	+1·04	0·12
8	1·41	+0·66	0·01

and negative directions are equally likely, and a jump to units 3 and 4 covers just under 0·7 nm along the helix direction, the effective migration distance (proportional to the square root of the number of jumps) is about 2·5 nm.

In the dynamic systems of interest here, or with polymers of indeterminate conformation, the calculation is carried out by considering various averaged possibilities for the orientation-distance factor. For example,[25] if poly(N-vinyl carbazole) is assumed to exist in conformations such that a majority of units resemble a 4:1 helix, the average separation between closest chromophores is 0·48 nm and the orientation factor, K^2, is 0·25. On the other hand when the orientation of the chromophores is completely randomized, the orientation factor is 0·67 if they are time-averaged (i.e. moving through many different orientations during the exciton residence time) and 0·48 if they are static. When the spectroscopic characteristics of 'monomer' singlet excitation of carbazole units are combined with these geometrical terms, migration coefficients of 1×10^{-8}, 2×10^{-8}, and $3 \times 10^{-8}\ \mathrm{m^2\,s^{-1}}$ are obtained for helical, random static, and random dynamic orientations, respectively. It can be seen that the calculation assuming an ordered helical geometry most closely fits the experimental observations reported in Table 6.1, as is reasonable for a molecule known to exhibit large energy barriers to segmental rotational conformation change. However, when similar data are applied to poly(vinylnaphthalene) the predicted migration coefficients are too small by a factor of ten, even when the freely · reorienting value of 0·67 is used for K^2. In this case experiment and calculation can be matched by reducing the average internaphthalene separation from 0·48 to 0·3 nm, an indication of much more facile segmental rotation and disappearance of helical structure in solutions.

It must be emphasized that in such considerations of dynamic systems, the average interchromophore separation is heavily weighted by the sixth-power dependence on small distances, and so must be smaller than distances based simply on population-number averages.

6.3.4. Phosphorescence as a probe of subgroup motions

The intensity of phosphorescence is a sensitive indication of both inter-system non-radiative processes and quenching by spurious impurities. This is partly a result of the relatively long lifetime of phosphorescence, which permits a variety of processes to compete with direct emission. Since intersystem processes are affected by molecular distortions, and quenching by molecular diffusion, both are sensitive to the scale and nature of local molecular movement.

In an early study of the temperature dependence of phosphorescence intensity in commercial polyethylene, Boustead[43] found that quenching processes in amorphous polymer increased markedly over temperature ranges centred at 110, 185, and 240 K, which temperatures also correspond to the onset of molecular motions observable in mechanical relaxation studies. In a related study of a variety of ethylene co-polymers, Somersall, Dan, and Guillet[44] found similar changes in the temperature coefficient of phosphorescence intensity at 110, 163, and 250 K. The very pronounced intermediate transition corresponds to the relaxation process which has been designated the γ-transition in a variety of dielectric and mechanical observations. It is associated with quite large-scale reorientations of the polyethylene chain. Whether this is the true 'glass transition', or is a subsidiary motion of a 'crankshaft' nature[45] is still not clear. The lowest temperature process appears to be the onset of group rotation at the keto-unit, and the highest temperature process is associated with motion of chain branches.

Related studies[44] of polystyrene, poly(methylmethacrylate), poly-(acrylonitrile), poly(methacrylonitrile) (Fig. 6.5) and their keto-polymers all clearly show the dependence of phosphorescence intensity on relaxation processes below the main glass transition.

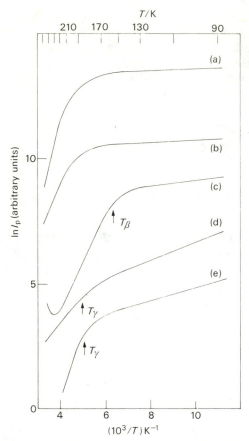

Fig. 6.5. Phosphorescence intensities: (a) polyacrylonitrile; (b) polymeth-acrylonitrile; (c) polyvinylchloride; (d) poly(methylmethacrylate); (e) polystyrene. Intensities in arbitrary units with each curve shifted vertically for clarity. Well characterized dielectric or mechanical β and γ transitions indicated.

6.4. Luminescence depolarization

6.4.1. Introduction

An important feature of a molecular electronic transition is that the transition moment bears a clearly defined geometrical relationship both to the molecular framework and to the electromagnetic vectors of radiation involved in the transition.[46] Consequently a study of the polarization characteristics of exciting and luminescence radiation can be a measure of the orientational properties of the chromophoric molecule. More specifically excitation using irradiation with specific polarization photoselects those chromophores in an appropriate orientation, and the polarization of the resulting luminescence reflects changes in the relative orientations of the absorption and emission vectors.

When the changes in polarization arise from molecular rotational phenomena, practicalities of observation require that the rotational-correlation time and the excited-state lifetime be of comparable magnitude. Consequently study of fluorescence depolarization is applicable to rotational processes with relaxation times between 10^{-10} and 10^{-8} s. These are encountered in the segmental rotation of dissolved flexible polymers, and the rotation of small molecules in viscous (though still fluid) environments. On the other hand phosphorescence depolarization is, in principle, capable of yielding information on rotational processes with relaxation times between 10^{-4} and 10^0 s, as might be encountered in gels and rubbers.

The basic principles of luminescence anisotropy,[47] their extension to molecular reorientation phenomena,[27] and the particular application to polymer systems[48,49] have all formed the subject of review articles.

The degree of polarization, p, of a partially polarized beam can be defined as

$$p = \frac{I_{\parallel} - I_{\perp}}{I_{\parallel} + I_{\perp}} \tag{6.14}$$

where $I_{\parallel}$ and $I_{\perp}$ are the intensities of the components of the beam resolved parallel and perpendicular to the reference direction of partial polarization. The basic geometry for studying depolarized emission perpendicular to the direction of a plane-polarized exciting beam is illustrated in Fig. 6.6. The

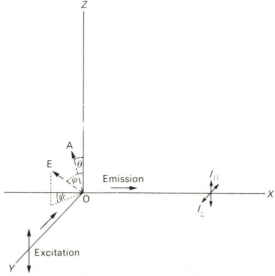

Fig. 6.6. Excitation and emission polarization geometries involving an absorption vector OA, forming an angle θ, with the Z-axis, and a subsequent emission vector OE, forming angles ϕ with the Z-axis and ψ with the X-axis in the XY-plane.

probability of absorption by a molecule with absorption vector oriented along OA is proportional to $\cos^2 \theta$. Correspondingly the polarized components of the emission are

$$I_{\parallel}(\phi, \psi) \propto \cos^2 \phi$$
$$I_{\perp}(\phi, \psi) \propto \sin^2 \phi \sin^2 \psi. \tag{6.15}$$

In a system of randomly oriented molecules, for all molecules with emission vector angle ϕ, the angle ψ will take all values between 0 and 2π, so that the average value $\sin^2 \psi$ is $\frac{1}{2}$ and

$$I_{\parallel}(\phi) \propto \cos^2 \phi$$
$$I_{\perp}(\phi) \propto \tfrac{1}{2} \sin^2 \phi. \tag{6.16}$$

The problem now reduces to a determination of the geometrical relationship between OA at zero time and OE at the moment of emission, followed by appropriately weighted integration over all possible values of the initial absorption angle θ.

The simplest case occurs when OA and OE are coincident, as is the situation for the $S_0 \leftrightarrow S_1$ transition (commonly observed in fluorescence) in immobile molecules.

Then

$$I_{\parallel} \propto \sum_{\theta=0}^{\theta=\pi/2} I_{\parallel}(\theta) \propto \sum_{\theta=0}^{\theta=\pi/2} f(\theta) \cos^2 \theta$$
$$I_{\perp} \propto \sum_{\theta=0}^{\theta=\pi/2} I_{\perp}(\theta) \propto \sum_{\theta=0}^{\theta=\pi/2} f(\theta) \tfrac{1}{2} \sin^2 \theta \tag{6.17}$$

where $f(\theta)$ is the fraction of molecules for which the absorption and emission vectors lie at the angle θ to the Z-axis, and is proportional to $\sin \theta$ for a random array. The parallel and perpendicular intensities are thus seen to be proportional to the average values of $\cos^2 \theta$ and $\frac{1}{2} \sin^2 \theta$ respectively. Substitution in eqn (6.14) yields

$$p = \frac{\frac{3}{2} \overline{\cos^2 \theta} - \frac{1}{2}}{\frac{1}{2} + \frac{1}{2} \overline{\cos^2 \theta}} \tag{6.18}$$

or

$$1/p - \tfrac{1}{3} = \frac{\frac{2}{3}}{\frac{3}{2} \overline{\cos^2 \theta} - \frac{1}{2}}. \tag{6.19}$$

Thus the problem reduces to calculation of $\overline{\cos^2 \theta}$. When this is weighted by the absorption probability (proportional to $\cos^2 \theta$) the final expressions are

$$\overline{\cos^2 \theta} = \frac{\int_0^{\pi/2} \cos^4 \theta \sin \theta \, d\theta}{\int_0^{\pi/2} \cos^2 \theta \sin \theta \, d\theta} = \tfrac{3}{5} \tag{6.20}$$

and

$$p = \tfrac{1}{2}. \tag{6.21}$$

So the angular dependence of absorption and emission probabilities determines that, even in this most favourable case, there must be some depolarization of luminescence. All processes which cause the emission vector to lie at an angle to the absorption vector must then cause further depolarization. When their angle is λ

$$1/p - \tfrac{1}{3} = \tfrac{5}{3} \left\{ \frac{2}{3 \cos^2 \lambda - 1} \right\}. \tag{6.22}$$

The angle λ may arise from intrinsic differences in the absorption and emission transitions. Under these circumstances depolarization is observed even in totally static molecules and eqn (6.22) defines the intrinsic depolarization p_0^{-1}. It can be seen that p_0 is $\tfrac{1}{2}$ for parallel moments (as is often observed in $0 \leftarrow 0$ $S_0 \leftarrow S_1$ fluorescence) but is $-\tfrac{1}{3}$ for perpendicular moments (as is sometimes the case in phosphorescence). In an extreme case the radiation is totally depolarized ($p_0 = 0$) when λ is 55°.

Alternatively the angular difference between the absorption and emission rates may arise by extrinsic processes occurring during the lifetime of the excited state. Two such are rotation of the excited chromophore and energy transfer to a chromophore with a slightly different orientation. It is the former process which is of interest in the present context.

6.4.2. Depolarization by molecular or segmental rotation

When rotation of a body is isotropic, it can be expressed in terms of a single rotation time defined by

$$\overline{(\cos \lambda)_t} = \overline{(\cos \lambda)_0} \exp(-t/\rho) \tag{6.23}$$

where the average values of $\cos \lambda$ change from the initial value (time zero) to that at time, t, according to an exponential rate law. The angular displacement occurring in a sequence of small jumps is related to the rotational diffusion coefficient, D_r, through the Einstein[50] relationship.

$$\overline{\lambda^2} = \frac{2t}{\rho} = 4D_r t = \frac{2RTt}{3\eta V} \tag{6.24}$$

where η is the solvent viscosity and V is the volume of the quasi-spherical rotor.

Small values of λ can be equated to $\sin \lambda$ whence combination with eqn (6.19) yields for small λ,

$$(1/p - \tfrac{1}{3}) = (1/p_0 - \tfrac{1}{3})(1 + 3\tau/\rho) \tag{6.25a}$$

$$= (1/p_0 - \tfrac{1}{3})(1 + 6D_r\tau) \tag{6.25b}$$

$$= (1/p_0 - \tfrac{1}{3})(1 + RT\tau/\eta V). \tag{6.25c}$$

These equations, applicable to the excitation of spherical molecules by vertically polarized radiation, were first derived by Perrin.[51]

Although these equations have been widely applied in the study of fluorescence depolarization in polymer systems, there are good reasons why their use is not as informative as is often implied. The principal difficulties arise in the assumption that the rotor is isotropic, and in the procedure adopted to evaluate the intrinsic depolarization p_0.

In many instances this second is achieved by extrapolation of depolarization, measured as a function of T/η, to conditions corresponding to an infinitely rigid environment ($T/\eta = 0$). However when the rotor is anisotropic, this extrapolation is not valid if it passes the region where the mean rotational and excited-state lifetimes are comparable,[52] quite apart from the inapplicability of the simple Stokes–Einstein relationships (and thus eqn (6.25c)) to such systems. In addition the principal contribution to the intensity of steady state fluorescence emission comes from molecules emitting in times less than the mean lifetime, and so the observation is 'weighted' by short-time events.

These deficiencies can be overcome by observation of the transient decay functions for the polarized components of emission. The functions for parallel and perpendicular emission differ because rotation intensifies one at the expense of the other. The two functions then contain sufficient information to yield the excited-state lifetime, and more than one rotational time, directly. This time-dependent decay of luminescence anisotropy is discussed in detail in a recent review by Wahl,[53] and only the most salient features are presented here.

In the study of anisotropy decay it is often convenient to replace the emission polarization by the emission anisotropy, which for vertically polarized exciting light is defined by

$$r = \frac{I_\| - I_\perp}{I_\| + 2I_\perp} = \frac{D}{S}. \tag{6.26}$$

The emission can be characterized in terms of these sum and difference intensities

$$S(t) = I_\| + 2I_\perp$$
$$D(t) = I_\| - I_\perp. \tag{6.27}$$

Thus, for luminescence characterized by a single lifetime, τ, and when $r(t)$, $D(t)$ are also single exponentials,

$$S(t) = S_0 \exp(-t/\tau)$$
$$r(t) = \tau_0 \exp(-t/\rho) \tag{6.28}$$
$$D(t) = D \exp(-t/\tau_D)$$

and

$$\rho = \frac{\tau \tau_D}{\tau - \tau_D}. \tag{6.29}$$

These equations yield the Perrin equation in terms of the static anisotropy Γ_0

$$r = \frac{\Gamma_0}{1 + \tau/\rho}. \tag{6.30}$$

The time-dependent equation for isotropic rotational motion is then

$$r(t) = r_0 \cdot \frac{\langle 3 \cos^2 \omega(t) \rangle - 1}{2} = \Gamma_0 \phi(t) \tag{6.31}$$

where $\omega(t)$ is the rotational angle defined by the positions of the emission vector between the excitation and emission times, and $\phi(t)$ is the orientation auto-correlation function. The rotational diffusion equation containing $\omega(t)$ can then be solved to yield again the familiar expression for the rotational correlation time,

$$r(t) = r_0 \exp(-t/\rho)$$

$$\rho = \frac{1}{6D_r}. \tag{6.32}$$

We note in passing that ρ describes relaxation of $\cos^2 \omega$, in contrast to the dielectric rotational correlation time, which measures relaxation of $\cos \omega$.

When the rotor is ellipsoidal, the emission anisotropy is expressed in terms of three exponentials

$$r(t) = A_1 \exp(-t/\rho_1) + A_2 \exp(-t/\rho_2) + A_3 \exp(-t/\rho_3) \tag{6.33}$$

in which the three correlation times, ρ_1, ρ_2, ρ_3, are functions of the two principal rotational diffusion coefficients D_1 round the major axis and D_2 round a transverse axis,

$$\rho_1 = \frac{1}{2S_1 + 4D_2}$$

$$\rho_2 = \frac{1}{5D_1 + D_2} \tag{6.34}$$

$$\rho_3 = \frac{1}{6D_1}.$$

A special case of such an anisotropic rotor is the chromophore in a polymer chain. Here rotation around one axis may be achieved by internal rotation, while rotation of the whole macromolecule is also possible.[54] Then

$$r(t) = \exp(-t/\rho_p)[A_1 \exp(-2t/\rho_c) + A_2 \exp(-t/6\rho_c) + A_3] \tag{6.35}$$

where ρ_p and ρ_c are the correlation times for rotation of the whole macromolecule (around three axes) and chromophore (around one axis) respectively.

While observations of fluorescence depolarization have been widely used to study the behaviour of dyes bound to biopolymers,[53] the technique has not been so widely applied to measurements of segmental motion in synthetic polymers. Some measurements, made by both continuous irradiation and emission anisotropy decay methods, which show that segmental rotation at the ends of dissolved macromolecules is much more rapid than in the chain interior,[55] are presented in Table 6.4. It can be seen that the rotational times generally agree with those observed by dielectric relaxation (Chapter 5).

Table 6.4

Rotational times of fluorescent chromophores on polymers dissolved in toluene at 298 K

Polymer chain	Chromophore and position	Rotational time (ns)
Poly(butylmethacrylate)	Anthracene—chain-end	0·8
	Anthracene—side-group in chain interior	4·0
Polystyrene	Anthracene—chain-end	0·8
	Anthracene—side-group in chain interior	6·0
	Naphthalene—side-group in chain interior	5·5
Poly(N-vinylcarbazole)	Anthracene—side-group in chain interior	26

Probably the most detailed studies of fluorescence depolarization by segmental rotation have been made on polystyrene in solution.[56] These confirm that the mean rotational relaxation time in a very mobile solvent is about 4 ns and show that the orientation-correlation function can be expressed as

$$\phi(t) = \exp(-t/\theta)\exp(-t/\rho)\,\mathrm{erfc}\,\sqrt{(t/\rho)} \tag{6.36}$$

where erfc is the error function complement and ρ and θ are relaxation times which, in this study, are considered to characterize respectively the diffusion of bond orientation along the chain and changes in orientation due to the fact that elementary motions are not confined to a tetrahedral lattice.

References

1. Birks, J. D. *Organic molecular photophysics, Vols. 1 and 2.* Wiley, New York and London (1973 and 1975).
2. Fox, R. B. *Pure appl. Chem.* **30**, 87 (1972).

3. Somersall, A. C. and Guillet, J. E. *Macromol. Rev.* (In press.)
4. Hirayama, F. *J. chem. Phys.* **42**, 3163 (1965).
5. Vala, Jr., M. T., Haebig, J. and Rice, S. A. *J. chem. Phys.* **43**, 886 (1965).
6. Klöpffer, W. In *Organic molecular photophysics* (ed. J. B. Birks), Vol. I, Chapter 7, p. 357. John Wiley, London (1973).
7. Longwort, J. W. and Bovey, F. A. *Biopolymers* **4**, 1115 (1966).
8. Phillips, D. *Photochemistry*, Vol. 5, p. 713. Spec. Periodical Reports, Chem. Soc. (London) (1974).
9. Fox, R. B., Price, T. R., Cozzens, R. F. and MacDonald, J. R. *J. chem. Phys.* **57**, 534 (1972).
10. Frank, C. W. and Harrah, L. A. *J. chem. Phys.* **61**, 1526 (1974).
11. Frank, C. W. *J. chem. Phys.* **61**, 2015 (1974).
12. Klöpffer, W. *J. chem. Phys.* **50**, 2337 (1969).
13. Birks, J. B. and Christophorou, L. G. *Proc. Roy. Soc. (Lond.)* **A277**, 571 (1964).
14. Heisel, F. and Laustriat, G. *J. chim. Phys.* **66**, 1881, 1895 (1969).
15. Klöpffer, W. and Liptay, W. *Z. Naturforsch.* **25a**, 1091 (1970).
16. David, C., Piens, M. and Geuskens, G. *Eur. polymer. J.* **9**, 533 (1973).
17. Harrah, L. A. *J. chem. Phys.* **56**, 385 (1972).
18. Förster, Th. *Ann. Phys.* **2**, 55 (1948); *Discuss. Faraday Soc.* **27**, 7 (1959).
19. Dexter, D. L. *J. chem. Phys.* **21**, 836 (1953).
20. Frenkel, Y. I. *Introduction to theory of metals* (3rd edn). Gosudarst. Izdatel. Fiz-Mat. Lit. Moscow (1958).
21. Klöpffer, W. *Kunstoffe* **61**, 533 (1971).
22. Johnson, G. E. *J. chem. Phys.* **62**, 4697 (1975).
23. Cozzens, R. F. and Fox, R. B. *J. chem. Phys.* **50**, 1532 (1969).
24. Somersall, A. C. and Guillet, J. E. *Macromolecules* **6**, 228 (1973).
25. North, A. M. and Treadaway, M. F. *Eur. polym. J.* **9**, 609 (1973) and unpublished results.
26. North, A. M. *Brit. polym. J.* **7**, 119 (1975).
27. Yokota, M. and Tanimoto, O. *J. phys. Soc. Jpn* **22**, 779 (1967).
28. North, A. M., Ross, D. A. and Treadaway, M. F. *Eur. polym. J.* **10**, 411 (1974).
29. North, A. M. and Ross, D. A. *J. polym. Sci. C* Symp. 55, 259 (1976).
30. Samedova, T. G., Karpocheva, G. P. and Davydov, B. E. *Eur. polym. J.* **8**, 599 (1972).
31. Terenim, A. N. and Ermolaev, V. L. *Trans. Faraday Soc.* **52**, 1042 (1956).
32. David, C., Demarteau, W. and Geuskens, G. *Eur. polym. J.* **6**, 537 (1970).
33. David, C., Naegelen, V., Piret, W. and Geuskens, G. *Eur. polym. J.* **11**, 569 (1975).
34. David, C., Demarteau, W. and Geuskens, G. *Eur. polym. J.* **6**, 1405 (1970).
35. Dan, E., Somersall, A. C. and Guillet, J. E. *Macromolecules* **6**, 228 (1973).
36. David, C., Demarteau, W., Lempereur, M. and Geuskens, G. *Eur. polym. J.* **8**, 409 (1972).
37. Golemba, F. J. and Guillet, J. E. *Macromolecules* **5**, 212 (1972).
38. Pearlstein, R. M. *J. chem. Phys.* **56**, 2431 (1972).
39. Lakatos-Lindenberg, K., Hemenger, R. P. and Pearlstein, R. M. *J. chem. Phys.* **56**, 4852 (1972).
40. Hemenger, R. P. and Pearlstein, R. M. *Chem. Phys. Lett.* **2**, 424 (1973).
41. Hemenger, R. P., Lakatos-Lindenberg, K. and Pearlstein, R. M. *J. chem. Phys.* **60**, 3271 (1974).
42. Ten Bosch, J. J. and Knapp, J. A. *Biochem. Biophys. Acta* **188**, 173 (1969).
43. Boustead, I. *Eur. polym. J.* **6**, 731 (1970).

44. Somersall, A. C., Dan, E. and Guillet, J. E. *Macromolecules* **7**, 233 (1974).
45. Schatzki, T. F. *Polym. Preprints* **6**, 646 (1965).
46. Liptay, W. In *Modern quantum chemistry* (ed. O. Sinanoglu), p. 45. Academic Press, New York (1965).
47. Weber, G. In *Fluorescence and phosphorescence analysis* (ed. D. M. Hercules), p. 217. Interscience, New York (1966).
48. Oster, G. and Nishijima, Y. *Fortschr. Hochpolym.-Forsch.* **3**, 313 (1964).
49. Nishijima, Y. *J. polym. Sci. C*, **31**, 353 (1970).
50. Einstein, A. *Ann. Phys.* **19**, 371 (1906).
51. Perrin, F. *Ann. Phys. (Paris)* **12**, 169 (1929).
52. Weber, G. *J. chem. Phys.* **55**, 2399 (1971).
53. Wahl, P. In *Biochemical fluorescence* (ed. F. Chen and H. Edelhoch), Vol. 1, Chapter 1, p. 1. Marcel Dekker, New York (1975).
54. Gottlieb, Y. and Wahl, P. *J. chim. Phys.* **60**, 849 (1963).
55. North, A. M. and Soutar, I. *J. chem. Soc., Faraday Trans I* **68**, 1101 (1972).
56. Valeur, B. and Monnerie, L. *J. polym. Sci. (Polymer Physics Edition)* **14**, 11 (1976).

VISCOELASTIC RELAXATION IN POLYMER SOLUTIONS AND MELTS

7.1. Introduction

Macromolecular species have the property of causing a marked increase in the viscosity of a solvent, even when present as less than one per cent of the solution. The increased viscosity is associated with the extra dissipation of energy in flow as the long thread-like polymer molecules swirl around in the sheared fluid.

Theoretical consideration of the collective motions of a chain when subjected to hydrodynamic shear forces have been discussed in Chapters 3 and 4. The majority of such theories refer to the motion of a polymer chain in isolation. However in practice, the motions are influenced not only by intra-, but also by intermolecular interactions.

Experimental efforts to observe the 'infinite' dilution limit reflect the precision with which the phase difference between stress and strain can be measured over an extended frequency range. In 1942 the concentrations studied were typically of the order of 15 per cent, in 1952 this had been reduced to 4 per cent, and by 1961 it was down to 1 per cent. Even these data, however, did not allow precise extrapolation to infinite dilution as required by theory. The first successful observations of the viscoelastic properties of a polymer in the infinite dilution limit involving concentrations as low as 0·1 per cent were performed in 1966.[1] Numerous such studies have been reported since that date.

7.1.1. Regions of hydrodynamic behaviour

The behaviour of polymer solutions can be classified broadly according to the degree of interaction of the polymer with its environment.

(i) *Infinite dilution limit.* This is the ideal condition described theoretically in Chapter 4. The movement of a polymer chain is analysed in terms of the superposition of a number of collective motions of appropriate subunits. The hydrodynamic flow field is assumed to be localized within the region around the subunit. According to the classification of Frisch and Simha[2] the infinite dilution limit exists for concentrations below $c[\eta] \sim 1$, where $[\eta]$ is the intrinsic viscosity of the polymer solution.

(ii) *Hydrodynamic screening limit.* A polymer subunit, in the infinite dilution limit, experiences the presence of other subunits in the same chain through hydrodynamic interactions with solvent. If the concentration of polymer in solution is raised slowly, a point will be reached when the subunit experiences hydrodynamic effects not only originating in its own chain, but also from neighbouring macromolecules. Perturbation of the subunit–solvent hydrodynamic interactions by this mechanism is expected to occur above a concentration defined by $c[\eta] \geqslant 1$. The effects may be expected to be cumulative, and will increasingly influence the transport properties up to concentrations corresponding to some limit of close-packing of polymer coils, $c[\eta] \geqslant 3-4$. The precise value of the concentration at which this condition is fulfilled is ill-defined, and depends on the flexibility of the polymer and the nature of solvent.

(iii) *Polymer–polymer contact region.* Once the critical packing concentration has been exceeded, the motion of the polymer will be dominated by the presence of direct polymer coil–polymer coil interactions. As a consequence the shear viscosity rises markedly and the dynamic behaviour of the polymer is modified significantly.

(iv) *Polymer chain entanglement regions.* Although direct polymer–polymer interactions are observed in region (iii), their effect becomes much increased in magnitude above concentrations of the order of $c[\eta] \geqslant 10$. This is a consequence of the interpenetration of the polymer coils, and is associated with chain entanglements. For entanglement to occur the polymer must possess a molecular weight which is greater than a critical value, M_c. The precise value of M_c is a function of the chemical nature of the polymer chain and particularly of its flexibility. The probability of an entanglement being formed will be a function of the number of chain contacts, their lifetimes, and topography. Dynamic entanglements may be considered as acting rather like 'hooks', impeding the translational motion of the polymer and producing drastic modification of its relaxation behaviour. In the extreme limit strong entanglements will effectively immobilize the molecule and the system will form a gel. A non-cross-linked system will, however, exhibit long-time creep behaviour as a consequence of the finite lifetime of the tie-points under prolonged stress.

(v) *The melt.* The extreme in a concentration study is the pure polymer which, if raised above its melt temperature, will exhibit creep or viscoelastic behaviour. The relative motion of a polymer molecule in a non-cross-linked matrix has been likened to the motion of a thread in a pipe.[3]

In Fig. 7.1 a schematic plot of the variation of the viscosity of a polymer solution with concentration shows the changes associated with the regions identified above.

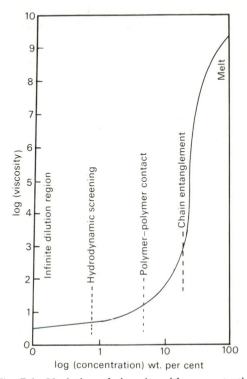

Fig. 7.1. Variation of viscosity with concentration.

7.1.2. The viscoelastic restoring force

It is not very difficult to visualize an elastic restoring force arising in the presence of polymer–polymer tie-points. This would have its origin in the entropic nature of polymer (rubber) elasticity. However it is a little more difficult to understand the origins of a similar force when the macromolecules are completely isolated.

The presence of such a restoring force was demonstrated experimentally by oscillatory shear studies at 10^4 Hz on 0·1 per cent polymer solutions as early as 1948.[4] Theoretical normal-mode analyses as discussed in Chapter 4 show how a coiled macromolecule is distorted by the hydrodynamic torque exerted on the subunits of the chain. The distribution of possible configurations adopted by the macromolecules is then less probable than the unperturbed value. This corresponds to a decrease in entropy and causes an increase in free energy which is stored, elastic energy.

7.2. Basic concepts in viscoelasticity of polymers

Theoretical and experimental studies of the viscoelastic behaviour of both dilute and concentrated polymer systems have proliferated over the last three decades. Certain of these have applied the concepts of continuum mechanics to dilute[5-13] and concentrated (including bulk)[14-20] polymer systems. Detailed consideration has been given to such specific aspects as chain entanglement in bulk and concentrated solutions,[21,22] hydrodynamic interactions,[23] and the theoretical description of the dynamics of conformational change.[24]

In the theory of elasticity,[25] the ratio of shear stress to strain for small deformations is the shear modulus, G. The ratios of the in-phase and out-of-phase stress to the strain are, respectively, the storage modulus, G', and loss modulus, G''. The former is a measure of energy stored and recovered in a cycle of deformation, the latter of energy dissipated as heat. In mathematical form, if the strain is a sine function of time

$$\gamma = \gamma_0 \sin \omega t. \tag{7.1}$$

The stress, σ, is given by

$$\sigma = \gamma_0 (G' \sin \omega t + G'' \cos \omega t) \tag{7.2}$$

where γ_0 is the peak strain and ω the radian frequency (2π times the frequency in Hertz). Both moduli G' and G'' depend on ω.

The discussion presented in Chapter 3 indicates that models based on the bead-spring concepts[26-8] lead to storage and loss moduli having the following frequency dependence

$$[G'] = (RT/M) \sum_{p=1}^{N} \omega \tau_p^2 / (1 + \omega^2 \tau_p^2) \tag{7.3}$$

$$[G''] = (RT/M) \sum_{p=1}^{N} \omega \tau_p / (1 + \omega^2 \tau_p^2). \tag{7.4}$$

In these equations, the square brackets represent limiting ratios when extrapolated to infinite dilution: thus

$$[G'] = \lim_{c \to 0} G'/c \tag{7.5}$$

$$[G''] = \lim_{c \to 0} (G'' - \omega \eta_s)/c. \tag{7.6}$$

In eqn (7.6), c is the concentration of polymer in solution and η_s is the solvent viscosity; the solvent contribution to the loss modulus is subtracted from the total to give the polymer contribution. No subtraction is necessary in (7.5) because the solvent contributes no elasticity.

It will be recalled that the principal factors determining the magnitude of the relaxation time in the above equations are the molecular weight, M, the size of the polymer and the strength of the 'hydrodynamic interaction parameter' h^*. This latter parameter is a measure of the drag produced between subunits of the chain, and may be thought of as the effective ratio of bead size to the average distance between neighbouring beads.

The longest relaxation time τ_1 is related to measurable quantities by

$$\tau_1 = [\eta]\eta_s M / RTS_1 \tag{7.7}$$

in which the intrinsic viscosity $[\eta]$ is defined by the familiar equation

$$[\eta] = \lim_{c \to 0} (\eta - \eta_s)/\eta_s c \tag{7.8}$$

where η_s is the viscosity of the solvent and η is the low-frequency limiting value of $G''/\omega\eta_s$. Two summations of importance are

$$S_1 = \sum_{p=1}^{N} \tau_p/\tau_1; \qquad S_2 = \sum_{p=1}^{N} (\tau_p/\tau_1)^2. \tag{7.9}$$

Table 7.1
Techniques for the study of viscoelastic behaviour

Technique	Frequency range	Comment	Ref.
Shear wave propagation	4–5×10^3 Hz	Poor accuracy—error typically of the order of 15 per cent in the dynamic viscosity.	175
Direct stress–strain measurements	10^{-4}–10^2 Hz	Very successful for studying the dynamic viscosity in the range 10^{-1}–10^5 N s m^{-2}.	176
Transducer measurement of stress and strain (Birnboim resonator)	10^{-2}–$1 \cdot 5 \times 10^3$ Hz 10^{-1}–1×10^5 Hz	Can be used to measure dynamic viscosities as low as 10^{-2} N s m^{-2} with high precision.	177–80
Impedance measurements	10^2–10^4 Hz	Successfully used to study solutions with viscosities of the order of 10^{-3} N s m^{-2}.	181
Torsional quartz crystals	10^4–$1 \cdot 2 \times 10^5$ Hz	Only applicable to liquids with viscosities below 1 N s m^{-2} accuracy ± 1 per cent.	182
Travelling-wave technique	10^4–2×10^5 Hz	Useful for liquids down to viscosities of 10^{-3} N s m^{-2}; accuracy 1 per cent for a value of 10^{-2} N s m^{-2}.	183–5
Normal and inclined incidence	5×10^6–10^8 Hz	Widely used for studies of viscous fluids.	186
Normal incidence technique	3×10^8–$1 \cdot 6 \times 10^9$ Hz	Viscous fluids only.	187–91
Light scattering	>1 GHz		192

The longest time τ_1 corresponds to a mode of motion in which the ends of the molecule are going in opposite directions, the others correspond to more complicated modes with increasing number of nodal points.

Equilibrium measurements of the viscosity are obviously a very valuable indication of the time-averaged interactions which occur in a polymer solution. However the real interest lies in the way in which G' and G'' vary with frequency. The techniques used for these studies have been discussed in detail elsewhere.[29,30] Broadly speaking, the measurements are concerned with observations of the phase difference between the stress and the strain and the mechanical impedance changes which occur when a surface undergoing oscillatory motion is in contact with a viscoelastic fluid. A survey of the techniques available (Table 7.1) indicates that whilst in theory viscoelastic measurements can be performed over a wide frequency range, the practical range for the study of dilute solutions in a solvent of low viscosity is somewhat more limited.

7.3. Viscoelasticity of dilute solutions

The importance of hydrodynamic interactions at low concentrations is well established. In steady flow the intrinsic viscosity in the region where polymer coils do not interpenetrate is

$$[\eta] = \lim_{c \to 0} \frac{(\eta - \eta_s)}{\eta_s c} = \Phi \langle r^2 \rangle^{\frac{3}{2}} / M \qquad (7.10)$$

where Φ is a 'universal' constant. This relates the intrinsic viscosity, molecular dimensions, and molecular weight for polymers with $M_w > 10^6$, and holds for a wide variety of polymer–solvent systems.[31] In theta solvents (when the second virial coefficient is zero and the chain is 'unperturbed' by polymer–solvent interactions) the experimental values of $[\eta] M / \langle r^2 \rangle^{\frac{3}{2}}$ remain very nearly constant, even down to rather low molecular weights ($M \sim 10^4$), while in good solvents the ratio begins to decrease as the molecular weight falls below 10^5. The observed value of Φ, agrees well with the theoretical limiting value ($h^* \to \infty$) calculated from the original Zimm theory ($\Phi = 0 \cdot 47$) or the Pyun/Fixmann modification ($\Phi = 0 \cdot 45$). The decrease of Φ with decreasing molecular weight is interpreted as a trend toward free-draining behaviour, and is predicted by theory since the hydrodynamic interaction parameter, h^*, itself decreases with decreasing molecular weight, being proportional to $M / \langle r^2 \rangle^{\frac{1}{2}}$.

Data on several polymers in both good and theta solvents are available[32-4] and indicate that if h^* is used as a reducible parameter, $[G'] M / RT$ and $[G''] M / RT$ can be placed in essentially quantitative agreement with moduli calculated from the Zimm eigenvalues.[35,36] Accurate calculations have been carried out for polymer molecules with $N = 200$ subunits. The values of h^*_{200} vary somewhat with solvent, but results for different polymers, molecular weights, and solvent powers are brought together by the correlation

$$h^*_{200}\alpha_\eta = 0.21 \pm 0.02 \tag{7.11}$$

in which α_η is the viscometric coil expansion factor relative to the theta solvent $([\eta]/[\eta]_\theta)^{\frac{1}{3}}$. This form is suggested by the $\langle r^2 \rangle^{\frac{1}{2}}$ term in the denominator of the hydrodynamic interaction parameter,

$$h^* = \frac{n\zeta_0}{(2\pi^3)^{\frac{1}{2}}\eta_s\langle r^2 \rangle^{\frac{1}{2}}} \tag{7.12}$$

in combination with eqn (7.10). ζ_0 is the friction coefficient for each of n main chain atoms. Thus the inference is drawn that coil expansion in good solvents shifts the dynamic properties toward free draining behaviour (reduced h^*). In the above equations, agreement between experiment and theory is only to be expected in the infinite dilution limit.

Detailed comparison[38] of the frequency-dependent properties predicted by the theories described in Chapter 4 and found in experiment may be performed by comparing the plots of reduced intrinsic complex modulus $[G^*]_R$ as a function of reduced angular frequency $\omega_R = \omega_1$.

In general the $[G']_R$ varies as ω_R^2 and $[G'']_R$ as ω_R in the low-frequency limit, $\omega\tau_1 < 0.5$, regardless of the magnitude of the hydrodynamic interaction parameter. In this low-frequency range $[G']_R$ is insensitive to h^* whereas $[G'']_R$ becomes larger as h^* increases. This difference is described quantitatively by the quantity S_2/S_1^2, which may be called the reduced steady shear compliance, since it is related to the steady-state compliance J_e^0 by

$$S_2/S_1^2 = \lim_{c \to 0} cRTJ_e^0\eta^2/M(\eta - \eta_s)^2. \tag{7.13}$$

This quantity is equal to 0.4 if $h^* \to 0$ and decreases monotonically to 0.206 as $h^* \to \infty$.[39]

In the high-frequency range $[G']_R$ and $[G'']_R$ are proportional to ω_R^a where a is $\frac{1}{2}$ if $h^* \to \infty$ and increases as h increases to a value of $\frac{2}{3}$ when $h^* \to \infty$. The ratio $[G'']_R/[G']_R$ also increases from 1 to $3^{\frac{1}{2}}$ as h^* increases from 0 to ∞. The dynamic behaviour in the limit $h^* \to 0$ is often referred to as 'Rouse-like' or 'free draining', and, as $h^* \to \infty$, as 'Zimm-like' or 'non-free draining'.

7.3.1. Solutions of linear polymers in theta conditions

A number of such systems have now been studied in the infinite dilution limit prescribed by theory,[38] and include polystyrene,[40] poly(α-methylstyrene),[41] polybutadiene,[42] and poly(dimethylsiloxane)[38]. The results for a high molecular-weight polystyrene sample in two theta solvents with quite different viscosities (Fig. 7.2) are typical of those obtained for these polymers, and are presented using the reduced plots discussed above. The use of universal abscissae does indeed reduce the plots for the two solutions on to a common curve. The heavy lines correspond to computations based on the Zimm theory with $h^* = 0\cdot25$. Excellent agreement is obtained between experiment and theory without the use of any additional adjustable parameters. Evidently the simple bead-spring model is quite satisfactory so long as the frequency does not go much higher than $10\tau_1^{-1}$ for molecular weights greater than 10^5 (the higher the molecular weight, the broader the satisfactory frequency range).

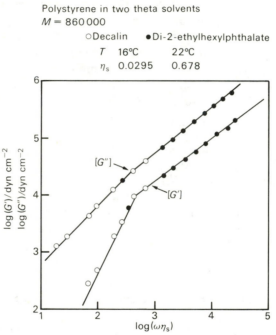

Fig. 7.2. Variation of modulus against reduced frequency for polystyrene in two theta solvents.

Two important conclusions can be drawn from these results. First the frictional resistance to motion comes entirely from the surrounding solvent, so that the molecule is behaving like a limp thread without any internal stiffness, and secondly the detailed chemical structure of the polymer has little influence on its hydrodynamic behaviour. For the polymers measured so far the observed reduced viscoelastic curves all have the same shape.

At high oscillatory shear frequencies, or with low molecular-weight samples, differences due to chemical structure are observed, and may be attributed to the intrinsic segmental mobility of the polymer influencing the over-all chain dynamics.

7.3.2. Non-theta solutions of linear polymers

A number of studies have been performed on polymers in good solvents: poly(isobutylene) in cyclohexane,[43-5] poly(methylmethacrylate)in chloroform,[43,46] poly(α-methylstyrene) in benzene and toluene,[47,48] and a few copolymers of styrene-butadiene in toluene and cyclohexane.[49] Measurements have been extended to more viscous solvents such as α-chloronaphthalene, 'arochlor', and decalin in the cases of polystyrene,[50] polybutadiene,[51] and poly(α-methylstyrene).[52] In all cases the frequency dependence of the intrinsic complex modulus is quantitatively more 'Rouse-like' than observed in the theta solutions. It follows that appropriate changes in the value of the hydrodynamic screening parameter should enable data from different solvents to be displayed on a common plot.

A recent extension of the simple bead-spring model discussed in Chapter 3 is rather successful for good solvents[53] and demonstrates the effectiveness of this approach. The model is basically that of Zimm,[54] with the value of h^* different from its ideal value, as indicated in eqn (7.11). The linear expansion factor α_n which appears in this equation is associated with excluded volume and deserves special comment.

7.3.3. Excluded volume

The idea of excluded volume arises from the restrictions which must be placed on an ideal flexible chain due to the physical requirement that the chain cannot cross itself, or occupy the same point in space. These restrictions on allowed conformations lead to an increase in the effective size of the polymer, and consequently to modification of the relaxation behaviour.

Traditionally, the excluded volume problem has been concerned with the limiting values of the exponents ν and ρ in the formulae for the mean square end-to-end distance $\langle R_n^2 \rangle$, and the probability $\langle P_n \rangle$ of ring closure after n steps in the formulation of the theory of the Gaussian coil molecule

$$\langle R_n^2 \rangle = An^\nu \tag{7.14}$$

$$\langle P_n \rangle = Bn^{-\rho}. \tag{7.15}$$

Numerous papers[55-74] have been published on this topic; for a chain on a three-dimensional lattice these exponents appear generally to have values of $\nu = \frac{6}{5}$ and $\rho = \frac{23}{12}$. These predictions are in agreement with Monte Carlo calculations[59,60] and analogous numerical studies based on a three-dimensional Ising model.[58] Experimentally the estimation of the 'excluded' volume[74-8] depends critically on the errors in the technique used for the measurement of the mean square end-to-end distance but seem to confirm that ν lies between $\frac{6}{5}$ and $\frac{4}{3}$ and ρ is $\frac{23}{12}$.

Recent frequency-dependent viscoelastic experiments have been evaluated in terms of an extension of the bead-spring model[79-82] which allows for varying hydrodynamic interaction, h^*, and excluded volume. This is introduced as an effective quadratic potential, Π. It has been concluded that Π increases with increasing molecular weight[83] or coil expansion.[81] Again h^* decreases with increasing coil expansion[85,86] or increasing molecular weight.[87-9]

7.3.4. Branched-chain polymers

The macromolecules discussed so far do approximate to linear chains of bead-like subunits. The viscoelastic behaviour of such systems in theta solutions is more influenced by the long-range connections between subunits of the same molecule than by the actual chemical structure of the molecule. The effects of these long-range interactions are particularly evident in the behaviour of branched, star (arms radiating from a central point), and comb (equal length arms spaced along a central backbone) polymers.[38]

The basic theory of star-shaped polymers is an extension of the simple formalism presented in Chapter 3, and involves placing restrictions on the activity of certain of the normal modes.[89] A branch is considered to place an effectively infinite mass at the branch point of the chain, and so modifies all modes not possessing a node at that point. The modified form of the steady-state compliance for star and comb polymers of f arms is obtained from

$$S_i = (f-1) \sum_{p \, \text{odd}} (\tau_p/\tau_1)^i + \sum_{p \, \text{even}} (\tau_p/\tau_1)^i \qquad (7.16)$$

and the reduced compliance is related to S_2/S_1^2 as described in (7.13). J_{eR}^0 decreases rapidly with increased branching, a prediction confirmed for star polymers with 4 and 9 arms.[90,91] The agreement in the case of a comb polymer[92] is not as good, and in both cases observed differences between experiment and theory have been attributed to the effects of a higher than normal concentration of polymer segments near the centre of the molecule. Experimentally it is found that branched polystyrene polymers exhibit an effective value of h^* approaching 0.40, which is significantly higher than the value of 0.25 usually attributed to the linear system.

The effects of branching on the 'universal' plots are presented in Fig. 7.3. A shoulder appears in both curves at frequencies a little above τ_i^{-1}, the storage modulus G' of the same molecular weight. The longest relaxation time is found to be larger than predicted by theory in the highly branched polymers.

7.3.5. High-frequency behaviour

At high frequencies the bead-spring model fails to describe the observed viscoelastic behaviour. As was pointed out in § 3.5 this is because the

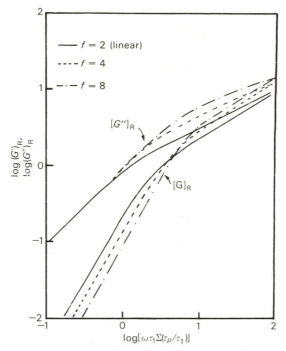

Fig. 7.3. Effect of chain branching on the dynamic modulus.

wavelength of the high frequency modes becomes comparable with the length of chains necessary to form the Gaussian subunit, and the motion of the chain segment becomes intimately connected with the specific conformational changes of the covalent bonds in the backbone. Such effects usually become apparent at frequencies two decades above τ^{-1}. Unfortunately most of these high-frequency measurements have been made on solutions of fairly high viscosity, and certain of these solvents are known to exhibit viscoelastic relaxation in the megahertz frequency range. The use of high-viscosity solvents cannot easily be avoided since the time–temperature superposition principle is usually invoked to extend the effective frequency range.[93]

Even with all these problems, it is well established that deviations from the simple Zimm- or Rouse-like behaviour do occur at high frequencies and this is illustrated by the behaviour of a 1·5 per cent polystyrene solution (Fig. 7.4). The broad frequency coverage is obtained by combining data from different temperatures using solvent viscosity as a reducing variable. At high frequencies, the polymer contribution to the loss modulus, $G'' - \omega \eta_s$, becomes directly proportional to frequency and the storage modulus appears to approach a limiting value. Two characteristic parameters are

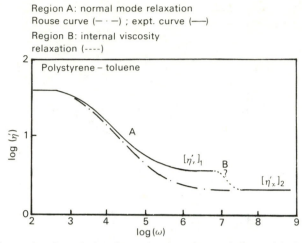

Fig. 7.4. Schematic plot of the dynamic viscosity variation with frequency for polystyrene in toluene.

used to describe this region: the ratio G'/v_2, (where v_2 is the volume fraction of polymer) and the ratio $(G'' - \omega\eta_s)/\omega\eta_s c$ extrapolated to infinite dilution—the high-frequency limiting intrinsic viscosity. Both of these quantities appear to be independent of molecular weight but depend on the specific chemical structure of the polymer (Table 7.2). Measurements have been

Table 7.2

Values of the limiting high-frequency intrinsic viscosity

Polymer	$M_w \times 10^{-3}$	$10^3[\eta'_\infty]\,\mathrm{m}^3\,\mathrm{kg}^{-1}$	Ref.
Polystyrene	98·2	15–16	94–7
	153	14–15	
	411	14–16	
	867	13	
	2850	14	
Poly(α-methylstyrene)	105	18	98, 99
	386	18	
	1400	19	
Poly(1,4-butadiene)	92	8	98
	244	7	
	910	8	
Polyisobutylene	400	7	98

$[\eta'_\infty]$ is the limiting value of $(G'' - \omega\eta_s)/\omega\eta_s c$ as $c \to 0$.

reported on a variety of polymers including polystyrene,[94-7] poly(α-methylstyrene),[98,99] poly(1,4-butadiene),[98] and polyisobutylene.[98] The physical interpretation of these measurements is not yet clear. Early theories treating the high-frequency behaviour as resembling that of rigid spheres[94,100,101] have now been superseded by those introducing internal viscosity[102-6] as described in Chapter 4. Even these theories, however, fail to predict the additional relaxation character now thought to exist at the highest frequencies.[107,108] How these high-frequency processes take place, and their significance for the description of the lower-frequency dynamic behaviour of the polymer molecules are, as yet, not fully understood. However it is clear that they are intimately connected with the local or 'segmental' motion of the polymer, and the implications of this have been discussed in a recent comparison[109] of the 'segmental' (non-correlated motions of localized chain segments) and 'internal viscosity' (correlated strongly-coupled short-wavelength modes) models.

7.3.6. Rigid rod and 'stiff' chain molecules

The classical system is that of poly(γ-benzyl-L-glutamate) which has been studied[110] in helicogenic solvents with different viscosities, dimethylformamide and m-cresol. The intrinsic moduli, extrapolated to infinite dilution, compare well with the predictions of Ullman[111] for rigid rods, agreement being best at the lowest frequencies. The data over the entire frequency range of three logarithmic decades can be described by a relaxation spectrum consisting of one terminal relaxation time separated from a sequence of relaxation times spaced as in the Zimm theory. The terminal time can be ascribed to end-over-end rotation of a rigid rod, and the additional effects observed at high frequency are attributed to flexing deformation of the helix. These measurements are paralleled by the behaviour reported for poly(γ-methyl-D-glutamate)[112,113] and deoxyribonucleic acid,[39] paramyosin,[114] and bovine serum albumin[114] in glycol–water mixtures.

For finite concentrations, it was found that the relaxation time could be represented by

$$\tau = m(\eta - \eta_s)M/cRT. \qquad (7.17)$$

Comparison of the values obtained from these experiments with those predicted by theory[115] indicated that the 'effective' molecular weight of the polymer was slightly lower than that determined by light scattering techniques. The apparently lower value is assumed to arise as a consequence of a slight flexibility of the rod-like polymer.

An alternative explanation for the viscoelastic behaviour of 'stiff' chains has been based on theories of the worm-like coil.[116,117] This approach has been shown to explain the relaxation observed in poly(N-vinylcarbazole)[118] and polyisocyanates.[119] The results of these experiments are in good agreement with other physicochemical observations.

It should be appreciated that the polymers described above represent an extreme of rigidity, and that theories used for a semi-flexible polymer should approximate to the rigid case in the low molecular-weight limit.

7.4. Viscoelastic behaviour in the hydrodynamic screening and coil contact regimes

7.4.1. Concentration effects and intermolecular interactions

While viscoelastic behaviour at low concentrations can be described in terms of the hydrodynamic behaviour of discrete particles,[120,121] at higher concentrations interchain effects become manifest, first through interaction with the flow field[122] and then by direct coil overlap. A number of coil-overlap criteria have been developed based on the hydrodynamic volume occupied by the expanding, rotating coils.[123-8] This depends on molecular weight and concentration, of course, and is reflected approximately in the quantity $c[\eta]$ referred to earlier.

A systematic approach[129] to segment-flow field-segment interactions[130] is to introduce a hydrodynamic screening factor. The inclusion of inter-molecular interactions leads to a hydrodynamic interaction tensor which, after averaging, differs from that of Kirkwood by a concentration-depen-dent correction term. The modified theory correctly predicts the transition from the non-free-draining to free-draining behaviour which has been observed in chain molecules as their concentrations are increased. Two parameters are introduced, h^* and b. The first is essentially the hydro-dynamic interaction parameter discussed previously and b is a screening parameter. Their definitions are respectively

$$h^* = N^{\frac{1}{2}}\zeta_0/(12\pi^3)^{\frac{1}{2}}b\eta \tag{7.18}$$

$$b = (\nu\zeta_0 Nb^2/12\eta)^{\frac{1}{2}} \tag{7.19}$$

where b is defined in terms of the mean square end-to-end distance, L, of the chain, of N springs,

$$L^2 = (N+1)b^2. \tag{7.20}$$

The other parameters are the friction coefficient, ζ_0, and the number of stationary beads per unit volume, ν.

The relaxation times, τ_k are given by

$$\tau_k = \frac{b^2 S_0}{6kT\lambda_k} = \frac{M\eta\eta_{\rm sp}(0)}{N_A kTC\lambda_k(\sum_{k=1}^{N} 1/\lambda_k)} \tag{7.21}$$

where $\eta_{\rm sp}(0)$ is the specific viscosity at $\omega = 0$ and k and T are the Boltzmann constant and the absolute temperature, respectively.

The form of specific viscosity can be shown to be

$$\eta_{\rm sp} = \frac{\eta_r - {\rm i}\eta_i}{\eta} = \frac{\pi^3/2N_A L^3 c}{4 \cdot 3^{\frac{1}{2}}M} \sum_{k=1}^{N} \frac{1 - {\rm i}(\omega\tau_1)(\lambda_1'/\lambda_k')}{\lambda_k'[1 + (\omega\tau_1)^2(\lambda_1'/\lambda_k')^2]} \tag{7.22}$$

with

$$L = N^{\frac{1}{2}}b$$

where η_r and η_i are defined by the relation

$$\eta_s = (\eta_r + \eta) - i\eta_i \tag{7.23}$$

and τ_1 is defined by (9.21).

The important difference between this and the previous theories (Chapter 4) is that the eigenvalues are concentration-dependent and the form of the equation is a function of the size of the polymer. This in turn depends on the solvent. Therefore the approach should apply to both 'good' and theta solvents, and also to the change from infinite dilution to these more concentrated solutions.

Application[131] of this approach to polystyrene in nitropropane (i.e. near-theta solvent), toluene (a good solvent), decalin, and dioctylphthalate (viscous solvents) has shown that theory and experiment are in agreement, the behaviour changing from non-free-draining to free-draining as the concentration is increased.

7.4.2. Viscosity correlations at intermediate concentrations

Fundamental theories of transport properties for systems of finite concentration are still rather tentative.[132] The difficulties in modelling this type of system are accentuated by the effects of concentration on equilibrium properties such as the coil dimensions. As the concentration is increased, intermolecular interactions modify the thermodynamic nature of the surrounding environment, and ultimately lead to the polymer collapsing to its theta dimensions.[133-5] Once the theta limiting value is reached the polymer dimensions appear to stay essentially constant right up to the melt.[136,137]

Viscosity is often used[138] to correlate transport data, and the effects of intermolecular interactions appear as the Huggins constant, k', in the equation for the zero shear viscosity,

$$\eta_0 = \eta_s(1 + [\eta]c + k'[\eta]^2c^2 + \cdots). \tag{7.24}$$

Experimentally k' is independent of molecular weight for a long chain, with values of roughly $0 \cdot 3 - 0 \cdot 4$ in good solvents and $0 \cdot 5 - 0 \cdot 8$ in theta solvents.

In dynamic observations intermolecular contributions appear first as an increase in the longest relaxation time, the more rapid relaxations remaining unaffected.[139] Apparently the largest-scale molecular motions are slowed as peripheral segments begin to impinge upon those of neighbouring chains. At somewhat higher concentrations, the coils begin to overlap appreciably and the faster relaxations are shifted to longer times. The relative spacings at long times increasingly resemble the Rouse spectrum.[140-2] This behaviour is believed to reflect an increasing free-draining character in the local flow pattern, caused by a gradual cancellation of intra-molecular hydrodynamic effects as the solution becomes uniformly filled with polymer segments.

Although a number of possible correlation parameters have been proposed,[143-9] there is no widely accepted convention. In an attempt to establish the factors which control the modifications to the viscoelastic spectrum in the region $c[\eta] = 1-10$, solvent efficiency and polarity have been studied.

Studies of a variety of polymers, including polystyrene,[138,146,147,149] polymethylmethacrylate,[138,150] polyvinylacetate,[145] and polyisobutylene[151] in a range of solvents have been quoted as evidence for a correlation between the viscosity at high concentrations and the glass transition temperature. In this region the viscosity loses its simple concentration dependence and appears to become proportional to a local frictional coefficient ζ_0. The value of ζ_0 depends on the nature of both solvent and polymer, being dependent on the free volume, as is the glass temperature of the mixture. Indeed such a correlation would follow automatically from the treatment of the rubber–glass transition as an isoviscous or isofree volume phenomenon.

7.5. Concentrated solutions and melts

An extensive literature[21,93,150,151] exists on the rheological properties of solutions with $c[\eta] > 10$ and of melts. The following discussion will be concerned with three principal characteristics of these systems, namely the viscosity–molecular weight relationships, the plateau modulus, and the steady-state compliance.

7.5.1. The zero shear viscosity[152,153]

In these highly concentrated solutions the viscosity loses its dependence on that of the solvent, and becomes a function of the product of two parameters. These are a friction factor, ζ, which is controlled solely by local features such as the free volume or the segmental jump frequency, and a structure factor, F, which is controlled by the large-scale structure and configuration of the chains.[21]

$$\eta_0 = \zeta \cdot F. \tag{7.25}$$

The friction factor depends on those features that govern the viscosity of small molecular fluids. At low temperatures it will depend on the proximity of the glass transition, and at high temperatures on an appropriate activation energy for flow.

The structure factor, on the other hand, depends on the number of chains per unit volume and on their molecular weight and dimensions. It is virtually insensitive to chemical structure and shows a strong correlation with the behaviour of the isolated chain extrapolated to high concentrations. In fact, for short chains, the Rouse equations adequately describe experimental data obtained in this region.

For longer chains the structure factor changes and is distinctly molecular weight-dependent. At high molecular weights the viscosity–molecular weight exponent changes from one to approximately 3·4. This occurs smoothly over a relatively narrow range of molecular weights. The characteristic value, M_c, for a polymer is obtained from the intersection of straight lines drawn through the two branches of the curve.

These trends are highlighted in the viscosity dependence of the segment contact parameter, M_c (Fig. 7.5). In many systems the transition in the slopes

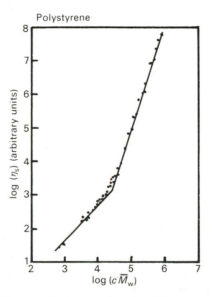

Fig. 7.5. Variation of the zero-frequency viscosity with concentration and molecular weight.

of the viscosity–molecular-weight curve remains fairly well defined, and within the accuracy of the data follows

$$(M_c)_{soln} = (\rho/c)M_c = M_c/\phi \qquad (7.26)$$

where M_c is the value for undiluted polymer, ρ is its density, and ϕ is the volume fraction of polymer in solution. This type of behaviour points towards chain entanglements as being responsible for the changes in the viscosity of the solutions above a certain segment contact value. The insensitivity of M_c to temperature and solvent power, and the fact that M_c

for different polymers corresponds to roughly the same number of main chain bonds, lead to the suggestion that the onset of the intermolecular interaction responsible for the effects in the viscosity is based on a geometric rather than a chemical characteristic of the polymer. Some typical values for various polymers are presented in Table 7.3.

Table 7.3
Characteristic molecular weights for undiluted linear polymers

Polymer	M_c
Polyethylene	3800
1,4-polybutadiene	5900
cis-polyisoprene	10 000
Polyisobutylene	15 200
Poly(vinylacetate)	24 500
Poly(dimethylsiloxane)	24 400
Poly(methylmethacrylate)	27 500
Poly(α-methylstyrene)	28 000
Polystyrene	31 200

The values of M_c were taken from refs. 21 and 93.

7.5.2. The plateau modulus

As indicated for moderately concentrated solutions, the effect of segment–segment interactions is to split the spectrum of relaxation into two or more features. The long-time mode moves to lower frequency whilst the higher-frequency modes are left virtually unchanged. This modification of the relaxation spectrum is demonstrated[154] by the plots of loss modulus versus frequency for narrow molecular-weight distribution polystyrene (Fig. 7.6). Similar behaviour is observed in the storage modulus. The width of the plateau depends strongly on molecular weight, while the characteristic modulus of the plateau region, G_N^0, is independent of molecular weight. The presence of the plateau confers rubber-like elastic properties over an intermediate range of frequencies or times. This plateau region is found to appear only when the molecular weight exceeds M_c and is taken as a direct indication of chain entanglement. Conceptually the solution must contain a number of rope-like interlooping polymers which can form a dynamic network. Rubber-like response to rapid deformation is obtained because the strands between coupled points can adjust rapidly whilst slip of the whole molecule, which involves contouring its entangled neighbours, requires a considerably longer time.

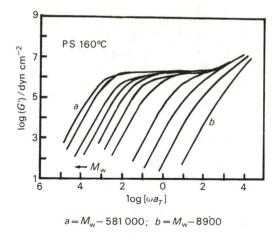

$a = M_w - 581\,000; \quad b = M_w - 8900$

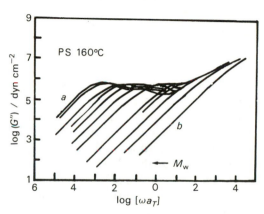

Fig. 7.6. Frequency dependence of the dynamic modulus for narrow-distribution polystyrene melts.[174]

The plateau value of the modulus can provide an apparent average molecular weight between coupling junctions

$$M_e = RT/G_N^0. \tag{7.27}$$

The number of junction points per molecule is therefore $E = M/M_e - 1$. Values of M_e for a number of polymers have been tabulated[155] and are summarized in Table 7.4.

The relaxation spacings in the terminal region are quite different from those of the Rouse model. They are usually observed to be much more tightly grouped around the mean, but with a tail of faster relaxations which continues into the plateau region.

Table 7.4

*Entanglement molecular weights for
undiluted linear polymers*

Polymer	M_e
1,4-polybutadiene	1900–2200
cis-polyisoprene	4500–5800
Poly(methylmethacrylate)	5900–10 000
Poly(dimethylsiloxane)	8100
Polyisobutylene	8900
Poly(vinylacetate)	12 000
Poly(α-methylstyrene)	13 500
Polystyrene	18 100

The values of M_e are defined as the apparent average molecular weight between coupling junctions: $M_e = \rho RT/G_N^0$.[93] Values quoted were taken from refs. 21 and 93, except polystyrene, (ref. 154).

7.5.3. Molecular entanglement theories of linear viscoelastic behaviour

The order-of-magnitude agreement between viscosities for undiluted polymers below M_c and calculated values from the Rouse theory suggests that behaviour at high molecular weights is modified by an interaction between chains. This interaction is a function of the product cM for concentrated solutions and melts, and appears to be an intrinsic property of such systems. The problem is to convert the naive picture of entangled rope-like molecules into an acceptable theory.[156]

For $M \gg M_c$ the low-frequency (affected by entanglement) relaxations are separated from the so-called transition (unaffected high-frequency) relaxations. The number and weight average relaxation times of the terminal time distribution are given by

$$\tau_n = \eta_0/G_N^0 = \eta_0 M_e \rho/c^2 RT \tag{7.28}$$

$$\tau_w = \eta_0 J_e^0 = 0.4 M_c' \rho/c^2 RT \tag{7.29}$$

where M_c' is molecular weight at which the compliance becomes constant (Fig. 7.6) and the breadth of the relaxation curves is described by

$$\tau_w/\tau_n = J_e^0 G_N^0 = 0.4 M_c'/M_e = 0.4 (M_c')_{\text{soln}}/(M_e)_{\text{soln}}. \tag{7.30}$$

A number of approaches in which the concept of entanglement is treated theoretically[157–69] have been formulated. The behaviour of gels and networks is currently an area of considerable theoretical interest.

An encouraging formulation of the behaviour of polymer chains in the melt and in concentrated solutions has been proposed by De Gennes[170] and Doi and Edwards[171]. This model assumes that the motion of an individual chain in the macromolecular matrix corresponds closely to that of an equivalent chain in a tube. The entangled matrix forms molecular loops around the reference chain and it is these which effectively constrain its motion to within the tube, the axis of which coincides with the persistence contour of the chain. The effective overall motion of the chain is then rather like that of a snake and so the term reptation has been used.

The application of a stress to the matrix will lead to the chain being compressed in certain regions and expanded in others. If the deformation is affine, then the motions executed by the chain in returning to equilibrium can be formulated in terms of rubber elasticity theory. The overall behaviour of the system can be separated into two parts: first the movement of the chain leading to an equilibrium conformational distribution, and second a rearrangement of the matrix to adjust to the new configuration of the chain (Fig. 7.7).

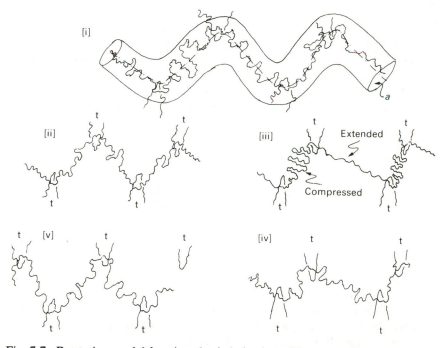

Fig. 7.7. Reptation model for viscoelastic behaviour. The upper diagram (i) illustrates the concept of the constraining tube radius a in which reptation motion occurs. The polymer molecule initially at equilibrium (ii) is stressed (iii) and certain elements of the chain are compressed others being extended. The polymer chain responds to the stress by wriggling to a new equilibrium configuration (iv). This situation is not a true equilibrium of the matrix and further translation occurs with the formation of new entanglements (v).

The contours of the chain are defined by the step size required for the chain to disentangle itself. The tube diameter is assumed to be independent of the molecular weight and the tube length directly proportional to it. The diameter of the tube, a, is a function of the matrix composition and is defined in terms of the polymer mean square end-to-end distance, $\overline{R^2}$, and contour length, L, $a = \overline{R^2}/L$. The self-diffusion coefficient for the chain is

$$D = kT/3N\zeta_r \tag{7.31}$$

where k is Boltzmann's constant, T is the temperature, and ζ_r is the molecular friction coefficient of a Rouse chain. N defines the number of random walk steps, each of length a, which can occur in the tube; $Na^2 = \overline{R^2}$. The analysis of the system leads to a characteristic time for the disentanglement of the chain from the originally defined tube as

$$\tau_d = L^2\zeta_r/\pi^2 kT \tag{7.32}$$

The model assumes that the number of contacts at equilibrium remains constant and therefore diffusion out of the original tube involves diffusion into a new tube. Since $\tau_d \propto M^3$, the processes of conformational equilibration and disentanglement will be widely separated in time so long as the chains are sufficiently long to form a matrix. Hence we find the origin of the so-called rubbery plateau characterized by linear viscoelastic stress relaxation. The stress relaxation modulus of the long time terminal region is given by

$$G(t) = \tfrac{4}{5}\nu NkT \frac{8}{\pi^2} \sum_{\text{odd } n} \frac{1}{n^2} \exp\left(-\frac{n^2 t}{\tau_d}\right) \tag{7.33}$$

where ν is the frequency of formation of contacts within the Gaussian network. This leads to the steady state viscosity and the rigidity modulus

$$\eta_0 = \frac{kT}{20} \frac{cN^2}{D^0} \frac{b^6}{a^4} \tag{7.34}$$

$$G_\infty = \tfrac{3}{5} cN_0 kT \, (b/a)^2$$

where c is the concentration of polymer (wt/vol.) and b is the effective bond length of the real chain. If we define the plateau modulus as G_N^0 it is possible to rearrange the above equations into the form[172]

$$D = \frac{G_N^0}{135} \left(\frac{cRT}{G_N^0}\right)^2 \left(\frac{\overline{R^2}}{M}\right) \frac{M_c}{M^2\eta_0(M_c)} \tag{7.35}$$

$$\tau_d = \frac{45}{2} \left(\frac{G_N^0}{cRT}\right)^2 \frac{M^3\eta_0(M_c)}{G_N^0 M_c} \tag{7.36}$$

$$\eta_0 = \frac{15}{4} \left(\frac{G_N^0}{cRT}\right)^2 \frac{M^3 \eta_0(M_c)}{M_c} \tag{7.37}$$

$$J_0^0 = \tfrac{6}{5} G_N^0 \tag{7.38}$$

The tube diameter can be expressed in terms of observables

$$a^2 = \frac{4}{5} \left(\frac{\overline{R^2}}{M}\right) \left(\frac{cRT}{G_N^0}\right). \tag{7.39}$$

The complex viscosity is given by

$$\eta^*(\omega) = \frac{G_N^0}{5} \sum_{p \, \text{odd}} \frac{8}{p^4 \pi^2} \frac{T_d}{1 + i(\tau_d/p^2)}. \tag{7.40}$$

This theory describes the stress strain curves which are observed in concentrated solutions of polystyrene in diethyl phthalate, and so forms the basis of a logical treatment of the chain-entangled state. However, there do remain one or two points of contention. First the predicted dependence of η_0 on molecular weight is M^3 whereas practically the relation $\eta_0 \propto M^{3.4-3.6}$ is observed. The dependence of τ_d on molecular weight is more satisfactory being M^3 from theory and $M^{3.1}$ from experiment. The dimensions of the tube predicted by the theory[3] are approximately 83 Å for polystyrene and 34 Å for polyethylene. These sizes reflect the distance between neighbouring entanglements, and as such are not unreasonable.

7.6. Technological aspects of polymer viscoelasticity

The viscoelastic properties of polymer systems and the consequences of the appropriate molecular motions are of great technological significance. A brief indication of three important phenomena is presented below, and the interested reader is referred to the appropriate detailed technical texts and journals.

7.6.1. Lubricants

Polymers are widely used as solutions in hydrocarbon lubricants as viscosity-index improvers.[173,174] In this application the additive reduces the variation of viscosity with temperature. For most uses this requires the polymer to have similar intrinsic viscosities at 40 and 100 °C. The theory behind viscosity-index improvers is that the size of the polymer coil increases with temperature, so offsetting the decrease in solvent viscosity. Copolymers of oil-insoluble units with oil-soluble units (e.g. methyl- and stearylmethacrylate copolymers) are particularly effective, and also confer other advantageous properties on the system. Particular phenomena which have to be dealt with include overcoming precipitation of the 'insoluble' polymer at low temperatures and dispersion of combustion products or other impurities in the oil.

7.6.2. Polymer melt flow

Many polymer manufacturing processes involve either the injection or the extrusion of molten polymer under conditions of high shear. The response of the melt to shear stress is controlled by the same factors which have been discussed in the frequency-dependent studies. It is now well known that the viscoelastic properties give an elastic memory to the fluid. For instance if a polymer melt is extruded through a cylindrical orifice, the flow lines which build up before the orifice can have a profound effect on the flow pattern after extrusion. At low extrusion rates the thread is ideally cylindrical. Increasing the flow rate produces an initial narrowing of the thread which then widens to slightly greater than its initial width some distance from the orifice. At high extrusion rates the polymer will neck dramatically and appear from the orifice in spurts. In certain conditions a spiral motion leads to a 'shish kebab' effect. These effects have been reviewed[175] and although understood qualitatively, in many respects have not yet been treated quantitatively.

7.6.3. Frictional drag reduction[176]

In very dilute solutions of certain high molecular-weight polymers, the onset of turbulence occurs at higher shear stresses than in the pure solvent. Consequently the frictional drag associated with flow at high shear rates is reduced. The explanation is that the polymer effectively suppresses incipient eddies which are necessary for the formation of the vortices leading to turbulence. The mechanism whereby this is achieved is not fully understood, although a number of definite correlations have been identified.

It is noted that there is a very good correlation between the characteristic time for normal mode or segmental relaxation and the time expected for the formation of the incipient eddies. When these times are similar the polymer can absorb energy from the fluid and in so doing can break the kinetic spectrum required for the build up of vortices.

It has been noted too that the effectiveness of drag reduction by undissolved polymer is often a function of time, and this suggests that polymer aggregates play an important part in the phenomenon. A definite correlation appears to exist also between the end-to-end distance of the polymer and the diameter of the eddies in the fluid.

The uniqueness of these correlations has been questioned, but it would appear that the effectiveness of a particular polymer in achieving drag reduction can be understood qualitatively by matching the time and wave length spectrum of the incipient eddies with those of the polymer coils or aggregate particles.

References

1. Tanaka, H., Sakanishi, A., Kaneko, M. and Furuichi, J. *J. polym. Sci.* **C15**, 317 (1966).
2. Frisch, H. L. and Simha, R. In *Rheology* (ed. F. R. Eirich), Vols. I and II. Academic Press, New York (1956).
3. Katz, R. A. and Edwards, S. F. *J. Phys. A* **5**, 674 (1972).
4. Baker, W. O., Mason, W. P. and Heiss, J. H. *Bull. Amer. phys. Soc.* **24**, 29 (1949).
5. Moore, W. R. *Prog. polym. Sci.* **1**, 1 (1967).
6. Meyerhoff, G. *Advan. polym. Sci.* **3**, 59 (1961).
7. Kurata, M. and Stockmayer, W. H. *Advan. polym. Sci.* **3**, 196 (1963).
8. Isihara, A. and Guth, E. *Advan. polym. Sci.* **5**, 233 (1967).
9. Berry, G. C. and Casassa, E. F. *J. polym. Sci. Macromol. Rev.* **4**, 1 (1970).
10. Bruce, C. and Schwarz, W. H. *J. polym. Sci. A-2* **7**, 909 (1969).
11. Moore, R. S., McSkimin, H. J., Gieniewski, C. and Andreatch, P. *J. chem. Phys.* **50**, 5088 (1969).
12. Node, I., Yamada, Y. and Nagasawa, M. *J. phys. Chem.* **72**, 2890 (1968).
13. Truesdall, C. and Noll, W. The non-linear field theories of mechanics. In *Handbuch der Physik* (ed. S. Flügge), Vol. 3, p. 3. Berlin: Springer Verlag, Berlin (1953).
14. Phillipoff, W. *Phys. Acoust.* **2B**, 1 (1965).
15. Bogue, D. C. and Doughty, J. O. *Ind. Engng chem. Fundam.* **5**, 243 (1966).
16. Semjonow, V. *Advan. polym. Sci.* **5**, 387 (1968).
17. Marvin, R. S. and Kinney, J. E. *Phys. Acoust.* **2B**, 165 (1965).
18. Hopkins, I. L. and Kurkjian, C. R. *Phys. Acoust.* **2B**, 91 (1965).
19. Eisenberg, A. *Advan. polym. Sci.* **5**, 59 (1967).
20. Tamura, M., Kurata, M., Osaki, K. and Tanaka, K. *J. phys. Chem.* **70**, 516 (1966).
21. Berry, G. C. and Fox, T. G. *Advan. polym. Sci.* **5**, 261 (1968).
22. Porter, R. S. and Johnson, J. F. *Chem. Rev.* **66**, 1 (1966).
23. Bloomfield, V. A. *Science* **161**, 1212 (1968).
24. Fixman, M. and Stockmayer, W. H. *Ann. Rev. phys. Chem.* **21**, 407 (1970).
25. Yamakawa, H. *Modern theory of polymer solutions*, Chapter 6, p. 214. Harper and Row, New York and London (1971).
26. Rouse, P. E. *J. chem. Phys.* **21**, 1272 (1953).
27. Zimm, B. H. *J. chem. Phys.* **24**, 269 (1956).
28. Tschoegl, N. W. *J. chem. Phys.* **40**, 473 (1964).
29. Matheson, A. J. *Molecular acoustics*, Chapter 6. Wiley-Interscience, London, New York (1971).
30. Osaki, K. *Advan. polym. Sci.* **12**, 11 (1973).
31. Berry, G. C. and Casassa, E. F. *J. polym. Sci., Part D Macromol. Rev.* **4**, 1 (1970).
32. Johnson, R. M., Schrag, J. L. and Ferry, J. D. *Polym. J.* **1**, 742 (1970).
33. Osaki, K., Mitsudi, Y., Johnson, R. M., Schrag, J. L. and Ferry, J. D. *Macromolecules*, **5**, 17 (1972).
34. Osaki, K., Schrag, J. L. and Ferry, J. D. *Macromolecules*, **5**, 144 (1972).
35. Lodge, A. S. and Wu, J. Y. Report 16. Rheology Research Centre, University of Wisconsin (July 1972).
36. Osaki, K. *Macromolecules*, **5**, 141 (1972).
37. Osaki, K. and Schrag, J. L. *Polym. J.* (Jpn) **2**, 541 (1971).
38. Tschoegl, N. W. *J. chem. Phys.* **39**, 149 (1963).

39. Ferry, J. D. *Accounts chem. Res.* **6**, 60 (1973).
40. Johnson, R. M., Schrag, J. L. and Ferry, J. D. *Polym. J.* **1**, 742 (1970).
41. Osaki, K., Schrag, J. L. and Ferry, J. D. *Macromolecules* **5**, 174 (1973).
42. Osaki, K., Mitsuda, Y., Johnson, R. M. and Schrag, J. L. *Macromolecules* **5**, 17 (1972).
43. Tanaka, H., Sakanishi, A., Kaneko, M. and Furuichi, J. *J. polym. Sci., Part C* **15**, 317 (1966).
44. Tanaka, H., Sakanishi, A., Kaneko, M. and Furuichi, J. *Zairyo*, **15**, 447 (1966).
45. Sakanishi, A. *J. chem. Phys.* **48**, 3850 (1968).
46. Tanaka, H., Sakanishi, A. and Furuichi, J. *Zairyo* **15**, 438 (1966).
47. Tanaka, H. and Sakanishi, A. *Zairyo* **16**, 528 (1967).
48. Tanaka, H. and Sakanishi, A. *Proceedings of the International Congress on Rheology* (ed. S. Onogi), p. 243. University of Tokyo Press, Tokyo (1970).
49. Sakanishi, A. and Tanaka, H. *Proceedings of the 4th International Congress on Rheology* (ed. S. Onogi), p. 251. University of Tokyo, Tokyo (1970).
50. Johnson, R. M., Schrag, J. L. and Ferry, J. D. *Polym. J.* **1**, 742 (1970).
51. Osaki, K., Mitsudi, Y., Johnson, R. M., Schrag, J. L. and Ferry, J. D. *Macromolecules* **5**, 17 (1972).
52. Osaki, K., Schrag, J. L. and Ferry, J. D. *Macromolecules* **5**, 144 (1972).
53. Tschoegl, N. W. *J. chem. Phys.* **40**, 473 (1964).
54. Lodge, A. S. and Wu, Y. J. *Rheol. Acta* **10**, 539 (1971).
55. Domb, C. *Advan. chem. Phys.* **15**, 229 (1969).
56. Loftus, E. and Gans, P. J. *J. chem. Phys.* **49**, 3828 (1968).
57. Bruns, W. *Makromol. Chem.* **124**, 91 (1969).
58. Mazur, J. *Advan. chem. Phys.* **15**, 261 (1969).
59. Mazur, J. and McCrackin, F. L. *J. chem. Phys.* **49**, 648 (1968).
60. Stockmayer, W. H. *Makromol. Chem.* **35**, 54 (1960).
61. Yamakawa, H. *Theory of dilute polymer solutions.* Harper and Row, New York (1970).
62. Edwards, S. F. *Proc. phys. Soc.* **85**, 613 (1965).
63. Alexandrowicz, A. *J. chem. Phys.* **51**, 561 (1969).
64. Gallacher, L. V. and Windwer, S. *J. chem. Phys.* **44**, 1139 (1966).
65. Domb, C. and Hioe, F. T. *J. chem. Phys.* **51**, 1915 1920 (1969).
66. Flory, P. J. *Principles of polymer chemistry.* Cornell University Press, Ithaca, New York (1953).
67. Fixman, M. *J. chem. Phys.* **45**, 785 (1966).
68. Flory, P. J. and Frisk, S. *J. chem. Phys.* **44**, 2243 (1966).
69. Fujita, H., Okita, K. and Norisuye, T. *J. chem. Phys.* **47**, 2723 (1967).
70. Yamakawa, H. *J. chem. Phys.* **48**, 3845 (1968).
71. Fixman, M. *J. chem. Phys.* **23**, 1645 (1955); **24**, 174 (1956).
72. des Cloiseaux, J. *J. phys. Soc. Jpn*, **26**, 42 (1969).
73. Naghizadeh, J. *J. chem. Phys.* **48**, 1961 (1968).
74. Reiss, H. *J. chem. Phys.* **47**, 186 (1967).
75. Alexandrowicz, A. *J. chem. Phys.* **46**, 3789 (1967).
76. Koyama, R. *J. phys. Soc. Jpn*, **22**, 973 (1967).
77. Yamakawa, H. and Tanaka, G. *J. chem. Phys.* **47**, 3991 (1967).
78. Kron, A. K. and Ptitsyn, O. B. *Polym. Sci. USSR*, **4**, 1075 (1963).
79. Tschoegl, N. W. *J. chem. Phys.* **39**, 149 (1963).
80. Tschoegl, N. W. *J. chem. Phys.* **40**, 473 (1964).
81. Tschoegl, N. W. *J. chem. Phys.* **44**, 2331 (1966).

82. Bloomfield, V. A. and Sharp, P. *J. chem. Phys.* **48**, 2149 (1968).
83. Frederick, J. E., Tschoegl, N. W. and Ferry, J. D. *J. phys. Chem.* **68**, 1974 (1964).
84. Sakanishi, A. and Tanaka, H. *Rep. Progr. polym. Phys. Jpn*, **10**, 89 (1967).
85. Sakanishi, A. *J. chem. Phys.* **48**, 3850 (1968).
86. Tanaka, H., Sakanishi, A., Kaneko, M. and Furuichi, J. *J. polym. Sci. C* **5**, 317 (1966).
87. Moore, R. S., McSkimin, H. J., Gieniewski, C. and Andreatch, P. *J. chem. Phys.* **47**, 3 (1967).
88. Ferry, J. D., Holmes, L. A., Lamb, J. and Matheson, A. J. *J. phys. Chem.* **70**, 1685 (1966).
89. Zimm, B. H. and Kilb, R. W. *J. Polym. Sci.* **37**, 19 (1959).
90. Osaki, K., Mituda, Y., Schrag, J. L. and Ferry, J. D. *Macromolecules* **5**, 17 (1972).
91. Mitsuda, Y., Osaki, K., Schrag, J. L. and Ferry, J. D. *Polymer J.* **4**, 24, (1973).
92. Mitsuda, Y., Schrag, J. L. and Ferry, J. D. *Polym. J.* **4**, 668 (1973).
93. Ferry, J. D. *Viscoelastic properties of polymer solutions* (2nd edn), p. 200. John Wiley & Sons, New York, London, Sydney, Toronto (1970).
94. Lamb, J. and Matheson, A. *Proc. phys. Soc. A* **281**, 207 (1964).
95. Philippoff, W. *Trans. Soc. Rheology*, **8**, 117 (1964).
96. Massa, D. J., Schrag, J. L. and Ferry, J. D. *Macromolecules*, **4**, 210 (1971).
97. Osaki, K. and Schrag, J. L. *Polym. J.* **2**, 541 (1971).
98. Cooke, B. J. and Matheson, A. J. *J. Chem. Soc. Faraday Trans. II* **72**, 679 (1976).
99. Osaki, K. and Schrag, J. L. *Polym. J.* **2**, 541 (1971).
100. Harrison, G., Lamb, J. and Matheson, A. *J. phys. Chem.* **68**, 1072 (1964).
101. Ferry, J. D., Holmes, L. A., Lamb, J. and Matheson, A. J. *J. Phys. Chem.* **70**, 1685 (1966).
102. Peterlin, A. *Polym. Lett.* **10**, 101 (1972).
103. Cerf, R. *Compt. Rend.* **234**, 1549 (1952).
104. Cooke, B. J. and Matheson, A. J. *Faraday Symp. Chem. Soc.* **6**, 194 (1972).
105. Zimmerman, R. D. and Williams, M. C. *Trans. Soc. Rheol.* **17**, 23 (1973).
106. Bazua, E. R. and Williams, M. C. *J. Polym. Soc. A2* **12**, 825 (1974).
107. Osaki, K. *Advan. Polym. Sci.* **12**, 1 (1973).
108. Moore, R. S., McSkimin, H. J., Gieniewski, C. and Andreatch, P. *J. chem. Phys.* **47**, 3 (1967).
109. McInnes, D. and North, A. M. *Polymer* **18**, 505 (1977).
110. Warren, T. C., Schrag, J. L. and Ferry, J. D. *Biopolymers*, **12**, 1905 (1973).
111. Ullman, R. *Macromolecules* **2**, 27 (1969).
112. Tanaka, H., Sakanishi, A., Kaneko, M. and Furuichi, J. *Zairyo* **15**, 302 (1966).
113. Tanaka, H., Sakanishi, A. and Furuichi, J. *Rep. Prog. polym. Phys. Jpn*, **10**, 87 (1967).
114. Allis, J. W. and Ferry, J. D. *J. Amer. chem. Soc.* **87**, 4681 (1965).
115. Scheraga, H. A. *J. chem. Phys.* **23**, 1526 (1955).
116. Kratky, O. and Porod, G. *Rec. Trav. Chim.* **68**, 1106 (1949).
117. Broersma, S. *J. chem. Phys.* **32**, 1626 (1960).
118. North, A. M. and Phillips, P. J. *Brit. polym. J.* **1**, 76 (1969).
119. Dev, S. B., Lockhead, R. Y. and North, A. M. *Discuss. Faraday Soc.* **49**, 244 (1970).
120. Simha, R. and Someynsky, T. *J. colloid. Sci.* **20**, 278 (1965).

121. Batchelor, G. K. and Green, J. T. *J. fluid Chem.* **56**, 401 (1972).
122. Simha, R. and Frisch, H. L. In *Rheology* (ed. F. R. Eirich), Vol. 1, Chapter 12. Academic Press, New York (1956).
123. Fixman, M. and Peterson, J. M. *J. Amer. chem. Soc.* **86**, 3524 (1964).
124. Dusek, K. and Prins, W. *Advan. polym. Sci.* **6**, 1 (1969).
125. Maron, S. H., Nakajima, N. and Krieger, I. M. *J. polym. Sci.* **37**, 1 (1959).
126. Zakin, J. L. and Simha, R. *J. chem. Phys.* **33**, 1791 (1960).
127. Bueche, F. *J. chem. Phys.* **20**, 1959 (1952).
128. Cronet, C. *Polymer* **6**, 373 (1965).
129. Wang, F. W. and Zimm, B. H. *J. polym. Sci.* **12**, 1619 (1974).
130. Onogi, S., Kobayashi, Y., Kojima, Y. and Taniguchi, Y. *J. appl. polym. Sci.* **7**, 847 (1963).
131. Wang, F. W. and Zimm, B. H. *J. polym. Sci.* **12**, 1639 (1974).
132. Bixon, M., Annual Review of Physical Chemistry, **27**, 65 (1976).
133. Orofino, T. A. *J. chem. Phys.* **45**, 4310 (1966).
134. Kirkbaum, W. R. and Godwin, R. W. *J. chem. Phys.* **43**, 4523 (1965).
135. Bueche, F., Cashin, W. M. and Debye, P. *J. chem. Phys.* **20**, 1956 (1952).
136. Ballard, D. G. H., Wignall, G. D. and Schelten, J. *Eur. polym. J.* **9**, 965 (1973).
137. Kirste, R. G., Kurse, W. A. and Schelten, J. *Makromol. Chem.* **162**, 299 (1973).
138. Gandhi, K. S. and Williams, M. C. *J. polym. Sci. C* **35**, 211 (1971).
139. Johnson, R. M., Schrag, J. L. and Ferry, J. D. *Polym. J.* (*Jpn*) **1**, 742 (1970).
140. Ferry, J. D. *Viscoelastic properties of polymer solutions*, p. 250. Wiley, New York (1970).
141. Frederick, J. E., Tschoegl, N. W. and Ferry, J. D. *J. phys. Chem.* **68**, 1974 (1964).
142. Holmes, L. A., Kusamizu, K., Osaki, K. and Ferry, J. D. *J. polym. Sci. A2* **8**, 2002 (1971).
143. Gandhi, K. S. and Williams, M. C. *J. appl. Polym. Sci.*, **16**, 2721 (1972).
144. Ferry, J. D., Foster, E. L., Browning, G. V. and Sayer, W. M. *J. colloid Sci.* **6**, 377 (1951).
145. Dreval, V. Ye, Tager, A. A. and Fomina, A. S., *Polym. Sci. USSR* **5**, 495 (1964).
146. Simha, R. and Zakin, J. L. *J. colloid Sci.* **17**, 270 (1962).
147. Quadrat, O. and Podnecka, J. *Coll. Czech. Chem. Commun.* **37**, 2402 (1972).
148. Johnson, M. F., Evans, W. W., Jordan, I. and Ferry, J. D. *J. colloid Sci.* **7**, 498 (1952).
149. Ferry, J. D., Grandine, L. D. and Udy, D. C. *J. colloid Sci.* **8**, 529 (1953).
150. Toms, B. A. and Stawbridge, D. J. *Trans. Faraday Soc.* **49**, 1225 (1953).
151. Dreval, V. Yu., Malkin, A. Yu. and Vinogradov, G. V. *Eur. polym. J.* **9**, 85 (1973).
152. Osaki, K. and Einaga, Y. *Prog. polym. Sci. Jpn*, **1**, 321 (1971).
153. Odani, H., Nemoto, N. and Kurata, M. *Bull. Inst. Chem. Rev. Kyoto U.* **50**, 117 (1972).
154. Onogi, S., Masuda, T. and Kitagawa, K. *Macromolecules* **3**, 109 (1970).
155. Tobolsky, A. V., Aklonis, J. J. and Akovali, G. *J. chem. Phys.* **42**, 723 (1965).
156. Graessley, W. W. *Advan. polym. Sci.* **16**, 1 (1974).
157. Williams, M. C. *A. I. Chem. E. J.* **13**, 955 (1967).
158. Williams, M. C. *A. I. Chem. E. J.* **12**, 1064 (1966).
159. Williams, M. C. *A. I. Chem. E. J.* **13**, 534 (1967).
160. Fixman, M. *J. chem. Phys.* **42**, 3831 (1965).
161. Bueche, F. *J. chem. Phys.* **40**, 484 (1964).

162. Aharoni, S. M. *J. appl. polym. Sci.* **17**, 1507 (1973).
163. Chompff, A. J. and Duiser, J. A. *J. chem. Phys.* **45**, 1505 (1966).
164. Duiser, J. A. and Staverman, A. J. *Physics of non-crystalline Solids* (ed. W. Prins), p. 376. North-Holland, Amsterdam (1965).
165. Edwards, S. F. and Grant, J. W. V. *J. Phys. A* **6**, 1169 (1973).
166. Forsman, W. C. and Grand, H. S. *Macromolecules* **5**, 289 (1973).
167. Edwards, S. F. and Grant, J. W. V. *J. Phys. A* **6**, 1186 (1973).
168. Graessley, W. W. *J. chem. Phys.* **54**, 5143 (1971).
169. De Gennes, P. G. *J. chem. Phys.* **55**, 572 (1971).
170. De Gennes, P. G. *J. Chem. Phys.* **55**, 572 (1971).
171. Doi, M. and Edwards, S. F. *J. Chem. Soc. Faraday Trans. II* **74**, 1789, 1802, 818 (1978); II, 38 (1979).
172. Graessley, W. *J. polymer Sci., Polymer Phys.* **18**, 27 (1980).
173. Hutton, J. F. *Lubricants*, p. 1, Adlard and Sons, Bartholomew Press, London (1973).
174. Walters, K. *Lubricants*, p. 9, Adlard and Sons, Bartholomew Press, London (1973).
175. Metzner, A. B., White, J. L. and Davis, M. M. *Rubber chemistry and technology*, p. 1426, American Chemical Institute (1968).
176. Block, H. *Molecular behaviour and the development of polymeric materials* (ed. A. Ledwidth and A. M. North), p. 541, Chapman and Hall, London (1974).
177. Ashworth, J. N. and Ferry, J. D. *J. Amer. Chem. Soc.* **71**, 622 (1949).
178. Morrisson, T. E., Zapas, L. J. and De Witt, T. W. *Rev. sci. Instrum.* **26**, 357 (1955).
179. Birnboim, M. H. and Ferry, J. D. *J. appl. Phys.* **32**, 2305 (1961).
180. Lamb, J. and Linden, P. *J. acoust. Soc. Amer.* **41**, 1032 (1967).
181. Birnboim, M. H. and Elyash, L. J. *Bull. Amer. phys. Soc.* **11**, 165 (1966).
182. Schrag, J. L. and Johnson, R. M. *Rev. sci. Instrum.* **42**, 224 (1971).
183. Sittel, K., Rouse, P. E. and Bailey, E. D. *J. appl. Phys.* **25**, 1312 (1954).
184. Mason, W. P. *Trans. Amer. Soc. Mech. Eng.* **69**, 359 (1947).
185. McSkimin, H. J. *J. acoust. Soc. Amer.* **24**, 355 (1952).
186. Barlow, A. J., Harrison, G., Richter, J., Seguin, H. and Lamb, J. *Lab. Pract.* **10**, 786 (1961).
187. Glover, G. M., Hall, G., Matheson, A. J. and Stretton, J. L. *J. Phys. E* **1**, 383 (1968).
188. Barlow, A. J., Dickie, R. A. and Lamb, J. *Proc. Roy. Soc.* **A300**, 356 (1967).
189. Barlow, A. J., Erginsav, A. and Lamb, J. *Proc. Roy. Soc.* **A298**, 481 (1967).
190. Lamb, J. and Richter, J. *Proc. Roy. Soc.* **A293**, 479 (1966).
191. Lamb, J. and Seguin, H. *J. acoust. Soc. Amer.* **39**, 519 (1966).
192. Fabelinskii, I. L. *Uspekhi Fiz. Nauk.* **63**, 355 (1966).

ULTRASONIC RELAXATION

8.1. Ultrasound propagation in fluids

The preceding chapter described the relaxation behaviour of a polymer subjected to pure high-frequency (10 Hz–100 kHz) shear waves. The pressure wave produced in the ultrasonic experiment is a composite of both shear and compression, and so is sensitive to relaxation behaviour involving shear, volume, and thermal mechanisms. While viscoelastic observations have been used extensively for the study of normal-mode motions of the whole polymer chain, ultrasonic measurements are principally concerned with the shorter time motions of relatively small elements of the polymer chain.

In order to appreciate fully the results obtained for polymer systems it is instructive to consider first the acoustic behaviour of simple molecular fluids, and then to discuss observations of polymer solutions.

Ultrasonic studies of gases had their advent in the mid-1920s and assessments on liquid systems commenced before the Second World War. However, a lack of readily available commercial equipment resulted in a very slow growth of interest in this area of research. Then, after the war, experience in acoustic detection of submarines stimulated interest in ionic processes such as occur in sea water, and in vibrational interactions in solids.[1] However, it is only in the last twenty years or so that physical chemists have recognized the potential of ultrasonic measurements in the study of conformational dynamics of small molecules,[2] solute–solvent interations,[3] structural relaxation in viscous media,[4] and more recently relaxation in polymers.[5] Application of the technique to molecular systems has illustrated that this type of measurement may provide data in situations where other techniques are either insensitive or uninformative.

The 'relaxation' or dispersion in the propagation characteristics of a high-frequency sound wave in a molecular system is intimately connected with the time dependence of the energy distribution in that system. The sound wave propagates through a medium with a complex velocity c^*, the magnitude of which is determined by the energy absorption from the sound wave and the modulus of the material

A classical fluid, such as liquid argon, is one in which the principal source of energy dissipation from a sound wave arises from so-called viscothermal losses. A sound wave may be considered either in terms of an adiabatic pressure perturbation or alternatively as a propagating translational temperature. If it is assumed that γ, the ratio of the specific heat at constant pressure to that at constant volume, is not unity, either description is applicable. The wave alternately raises and lowers the translational temperature, and consequently provides an oscillatory perturbation to the kinetic energy distribution. For a monatomic system such as argon, perturbation of the translational energy can lead to dissipation only when the temperature wave becomes out of phase with the pressure wave. The relative movement of atoms from a higher density configuration to a lower one and vice versa, may be described in terms of a complex bulk viscosity or dynamic bulk modulus, and the transfer of momentum out of phase with the wave leads to a thermal loss. Molecular systems possess two important additional degrees of freedom which can give rise to further energy loss processes. These constitute vibration–rotation and conformational relaxation.

(a) *Vibration–rotation relaxation* arises as a consequence of an inelastic collision between two suitably activated molecules such that part of the kinetic energy of the pair is converted into internal vibrational and rotational energy. The energy is 'stored' for a period and may then be dissipated by radiation, or more usually returned to translation by a further inelastic collision.

(b) *Conformational relaxation* has a similar origin and is the consequence of inelastic collisions activating internal torsional modes and so promoting a perturbation of the distribution of molecular conformations. This process, too, represents a mechanism for a dynamic storage of energy and gives rise to a dispersion in the propagation parameters.

Of course, elastic collisions similar to those described for a monatomic system lead to 'classical' viscothermal losses in such systems. In a molecular system, atomic and molecular mass, molecular symmetry, strength of the intermolecular forces, and energy of the lowest vibrational mode are all important in determining the frequency and magnitude of the vibrational and viscostructural relaxations in the system. In a low-viscosity system of highly asymmetric molecules contributions from vibrational and viscothermal effects are generally small, unless strong dipolar interactions lead to configurational ordering at a molecular level, as in liquid crystals.[6] For example, the major contribution to the acoustic loss in a molecule such as 1,2-dibromo-1, 1-di-fluoroethane arises from rotational isomeric processes

(Fig. 8.1). The acoustic absorption in polymeric systems is somewhat more complex than that in such simple molecules, although the concepts involved are essentially the same.[5]

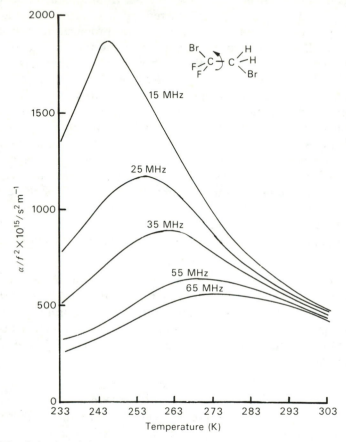

Fig. 8.1. Acoustic attenuation in 1,2-dibromo-1,1-difluoroethane.

8.2. Resumé of background theory

It was pointed out in Chapter 4 that the acoustic observation involves the collective motion of the particles in a system, and that this is reflected in the long-time tail of the velocity auto-correlation function. In this respect the technique is not very sensitive to the part of the velocity auto-correlation function which dominates spectroscopic and scattering observations. It means, too, that this low-frequency limit can adequately be described by the classical continuum approach.

8.2.1. Molecular fluid-continuum approach to relaxation

The theory of ultrasonic propagation has been described adequately elsewhere.[7-11] The continuum approach considers the ultrasonic wave propagating in a classical fluid obeying the Navier–Stokes equations, and treats relaxation as arising from the application of a perturbation to the thermodynamic variables of the system. For a non-relaxing fluid the hydrodynamic equations yield a square law dependence of the absorption on the frequency given by

$$\alpha/f^2 = \frac{2\pi^2}{\rho c^3}\left((\tfrac{4}{3}\eta_v + \eta_s) + \frac{(\gamma-1)}{C_p}\kappa\right) \tag{8.1}$$

where ρ is the density, c is the speed of the sound wave, η_s and η_v are respectively the shear and volume viscosities, γ is the ratio of the specific heat at constant pressure, C_p to that at constant volume, C_v, κ is the thermal conductivity and f is the frequency ($\omega = 2\pi f$) of the observation. At frequencies below approximately 10 GHz the absorption coefficient divided by the frequency squared is found to be independent of frequency for most non-associated liquids. Observations above 10 GHz deviate from this simple law because of the inadequacies of the continuum approach when the time constant approaches that of a molecular collision.

In a molecular system with internal degrees of freedom, dynamic storage of energy, which leads to absorption of sound, is possible through inelastic collisions. In the context of polymers the most important of the perturbations is that involving internal modes of the molecule and especially excitation of the torsional mode leading to conformational change. Studies of a large number of organic molecules[12-16] have indicated that in all systems, except those with either large differences in their internal vibrational energy distribution or with a high degree of symmetry, the principal contribution to the absorption arises from intermolecular rearrangement of the molecule. If the perturbation is slow, the molecules being promoted to and relaxed from the higher energy state maintain a dynamic equilibrium and any departure from the mean remains in phase with the temperature wave. As the exchange and perturbation times become comparable, the dynamic equilibrium becomes out of phase with the acoustic perturbation and an energy loss is observed. The dispersion for a simple rotational isomeric process (Fig. (8.2)) may be described by the equation

$$\alpha/f^2 = [A/(1+(f/f_c)^2)] + B \tag{8.2}$$

where A is the amplitude of the relaxation process, B is the classical non-relaxing absorption described by eqn (8.1). The characteristic frequency, $f_c (2\pi f_c \tau = 1)$, represents the mean time for the exchange process. In practice alternative equations, allowing for either a distribution of

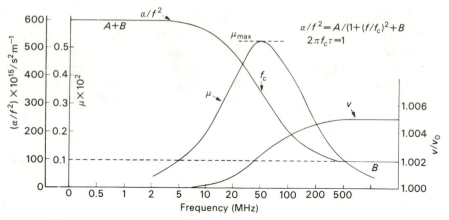

Fig. 8.2. Schematic representation of an ideal acoustic relaxation.

relaxation times and or a number of relaxation processes, may.be constructed.

The amplitude of the acoustic relaxation, for a single relaxation process is related to the thermodynamic quantities associated with the conformational equilibrium in the following manner

$$A = 2\mu_m/cf_c \tag{8.3}$$

$$2\mu_m/\pi = (\gamma - 1)\delta C_p/(C_p - \delta C_p)(1 - (\Delta V^\circ C_p)/(V^\circ \Delta H^\circ \theta))^2 \tag{8.4}$$

$$\delta C_p = R(\Delta H^\circ/RT)^2 \exp(-\Delta G^\circ/RT)/(1 + \exp(-\Delta G^\circ/RT))^2 \tag{8.5}$$

$$c^2 = (\gamma - 1)C_p/\theta^2 V^\circ T \tag{8.6}$$

where μ_m is the acoustic loss at the characteristic frequency, C_p and δC_p are respectively the specific heat at constant pressure and the relaxing specific heat, ΔH°, ΔV°, ΔG°, and ΔS° are respectively the standard state enthalpy, volume, free energy, and entropy differences associated with the conformational equilibrium, and θ is the expansion coefficient. The detailed derivation of the above relationships and discussion of the associated approximations is presented elsewhere.[2]

Very often there is no experimental information available over a range of pressure and it has to be assumed that $C_p \Delta V^\circ \ll \Delta H^\circ V^\circ \theta$ and that $\Delta V^\circ/V^\circ$ is small. The above equations can then be combined to yield

$$\mu_m = \frac{(\gamma - 1)R}{2C_p}\left(\frac{\Delta H^\circ}{RT}\right)^2 \frac{\exp(-\Delta G^\circ/RT)}{(1 + \exp(-\Delta G^\circ/RT))^2}. \tag{8.7}$$

It will be appreciated that the above equations do not yield unique solutions for ΔH° and ΔG°. However if the additional assumption is made that $\Delta G° > 3RT$, then the equations can be simplified to yield

$$\mu_m = \frac{(\gamma-1)R}{2C_p}\left(\frac{\Delta H°}{RT}\right)^2 \exp(-\Delta H°/RT)\exp(\Delta S°/R) \qquad (8.8)$$

with an error of less than 5 per cent. A plot of log $(T\mu_m/c^2)$ versus $1/T$ yields $\Delta H°$.

This procedure is inaccurate when $\Delta G° < 3RT$ and an alternative assumption is then required about the magnitude of $\Delta S°$. If $\Delta S° = 0$,

$$\delta C_p = R\left(\frac{\Delta H°}{RT}\right)^2 \frac{\exp(-\Delta H°/RT)}{(1+\exp(-\Delta H°/RT))^2} \qquad (8.9)$$

and

$$\mu_m = \frac{(\gamma-1)R}{2C_p} \cdot \frac{(\Delta H°)^2}{RT} \cdot \frac{\exp(-\Delta H°/RT)}{(1+\exp(-\Delta H°/RT))^2}. \qquad (8.10)$$

$\Delta H°$ can here be evaluated from a series of values of μ_m by an iterative process. The variation of δC_p with $\Delta H°/RT$ (the Schottky function) is shown in Fig. 8.3. It may be noted that the maximum, whilst being a function of $\Delta S°$, is located at approximately $\Delta H° \simeq 2\cdot4RT$.

The activation energy for a conformational change can be evaluated from a study of the temperature dependence of the relaxation frequency, which is related to the forward (k_f) and reverse (k_b) rate constants by

$$f_c = 1/2\pi\tau = (k_f+k_b)/2\pi \qquad (8.11)$$

where τ is the relaxation time of the equilibrium. The theory of rate processes[17] expresses the rate constant k_b of a reaction in terms of the increase in free energy $\Delta G^{\ddagger}$ on moving from the initial state to an intermediate activated state.

$$k_b = (kT/h)\exp(-\Delta G^{\ddagger}/RT). \qquad (8.12)$$

Hence

$$f_c = (1/2\pi)(kT/h)\{\exp(-\Delta G_f^{\ddagger}/RT)/1+\exp(-\Delta G_b^{\ddagger}/RT)\} \qquad (8.13)$$

where f and b refer to the forward and back reactions and k and h are the Boltzmann and Planck constants.

If we assume that equilibrium in this reaction lies well to the left, that is

$$\Delta G_b^{\ddagger} \ll \Delta G_f^{\ddagger}$$

$$f_c = (1/2\pi)(kT/h)\exp(-\Delta G_b^{\ddagger}/RT). \qquad (8.14)$$

Experimentally f_c is obtained over as wide a range of temperature as possible and $\Delta H_b^{\ddagger}$ and $\Delta S_b^{\ddagger}$ determined from the slope and intercept of a graph of $\log(f_c/T)$ against $1/T$, when $\Delta G_b^{\ddagger}$, $\Delta G_f^{\ddagger}$, k_b, and k_f are obtained

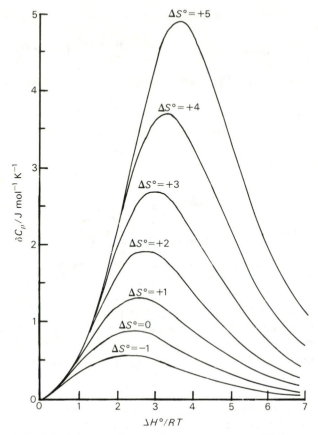

Fig. 8.3. Schottky diagram for a two-state equilibrium.

from eqn (8.11)

$$k_b \simeq 2\pi f_c. \tag{8.15}$$

This illustrates how the study of the acoustic attenuation, α/f^2, as a function of frequency and temperature can provide information on the activation energy of conformation interchange as well as the conformational equilibrium parameters.

8.3. Ultrasonic wave propagation and polymer solution dynamics

It will be evident from the brief summary of the theory presented above that the ultrasonic technique is ideally suited for the study of conformational dynamics in systems where there exists a significant difference in energy between conformational states and the barrier to interchange is relatively small. Most of the work reported to date has been concerned with the evaluation of the temperature dependence of the observed relaxation or with elucidating the nature of the molecular process responsible. Initial studies of polymer solutions were performed with pulsed techniques[16,19] and with interferometers[20,21] operating over the frequency range 5–50 MHz. For an ideal relaxation of width 1·4 decades, this frequency range would be expected to be hardly adequate for differentiation between ideal and non-ideal relaxation processes. Considerable efforts have been made over the past decade to extend the frequency range of the ultrasonic technique, and experimental observations are now possible between 100 kHz and 3 GHz (Table 8.1).

Table 8.1

Techniques available for ultrasonic studies of polymers

Method	Range (Hz)	Comment	Reference
Reverberation	10^4–10^6	Fairly accurate (5 per cent). Important for very low frequencies.	74
Streaming	10^5–10^6	Poor accuracy (10 per cent).	75
Optical methods	10^6–10^8	Classical visualization of sound field. Now largely obsolete in this form although fairly accurate.	76, 77
Resonance	10^5–10^7	Capable of high accuracy (1 per cent) has a low and high frequency limit.	78, 79
Interferometry	10^6–10^8	Poor accuracy for highly absorbing media. Very accurate velocity measurements.	80
Pulse methods	10^7–10^9	Principal present-day technique. Accuracy 1–3 per cent.	18, 81–9

Although measurements of the acoustic absorption of polymer solutions were made as early as 1953,[22] studies outside the narrow frequency range 1–10 MHz were not undertaken until after 1964.[23-7] The first definite quantification of acoustic relaxation in polymer solutions and its analysis in terms of a definite model was reported at the 6th International Congress on Acoustics held in Tokyo in 1968.[28,29] Subsequently measurements have been reported over a frequency range covering 100 KHz to 3 GHz, establishing the existence of relaxation in a wide variety of polymer systems and providing a fascinating picture of the dynamic behaviour of macromolecular systems.

For convenience the acoustic observations on polymer solutions will be discussed according to the influence of various factors on the relaxation spectrum.

8.3.1. Molecular-weight effects

It is not very surprising to find that the most widely studied polymer in solution is polystyrene. The early measurements of Hässler and Bauer[28] suggested that the acoustic relaxation was independent of molecular weight. More recently studies with narrow molecular-weight samples, have established that a molecular-weight effect does influence the relaxation spectrum.[30-7] Both the amplitude and frequency of the relaxation change with molecular weight (Fig. 8.4). The ultrasonic relaxation of polystyrene

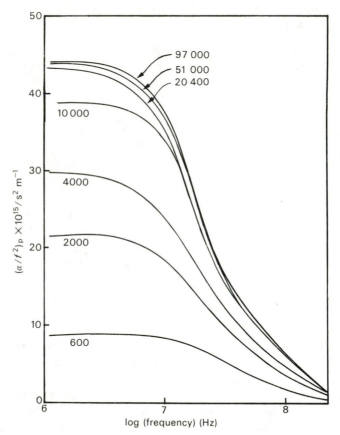

Fig. 8.4. Acoustic attenuation in narrow molecular-weight polystyrene solutions in toluene at 303 K.

solutions in the megahertz frequency range has the following characteristics:[35-7]

(i) The relaxation may be described in terms of either a 'single' process with a distribution of relaxation times or, at the most two processes.

Some discussion exists as to which of the above descriptions is correct, however it is doubtful whether the experiments are of sufficient precision for this question to be resolved satisfactorily. In practice a good representation of the relaxation curve can be obtained using the following empirical equation

$$\alpha/f^2 = A\{1 + (f/f_c)^\beta \cos(\pi\beta/2)]/\{1 + 2(f/f_c)\beta \cos(\pi\beta/2) + (f/f_c)^{2\beta}\}$$

$$(8.16)$$

where β is the distribution coefficient for a single non-ideal relaxation processes. The value of β obtained from the data presented in Fig. 8.4 lies in the range 0·8 to 0·93, indicating the near-ideality of the relaxation.

(ii) The relaxation frequency obtained from studies in various solvents is independent of molecular weight[30-4] provided that the molecular weight is greater than about 10 000. The variation of the characteristic frequency with molecular weight (Fig. 8.4) is very similar in form to that reported for the change in spin correlation relaxation times observed for electron spin probes[38,39] and ^{13}C nuclei.

(iii) The relaxation strength increases with increasing molecular weight up to a value of $M_n = 10\ 000$, above which it is independent of chain length.

Two possible mechanisms for a molecular-weight dependence of the acoustic absorption have been proposed. The first involves a frequency-dependent modification to the contribution to the 'classical' shear and volume viscosity absorption as a consequence of the viscoelastic properties of the macromolecules. Using the models described in the previous chapter it is found that whilst the contribution is significant, it does not explain the observed trends. Subtracting the viscoelastic contribution to the dispersion observed in high molecular-weight systems makes the residual relaxation almost ideally Debye, i.e. $\beta \sim 1$.

By analogy with small molecules[2] the acoustic relaxation arises principally as a result of conformational changes of the backbone structure of the polymer. A molecular-weight dependence of the amplitude and relaxation frequency would infer that the form of the potential profile governing the internal rotation is itself a function of molecular weight. Such an effect would be expected if the rotation about one bond was in some way coupled to the motion of its neighbours, or, alternatively, if the energies of the conformational states involved are influenced by other than nearest-neighbour interactions. Analysis of the variation of the acoustic amplitude with chain length[35] indicates that six to eight units are necessary before the motion of the chain becomes truly representative of internal segments uninfluenced by chain ends.

The relaxation behaviour of short-chain, normal alkanes indicates that although the detailed profile involved in a particular conformational change may be very complex (Fig. 8.5), the observed dynamics and equilibrium

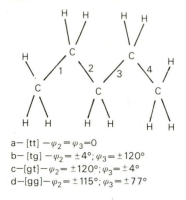

a— [tt] $-\varphi_2 = \varphi_3 = 0$
b— [tg] $-\varphi_2 = \pm 4°; \varphi_3 = \pm 120°$
c— [gt] $-\varphi_2 = \pm 120°; \varphi_3 = \pm 4°$
d— [gg] $-\varphi_2 = \pm 115°; \varphi_3 = \pm 77°$

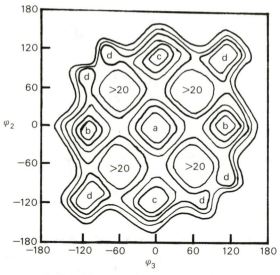

Fig. 8.5. Energy map calculated for internal rotations in n-pentane, with $\phi_1 = \phi_4 = 0$: contours are shown at intervals of 4 kJ mol^{-1}. Minima are indicated by a, b, c, and d.

behaviour may often approximate to a simple two-state model.[40] These predictions are in good agreement with the acoustic observations on normal alkanes with chain lengths varying from C-4 to C-14.[41,42]. The effect of coupling in the motion of neighbouring carbon atoms in the alkane chain can be seen in both acoustic and nuclear magnetic ^{13}C relaxation.

Specific conformational changes have been proposed for the relaxation in polystyrene.[30] However such detailed analysis must remain speculation because of the limited precision of relaxation data. Recently a study of the model compound 2,4-diphenylpentane indicated[43] that the relaxation behaviour of the meso and racemic forms are significantly different (Table 8.2). The single relaxation observed for each isomer was assigned to a

Table 8.2

Rotational isomeric parameters for 2,4-diphenyl pentane

Isomer	ΔH°/kJ mol^{-1}	ΔS°/JK^{-1} mol^{-1}	$\Delta H^\ddagger$/kJ mol^{-1}	$\Delta S^\ddagger$/JK^{-1} mol^{-1}
Meso	$5\cdot85 \pm 2\cdot9$	$-5\cdot4 \pm 4\cdot2$	$13\cdot4 \pm 7\cdot1$	$-25 \pm 30\cdot5$
Racemic	$5\cdot85 \pm 2\cdot9$	$13\cdot0 \pm 1\cdot2$	$13\cdot0 \pm 0\cdot5$	$-20\cdot5 \pm 8\cdot4$

rotational isomeric process between g − t and tt conformations for the meso isomer and between tt and gg conformations for racemic isomer. The racemic isomer exhibits a relaxation at approximately 220 MHz and the meso at 74 MHz at room temperature. It is clear that characterization of the relaxation of the higher oligomers is necessary before it is possible to discuss the detailed conformational changes which occur in the polymeric materials. Indeed the highly specific structures adopted by short-chain molecules may be not as stable as slightly distorted structures in longer chain systems. This distortion hypothesis predicts an increase in activation energy and relaxation amplitude with chain length, both of which effects are consistent with experimental observation.

Molecular weight effects similar to those found in polystyrene have been observed in poly(α-methylstyrene)[44] and poly(methylmethacrylate).[44]

8.3.2. Tacticity

As indicated in the previous section, the detailed stereochemistry of the polymer may modify the relaxation spectrum associated with rotational isomerism. Studies of solutions of poly(methylmethacrylate)[44] have indicated that the relaxation in the megahertz frequency range does vary with the tacticity of the polymer as well as with molecular weight. The two-state energy difference for the syndiotactic form is higher than that of the isotactic form (Table 8.3).

Studies of the glass transition temperatures of the tactic forms of this polymer indicate that for the isotactic form T_g is approximately 40 °C and for the syndiotactic for T_g is 120 °C.[45] If the 'flex energy' between different rotational states in the solid polymer is calculated from the Gibbs–DiMarzio theory[46] of the glass transition, this quantity compares well with the acoustic interstate energy difference (Table 8.3). The close correspondence of the

Table 8.3

Effect of structure on rotational isomeric parameters

Polymer	Tacticity	ΔH°/kJ mol^{-1}
Poly(methylmethacrylate)	Isotactic	6·3
	Atactic	6·3
	Syndiotactic	3·7
Poly(α-methylstyrene)	Syndiotactic	8·3
Polystyrene	Atactic	5·4

values is quite surprising since the acoustic measurements were performed on dilute solutions and the glass transitions occur in bulk solids. It is not yet clear how far this agreement is real, but it may indicate that important factors determining the rotational profile arises from intra- rather than intermolecular interactions.

8.3.3. Concentration effects

Very few measurements of the effects of concentration on the ultrasonic relaxation properties of polymer solutions have been reported. In the case of polystyrene in dibutyl phthalate[33] the acoustic absorption varies linearly with concentration from 1 to 8 weight per cent at all frequencies between 3 and 52 MHz as expected for an intrachain process. Similar conclusions can be drawn from studies of the acoustic absorption of polystyrene in toluene[47-9] (Fig. 8.6). It will be noted however, that above a concentration of 10 per cent, the relaxation amplitude deviates from a linear concentration dependence, although the general shape of the relaxation appears little affected. This is despite a change in viscosity of almost two decades.

Adopting the procedure of analysing the dispersion in terms of a thermal (rotational isomeric) and viscoelastic contribution leads to an almost ideal relaxation in the megahertz region (Fig. 8.6). The amplitude of the thermal process is then again a linear function of concentration above about 10 per cent. Extension of the analysis, with consideration of volume effects, has lead to the suggestion that the acoustic relaxation in the megahertz frequency range in the concentrated solutions has a contribution from entropic effects. This result is not really very surprising, since a major factor governing the behaviour of an entangled liquid or rubber is the entropy of deformation.

8.3.4. Solvent effects

Studies[30-7] of polystyrene in various solvents have indicated that although the amplitude of the relaxation depends on the solvent, the relaxation

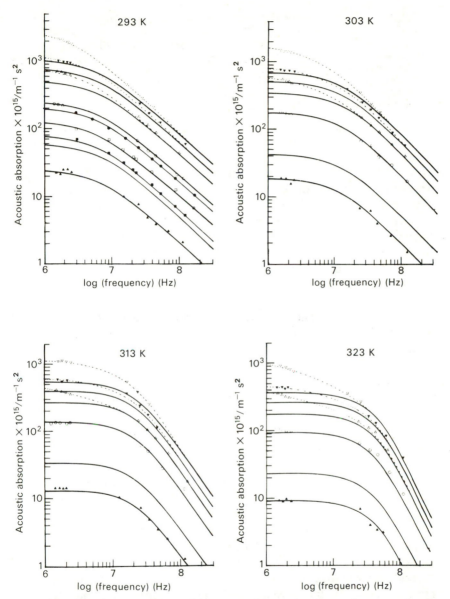

Fig. 8.6. Effect of temperature and concentration on the ultrasonic attenuation for polystyrene in toluene. Concentration (wt per cent): (▲) 1; (■) 5; (□) 7; (●) 9; (○) 10; (△) 20; (▼) 30; and (▽) 40. Full line: absorption after subtraction of viscoelastic loss.

Table 8.4

Effect of solvent on the relaxation in polystyrene

Solvent	Viscosity (cP)	$\Delta H^\circ/\text{kJ mol}^{-1}$	$\Delta H^{\ddagger}_{21}/\text{kJ mol}^{-1}$
Methylethylketone	0·44	7·1	1·2
Toluene	0·52	1·7	10·7
Decalin	2·4	4·5	10·7
Dibutylphthalate	20·7	7·0	27·6
Benzene	0·56	3·67	27·6
Carbon tetrachloride	0·84	3·67	27·6

frequency is only slightly sensitive to solvent composition and does not appear to correlate well with changes in the solvent viscosity (Table 8.4).

Two possible effects should be considered when varying the solvent: first, its effect on the 'effective' volume associated with the rotational relaxation process[50] and, secondly, changes in the conformational energy difference resulting from changes in dielectric constant.[51] Correlations based purely on a volume effect do not appear to provide the full answer and an improvement is obtained by allowing for dielectric constant changes.[52]

Extension of the experimental frequency range below 1 MHz has revealed that the solvent has a very significant effect on the 'viscoelastic' contribution to the acoustic absorption. Data have been obtained[53] between 10^4 and 10^5 Hz using reverberation techniques on solutions of polystyrene of varying molecular weight in xylene. The amplitude of this process is significantly greater than the higher frequency loss discussed earlier. This acoustic absorption appears to increase with molecular weight and solvent viscosity as expected for a viscoelastic process.[52,54]

8.3.5. Coupling in isomeric relaxation

The principal problem in the analysis of any relaxation data is choice of a suitable model. Part of this lies in determining what exactly is the unit undergoing, in this case, conformational charge. An estimation of the effective range of influence of one group on the relaxation of another can be gained from study of block copolymers with short chains (Fig. 8.7). Acoustic studies of copolymers of α-methylstyrene and styrene with alkane chains has indicated that as the length of the alkane block is increased to C_6 so the activation energy for rotation is decreased,[55] but there is no further effect between C_6 and C_{10}. It may be inferred that once the aromatic moieties have been separated by six carbon atoms, the relaxation process is virtually independent of neighbouring units.

A study of copolymers of vinylcarbazole[56] indicates that the relaxation spectrum is not significantly modified by the introduction of a proportion of more mobile monomer units. This observation would suggest that, in a

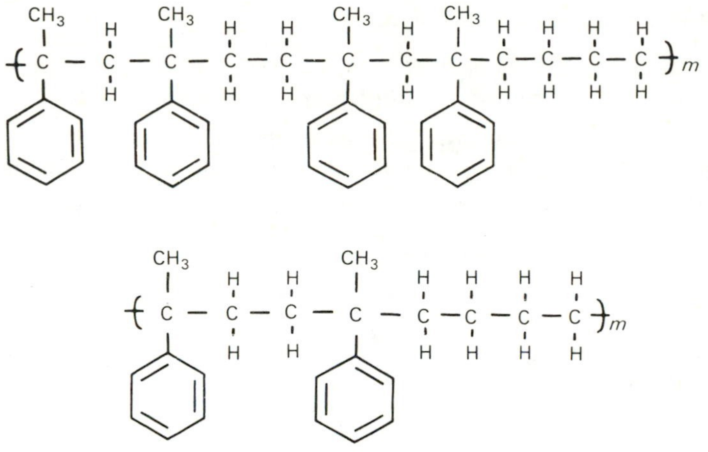

Fig. 8.7. Structures of α-methylstyrene-alkane co-polymers. Upper structure is that of the tetramer, the lower that of the dimer; m is the degree of polymerization.

random copolymer of this type, the effects of long sequences of the less mobile vinylcarbazole can significantly damp the plasticizing action of separated mobile monomer units.

8.3.6. *Polymer–solvent interactions*

It is well known that many simple liquids possess high acoustic losses as a result of vibrational relaxation, and this often occurs in the high megahertz frequency range. Examples of this type of behaviour are shown by benzene, carbon tetrachloride, and chloroform. The observation of a negative increment of the absorption in solutions of poly(methyl-methacrylate)[57–61] in benzene has been suggested as evidence for interaction between the vibrational relaxation of benzene and the intramolecular vibrations of the polymer molecule.[62] Similar effects have been observed in polycarbonate solutions in dioxane and chloroform.[63] Some support is given to this hypothesis by the fact that the Raman spectra of solutions of polycarbonate in chloroform contain lines at 367 and 364·2 cm^{-1} of which the intensity is concentration-dependent. Other thermodynamic[64] and spectroscopic[65] observations support the presence of intense solvent–solute interactions in this system.

A similar deduction has been drawn recently from high-frequency measurements of solutions of polybutadiene in cyclohexane.[66] It is clear that in solvents where there is a significant contribution to the loss from vibrational relaxation and in which solvent-solute interactions may occur, a modification of the relaxation spectrum of the solvent due to the polymer is to be expected.

8.4. Studies of some specific polymers

8.4.1. Ultrasonic relaxation in solutions of poly(vinylacetate), poly(vinyl-propionate), and poly(vinylbutyrate)—side-group effects

The effect of side-groups may be investigated by a comparison of the ultrasonic relaxation in poly(vinylacetate) poly(vinylpropionate) and poly(vinylbutyrate) (Fig. 8.8).[57-9] These polymers have been studied in both

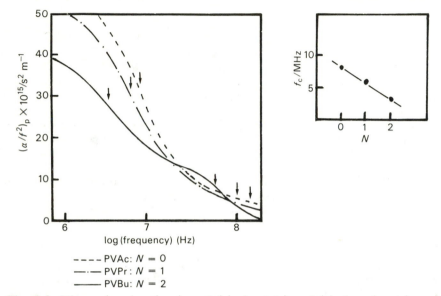

    ---- PVAc: $N = 0$
    —·— PVPr: $N = 1$
    ——— PVBu: $N = 2$

Fig. 8.8. Ultrasonic relaxation in poly(vinylacetate), poly(vinylpropionate), and poly(vinylbutyrate).

toluene and acetone.[60] In acetone the polymers exhibit a single relaxation, whereas in toluene two relaxations are easily identified. Two explanations appear to exist as to the origin of the relaxations. The first involves consideration of two local internal rotation processes in the backbone, one of wide range and the other a more local process.[61] Alternatively the relaxations may be ascribed to a local relaxation of the backbone and to relaxation of the side-chain. Some support for this second model is gained from the fact that the position of the low-frequency relaxation appears to correlate with the moment of inertia of the backbone segment, rather than chain molecular weight (Figure 8.8). The activation parameters associated with the relaxations are summarized in Table 8.5.

8.4.2. Ultrasonic relaxation in aqueous solutions

Certain characteristic features of relaxation in aqueous systems are illustrated by the three polymer types, poly(N-vinylpyrrolidone),

Table 8.5
Activation parameters for PVAc, PVPr, and PVBu

Polymer	Solvent	$\Delta H°/\text{kJ mol}^{-1}$	$\Delta H^{\ddagger}_{21}/\text{kJ mol}^{-1}$
Poly(vinylacetate)	Toluene	5·7	4·2
	Acetone	6·3	23·8
Poly(vinylpropionate)	Toluene	5·45	5·35
Poly(vinylbutyrate)	Toluene	5·35	9·4

poly(ethyleneoxide) and poly(saccharides). Generally strong solvation (hydration) interactions complicate separation of viscoelastic and conformational contributions to the relaxation. In addition the volume change associated with conformational rotation may be increased to very significant proportions.

(i) *Poly(N-vinylpyrrolidone)*. The polymer with pyrrolidone side chain soluble in water. It has been studied as a model substance for proteins since it contains the —N—C— group in the side chain. The polymer is strongly

$$\begin{array}{c} \| \\ O \end{array}$$

hydrated at low temperatures (room temperature and below), although on raising the temperature the polymer is virtually dehydrated at 45–50 °C when it shows the same behaviour as do synthetic polymers in non-aqueous solvents. This effect has been observed also using dielectric relaxation.[67]

The effect of hydration on the relaxation curves is clearly indicated in Fig. 8.9. The lower-frequency absorption is modified by the appearance of a shoulder at low temperatures.[68] The relaxation observed in dimethylformamide is similar in form to that observed in aqueous solution, the position- and temperature-dependence of the relaxation being similar to that of the dehydrated polymer.[69] The activation energy for backbone rotation is significantly modified by the hydration which occurs at low temperatures, (Table 8.6). Below the dehydration temperature $\Delta H^{\ddagger}$ is connected with the interaction between water of hydration and that 'free' in

Table 8.6
Activation parameters for poly(N-vinylpyrrolidone)

	$\Delta H^{\ddagger}_{21}/\text{kJ mol}^{-1}$		$\Delta H°/\text{kJ mol}^{-1}$	
Molecular weight	10–20°	35–40°	10–20°	35–40°
10 000	18	16·7	2·9	2·9
25 500	5·85	16·7	2·9	2·9
40 000	5·85	16·7	2·9	2·9
360 000	5·85	16·7	2·9	2·9

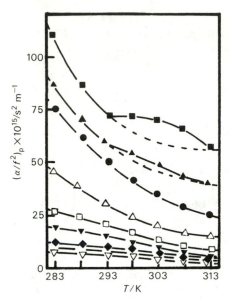

Fig. 8.9. Relaxation in poly(N-vinylpyrrolidone). Frequency of observation (MHz): (■) 5; (▲) 7; (●) 9; (△) 15; (□) 21; (▼) 30; (◆) 70; (▽) 130.

the surrounding solution. Above 25 °C it is related to the interaction between non-hydrated segments and the surrounding water.

In general, the lower-frequency relaxation spectrum in this system is independent of the solvent, whereas the higher-frequency process is significantly affected by the solvent.

(ii) *Poly(ethyleneoxide)*. The acoustic absorption coefficients of solutions of various molecular weights (in the range 62–20 000) have been studied in aqueous solution. Samples with molecular weights of 4500 to 20 000 give virtually identical dispersion curves. It has been argued that the relaxation in these systems arises from a combination of conformational[70] and viscoelastic processes.[71] Calculations based on a Zimm model appear to explain the observed results only partially. It has been suggested that solvation complicates interpretation of these data since the volume term associated with conformational change may not be negligible.

(iii) *Dextran and cellulose*. Studies of the attenuation in aqueous solution of dextran[72,73] (Fig. 8.10) indicate that the absorption coefficient α/f^2 for the solution is independent of molecular weight above 10 000. A study of solutions of cellulose triacetate in cyclohexanone indicates that the relaxation is of greater amplitude than in the case of dioxan–water mixtures. Although the relaxation has a similar form to that observed in poly(ethyleneoxide) the contribution from shear relaxation is negligible and in this case the relaxation must be ascribed to conformation changes in the polymer.

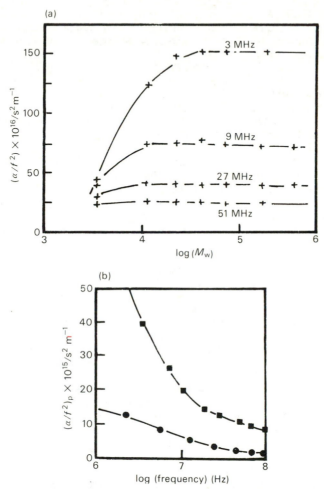

Fig. 8.10. Ultrasonic relaxation in aqueous solutions of dextran and cellulose triacetate in cyclohexane. (a) Plot of acoustic absorption of aqueous dextran versus molecular weight; (b) comparison of the acoustic absorption of similar molecular-weight samples of cellulose acetate in cyclohexane (■) and dextran in water (●).

References

1. Truell, R., Elbaum, C. and Chick, B. B. *Ultrasonic methods in the solid state.* Academic Press, New York (1969).
2. Wyn-Jones, W. and Pethrick, R. A. In *Topics in stereochemistry* (ed. E. L. Eliel and N. L. Allinger), Vol. 5, p. 205. Wiley Interscience, New York (1970).

3. Blandamer, M. J. *Introduction to chemical ultrasonics.* Academic Press, London (1973).
4. Harrison, G. *The dynamic properties of supercooled liquids.* Academic Press, London (1976).
5. Pethrick, R. A. *Rev. macromol. Chem.* **C9**, 91 (1973).
6. North, A. M. and Pethrick, R. A. In *Transfer and storage of energy by molecules* (ed. G. Burnett, A. M. North, and J. N. Sherwood), Vol. 4, p. 441. Wiley Interscience, London (1974).
7. Andrae, J. H. and Lamb, J. *Proc. Phys. Soc.* **68**, 814 (1956).
8. Lamb, J. In *Physical acoustics* (ed. W. P. Mason), Vol. 11A. Academic Press, New York (1956).
9. Litovitz, T. A. *J. acoust. Soc. Amer.* **31**, 681 (1959).
10. Matheson, A. J. *Molecular acoustics.* Wiley Interscience, London (1971).
11. Beyer, R. T. and Letcher, S. V. *Physical ultrasonics.* Academic Press, London (1969).
12. Naugle, D. G., Lunsford, J. and Singer, J. *J. chem. Phys.* **45**, 4669 (1966).
13. Jarzynski, J. and Litovitz, A. *J. chem. Phys.* **41**, 1290 (1964).
14. Jarzynski, J. *Proc. Phys. Soc.* **81**, 745 (1963).
15. Victor, A. E. and Beyer, R. T. *J. chem. Phys.* **52**, 1573 (1970).
16. Carnevale, E. H. and Litovitz, T. A. *J. acoust. Soc. Amer.* **27**, 547 (1955).
17. Glasstone, S., Laidler, K. J. and Eyring, H. *Theory of rate processes.* McGraw-Hill, New York (1958).
18. Pellam, J. R. and Galt, J. K. *J. chem. Phys.,* **14**, 608 (1946).
19. Pinkerton, J. M. M. *Proc. phys. Soc.* **B62**, 129, 286 (1949).
20. Hubbard, J. C. *Phys. Rev.,* **38**, 1011 (1931).
21. Greenspan, M. *J. acoust. Soc. Amer.* **22**, 568 (1950).
22. Wada, Y. and Shimbo, S. *J. acoust. Soc. Amer.* **25**, 549 (1953).
23. Gooberman, G. *Nature* **191**, 693 (1961).
24. Cerf, R., Zana, R. and Candau, S. Comptes Rendus, **252**, 681 (1961).
25. Cerf, R., Candau, S. and Zana, R. *Z. Phys. Chem.* **65**, 687 (1961).
26. Candau, S., Zana, R. and Cerf, R. *Comptes Rendus,* **252**, 2229 (1961).
27. Cerf, R. *Comptes Rendus* **270**, 1075 (1970).
28. Hassler, H. and Bauer, H. J. *Kolloid Z.* **230**, 194 (1969).
29. Nomura, H., Kato, S. and Miyahara, Y. *Nippon Kagaku Zasshi* **89**, 149 (1968).
30. Ludlow, W., Wyn-Jones, E. and Rassing, J. *Chem. Phys. Lett.* **13**, 477 (1972).
31. Ono, K., Shintani, H., Yano, O. and Wada, Y. Polymer J., **5**, 164 (1973).
32. Ott, H., Cerf, R., Michels, B. and Lemarechal, P. *Chem. Phys. Lett.* **24**, 323 (1974).
33. Nomura, H., Kato, S. and Miyahara, Y. *Nippon Kagaku Zasshi* **91**, 837 (1970).
34. Nomura, H., Kato, S. and Miyahara, Y. *J. Material Soc. Jpn,* **20**, 669 (1971).
35. Cochran, M. A., Dunbar, J. H., North, A. M. and Pethrick, R. A. *J. chem. Soc. Faraday Trans. II* **70**, 215 (1974).
36. Nomura, H. and Miyahara, Y. *Nippon Kagaku Zasshi* **88**, 502 (1967).
37. Nomura, H., Kato, S. and Miyahara, Y. *J. chem. Soc. Jpn,* (*Chem. and Ind. Chem.*) 1554 (1973).
38. Bullock, A. T., Cameron, G. G. and Smith, P. M. *J. phys. Chem.* **77**, 1635 (1973).
39. Allerhand, A. and Hailstone, R. K. *J. chem. Phys.* **56**, 3718 (1972).
40. Scott, R. A. and Scheraga, H. A. *J. chem. Phys.* **42**, 2209 (1965).
41. Cochran, M. A., Jones, P. B., North, A. M. and Pethrick, R. A. *J. chem. Soc. Faraday Trans. II* **68**, 1719 (1972).

42. Piercy, J. E. and Rao, M. G. S. *J. chem. Phys.* **46**, 3951 (1967).
43. Froelich, B., Noel, C., Jasse, B. and Monnerie, L. *Chem. Phys. Lett.* **44**, 159 (1976).
44. Dunbar, J. H., North, A. M., Pethrick, R. A. and Steinhauer, D. B. *J. chem. Soc. Faraday Trans. II* **70**, 1478 (1974).
45. Karasz, F. E. and MacKnight, W. J. *Macromolecules* **1**, 537 (1968).
46. Gibbs, J. H. and DiMarzio, E. A. *J. chem. Phys.* **28**, 373 (1958).
47. Mikhailov, I. G. and Safina, E. B. *Sov. Phys. Acoustics* **17**, 335 (1972).
48. Nomura, H., Kato, S. and Miyahara, Y. *J. Material Soc. Jpn*, **20**, 669 (1971).
49. Dunbar, J. I., North, A. M., Pethrick, R. A. and Steinhauer, D. B. *J. polym. Sci., Polym. Phys.* **15**, 263 (1977).
50. Kato, S., Kondo, H., Fujio, I., Nomura, H. and Miyahara, Y. *J. chem. Soc. Jpn*, (*Chem. and Ind. Chem.*) 1981 (1974).
51. Abraham, R. J. and Bretschneider, E. In *Internal rotation in molecules* (ed. W. J. Orville Thomas), p. 481. Wiley, New York, (1974).
52. Dunbar, J. H., North, A. M., Pethrick, R. A. and Poh, B. T. (Unpublished results.)
53. Okano, K., Shintani, H., Yana, O. and Wada, Y. *Polymer J.* **5**, 164 (1973).
54. Okano, K. *Rep. Prog. polym. Phys. Jpn* **5**, 67 (1962).
55. North, A. M., Rhoney, I. and Pethrick, R. A. *J. chem. Soc. Faraday Trans. II* **70**, 223 (1974).
56. Shanker Iyer, P. N., Steinhauer, D. B., North, A. M. and Pethrick, R. A. *Polymer* **16**, 797 (1975).
57. Nomura, H., Kato, S. and Miyahara, Y. *J. Material Soc., Jpn* **21**, 476 (1972).
58. Nomura, H., Kato, S. and Miyahara, Y. *J. chem. Soc. Jpn* (*Chem. and Ind. Chem.*) 1291 (1972).
59. Nomura, H., Kato, S. and Miyahara, Y. *J. chem. Soc. Jpn* (*Chem. and Ind. Chem.*) 2398 (1973).
60. Masuda, Y., Ikeda, H. and Ando, M. *J. Material Soc. Jpn* **20**, 675 (1975).
61. Funfschilling, O., Lemarechal, P. and Cerf, R. *Comptes Rendus* **270**, 659 (1970).
62. Tondre, C. and Cerf, R. *J. chim. Phys.* **65**, 1105 (1968).
63. Nomura, H., Kato, S. and Miyahara, Y. *Memoirs Faculty of Engng, Nagoya Univ.* **27**, 72 (1975).
64. Teramachi, S., Takahashi, A. and Kagawa, I. *Nippon Kogyo Kagaku Zasshi* **69**, 685 (1966).
65. Reeves, L. W. and Schneider, W. G. *Can. J. Chem.* **35**, 251 (1957).
66. Dunbar, J. H., North, A. M. and Pethrick, R. A. *Polymer* **18**, 577 (1977).
67. Nomura, H. and Miyahara, Y. *Bull chem. Soc. Jpn* **30**, 1599 (1966).
68. Kato, S., Kondo, H., Fujio, I., Honura, N. and Miyahara, Y. *J. chem. Soc. Jpn* (*Chem. and Ind. Chem.*) 1981 (1974).
69. Kato, S., Uehara, I., Kondo, H., Nomura, H. and Miyahara, Y. *J. chem. Soc. Jpn* (*Chem. and Ind. Chem.*) 1651 (1975).
70. Hames, G. G. and Lewis, T. B. *J. phys. Chem.* **70**, 1610 (1966).
71. Kessler, L. W., O'Brian, W. D. and Dunn F. *J. phys. Chem.* **74**, 4069 (1970).
72. Hawley, S. A. and Dunn, F. *J. chem. Phys.* **50**, 3523, (1969).
73. Kato, S., Uehara, I., Kondo, H., Nomura, H. and Miyahara, Y. *Report Prog. polym. Phys. Jpn* **18**, 131 (1975).
74. Ohsawa, T. and Wada, Y. *Japan J. appl. Phys.* **6**, 1351 (1967).
75. Hall, D. N. and Lamb, J. *Proc. phys. Soc.* **B73**, 354 (1959).
76. Debye, P. and Sears, F. W. *Proc. nat. Acad. Sci., US* **18**, 410 (1932).

77. Lucas, R. and Biquard, P. *J. Phys. Radium* **3**, 464 (1932).
78. Eggers, F. *Acoustica*, **19**, 323 (1968).
79. Pethrick, R. A. *J. Phys. E* **5**, 571 (1972).
80. Hubbard, J. C. *Phys. Rev.* **46**, 525 (1934).
81. Seki, H., Granato, A. V. and Truell, R. *J. acoust. Soc. Amer.* **28**, 230 (1956).
82. Papadakis, E. P. *J. acoust. Soc. Amer.* **40**, 863 (1966).
83. Redwood, M. and Lamb, J. *Proc. IEE B* **103**, 773 (1955).
84. Truell, R. and Oates, W. *J. acoust. Soc. Amer.* **35**, 1382 (1963).
85. Papadakis, E. P. *J. appl. Phys.* **35**, 1474 (1964).
86. Williams, J. and Lamb, J. *J. acoust. Soc. Amer.* **30**, 308 (1958).
87. Papadakis, E. P. *J. acoust. Soc. Amer.* **42**, 1045 (1967).
88. McSkimin, H. J. *J. acoust. Soc. Amer.* **37**, 864 (1965).
89. McSkimin, H. J. and Andreatch, P. *J. acoust. Soc. Amer.* **34**, 609 (1962).

NUCLEAR MAGNETIC AND ELECTRON SPIN RELAXATION STUDIES

9.1. Introduction

The nuclear magnetic resonance technique made its impact in polymer science, in the late 1950s, as a method for the analysis of composition, configuration, and conformation in macromolecules. Initially observations were performed almost exclusively using the hydrogen nucleus (^{1}H). However, in the last ten years the advent of the Fourier transform technique has made possible studies of nuclei present in small concentrations (either because of natural abundance or chemical composition) such as ^{13}C, ^{19}F, ^{29}Si, and ^{31}P. During the same period electron spin resonance experiments have been reported on polymers with chemically attached spin probes or on radicals generated on macromolecules. Much of the electron spin resonance literature has been concerned with a quantification of the atomic spatial organization around the unpaired electron. However the technique has recently been applied successfully to the study of the dynamics of molecular motion both in polymer solutions and solids.

In a text concerned with molecular motion it is inappropriate to consider in detail n.m.r. and e.s.r. analyses of conformational and sequence structure of macromolecules. However, an understanding of the factors which enable these techniques to provide structural information is necessary before the problems associated with the interpretation of relaxation data can be appreciated. The fine structure in the n.m.r. and e.s.r. spectra reflects not only the complexity of nuclear and electronic interactions but also those factors which control the relaxation behaviour. Fortunately a number of these interactions are resolvable and allow, in certain instances, a unique description of the relaxational motion of particular particles.

In order to model the motion of a particular particle we shall consider first the nature of the resonance process being observed. For convenience the following discussion will refer specifically to nuclear magnetic resonance, although a similar formalism can be obtained for electron spin resonance.

9.2. Concepts of nuclear magnetic resonance[1-7]

Application of a strong magnetic field to a system of nuclear spins leads to a loss in degeneracy of the energy of the spin states.

Transitions between the lower and upper energy levels can be induced by the application of a radio frequency field with an energy equal to the gap between spin states. The radio and applied magnetic fields are arranged to be perpendicular (minimizing mutual interaction). Transitions between spin states produce absorption of the electromagnetic radiation and a dispersion in the magnetic field (Fig. 9.1). The precise value of the energy gap between spin states is a function of the nuclei both chemically bonded to and spatially close to the resonating species, and of the form of the Fermi surfaces involved. A discussion of the factors influencing chemical shift and spin–spin coupling is to be found in a series of excellent reviews.[1-6] In the context of molecular motion in polymers it is important to realize that if a nucleus is strongly coupled to its neighbours, its relaxation, too, will reflect the nature of the coupling interactions.

9.2.1. Bloch equations

The resultant magnetic moment $\mathbf{M}$, of an assembly of nuclei (Fig. 9.1) precesses with a frequency ω_0 about an axis chosen such that the z-axis corresponds to the direction of the applied magnetic field $\mathbf{H}_0$. In the absence of relaxation effects, the projection of $\mathbf{M}$ on the z-axis, $\mathbf{M}_z$, will remain constant

$$d\mathbf{M}_z/dt = 0. \tag{9.1}$$

The magnitudes of the projections along the x- and y-axes will not remain constant but will vary with time. Further, when the projection along the x-axis, $\mathbf{M}_x$, is a maximum, that along the y-axis, $\mathbf{M}_y$, will be minimum and equal to zero, and vice versa. The projections will therefore be 180° out of phase. This time-dependent behaviour may be described by

$$d\mathbf{M}_x/dt = \omega_0\mathbf{M}_y = \gamma\mathbf{M}_y\mathbf{H}_0 \tag{9.2}$$

$$d\mathbf{M}_y/dt = -\omega_0\mathbf{M}_x = -\gamma\mathbf{M}_x\mathbf{H}_0 \tag{9.3}$$

where γ is the gyromagnetic ratio. In the nuclear magnetic resonance experiment a second magnetic field is applied at right angles to the main field and is usually several orders of magnitude smaller than the main field. The vector $\mathbf{H}_1$, rotates in the xy-plane with a frequency ω. If this frequency equals that of the Larmor precession frequency for the nucleus, ω_0, a transition can occur. The effect of $\mathbf{H}_1$ on the static magnetic field may be expressed as:

$$d\mathbf{M}_x/dt = \gamma\{\mathbf{M}_y\mathbf{H}_0 - \mathbf{M}_z(\mathbf{H}_1)_y\} \tag{9.4}$$

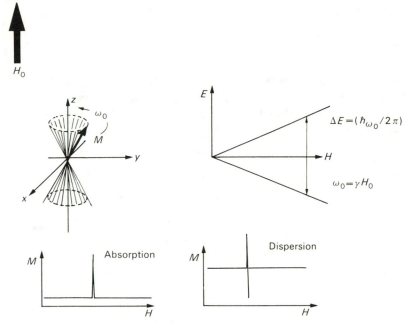

Fig. 9.1. Schematic representation of the nuclear magnetic resonance technique. Nuclei precess with a frequency equal to the Larmor frequency. Resonance occurs when the energy and field conditions are fulfilled.

$$dM_y/dt = \gamma\{M_x H_0 + M_z (H_1)_x\} \tag{9.5}$$

$$dM_z/dt = \gamma\{M_x (H_1)_y + M_y (H_1)_x\} \tag{9.6}$$

where $(H_1)_x$ and $(H_1)_y$ are the components of H_1 along the x- and y-axes, and are given by

$$(H_1)_x = H_1 \cos \omega t \tag{9.7}$$

$$(H_1)_y = H_1 \sin \omega t. \tag{9.8}$$

In the above equations no allowance has been made for the relaxation of the magnetic field components in the x-, y-, and z-directions. Without a relaxation mechanism it would be impossible to observe a spectrum. Phenomenologically the relaxation of the component in the z-direction may be expressed by

$$dM_z/dt = (M_z - M_0)/T_1 \tag{9.9}$$

where T_1 is referred to as the longitudinal relaxation time, M_0 is the equilibrium value of the magnetization. Similarly, the relaxation in the z- and y-directions may be described by a transverse relaxation time, T_2, using the equations

$$dM_x/dt = -M_x/T_2 \tag{9.10}$$

$$dM_y/dt = -M_y/T_2. \tag{9.11}$$

These relaxation processes differ in that the former (T_1) describes the decay of the magnetic field to an equilibrium value, the latter (T_2) the decay of the field to zero. Combination of the steady-state equations and the above yields the so called Bloch equations

$$dM_x/dt = M_y H_0 - M_z H_1 \sin \omega t] - M_x/T_2 \tag{9.12}$$

$$dM_y/dt = [M_x H_0 - M_z H_1 \cos \omega t] - M_y/T_2 \tag{9.13}$$

$$dM_z/dt = [M_x H_1 - \sin \omega t - M_y H_1 \omega \cos \omega t] - (M_z - M_0)/T_1. \tag{9.14}$$

In order to appreciate the significance of the relaxation terms in the above equations it is usual to change the frame of reference from a fixed set of axes, x, y, and z, to a set imagined to rotate with the applied field H_1. In the rotating frame, both H_0 and H_1 are fixed. It is usual to resolve the projection of M on the xy-plane into components u and v along the perpendicular to H_1 which are accordingly in-phase and out-of-phase with H_1. In transforming the original frame of reference to this new frame, the following expressions are obtained

$$M_x = u \cos \omega t - v \sin \omega t \tag{9.15}$$

$$M_y = -(u \sin \omega t + v \cos \omega t). \tag{9.16}$$

Using these relationships and $\gamma H_0 = \omega$: the above equations become equal to

$$du/dt + u/T_2 + (\omega_0 - \omega)v = 0 \tag{9.17}$$

$$dv/dt + v/T_2 - (\omega_0 - \omega)u + \gamma H_1 M_z = 0 \tag{9.18}$$

$$dM_z/dt + (M_z - M_0)T_1 - \gamma H_1 v = 0. \tag{9.19}$$

The term $(\omega_0 - \omega)$ is a measure of how far the rotating frame is from the resonance condition. Inspection of the above equation indicates that changes in M_z are associated only with v, the out-of-phase component of the macroscopic magnetic moment, and not with u. The absorption spectrum is therefore associated with the measurement of v and the dispersion with u.

The above equations can be approximated to give relationships between line widths and the relaxation times T_1 and T_2.[4,5] The reader is referred to the standard texts for the extension of this analysis to the equilibrium observations of the absorption and dispersion spectrum.[1-4] The above equations indicate that independent measurements of T_1 and T_2 are possible by the correct selection of observation conditions. Although the relaxation times T_1 and T_2 have been introduced in a purely phenomenological manner they also have very important mechanistic implications.

9.2.2. Spin-lattice relaxation time (T_1)

Contributions to T_1 arise from rapid fluctuations of the magnetic field at the nucleus due to the relative motion of the molecules, in which the nucleus is embedded, within the local field due to the other molecules. From studies of pure liquids it is clear that fluctuations in the local magnetic field have correlation times which are of the order of 10^{-11} s and hence are well removed from the Larmor precession time, which is typically of the order of 10^{-7} to 10^{-8} s. As a result the coupling between the local fluctuations and the precessing nuclear magnetic moment is very weak and the transfer of energy is very inefficient. Typically T_1 has a value ranging from 10^{-1} s for a diamagnetic liquid to 10 s for protons in organic liquids contaminated with dissolved paramagnetic oxygen. In the case of polymers the value of T_1 can vary between 10^{-2} and 10 s, depending on the nature of the nuclei involved and the form of the distribution of relaxation times at the Larmor frequency. It may be noted that segmental and local motions of a macro-molecular backbone often possess a significant fluctuation density at the Larmor frequency and hence constitute a dominant relaxation mechanism.

9.2.3. Spin–spin or spin–phase memory relaxation time (T_2)

The nuclei, before application of the radio frequency field, will possess random phases relative to their neighbours with similar energy. The radio frequency field $\mathbf{H}_1$ produces a phase coherence of the projected nuclear moments in the xy-plane and results in a signal being detected in the sensor coils. This phase coherence will decay with a characteristic time T_2, which is a measure of the strength of the nuclear coupling interactions. The dephasing is achieved by the mutual interaction of two spins. Pictorially, T_2 involves spin flips promoted by electron coupling with other nuclei, which are themselves undergoing relaxation. Chemical processes and rotational isomeric changes have been studied by this technique and the analysis used is discussed in detail elsewhere.[3,4]

In the case of polymer systems both T_1 and T_2 can provide information on the dynamics of the polymer chain. For precise studies of the n.m.r. relaxation times, pulse radiation rather than the conventional continuous-wave radio signal is usually employed.

9.3. Pulsed experiments

The experiments fall neatly into three classes—Carr–Purcell, rotating frame, and Fourier transform experiments. The choice of which experiment to use for the study of a particular system depends on the nucleus, its interaction with neighbouring spin systems, and the time scale of the molecular perturbation.

9.3.1. Carr–Purcell experiments

Pulsed experiments are usually described in terms of the rotations that a particular pulse sequence induces on the magnetic vector. In the typical experiment a pulse of duration t_ϕ rotates the magnetization vector through an angle ϕ. The first pulse is usually chosen so that the magnetization vector, initially in the z-direction, is rotated into the $-z$ direction. This is termed a π pulse, since it effectively rotates the magnetization through 180°. A second pulse of duration $\pi/2$ is used to measure the magnetization which remains aligned along the z-direction at a time t_p later. The steps in this process can be described (Fig. 9.2) in terms of the magnetization vector after each pulse,

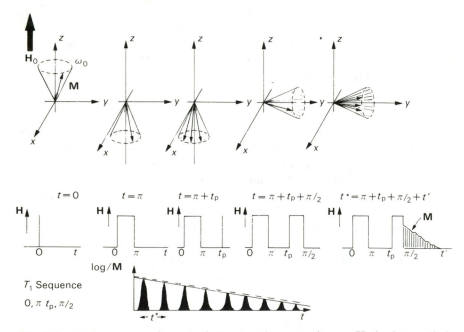

Fig. 9.2. Pulse sequence for a typical relaxation experiment. H_0 is the applied magnetic field, M is the resultant magnetic polarization.

and represented mathematically by a simple exponential decay with time constant T_1

$$M(t_p) = M_0(1 - 2\exp(-t_p/T_1)) \tag{9.20}$$

where $M(t_p)$ is the magnetization after a time t_p and M_0 is the initial equilibrium magnetization. It is important to note that this technique will give a mean relaxation time for all the nuclei in resonance. For experimental reasons this approach is unable to distinguish between the relaxation of strongly coupled or closely spaced resonances.

9.3.2. Pulsed experiments in the rotating frame

Of the possible pulse sequences which can be applied to the system one requires special mention, the 'rotating frame' sequence. As in the previous experiment, an intense radio frequency pulse is applied to the system. However it is of duration such that the vector is rotated through $\pi/2$. When this pulse is switched off a second pulse is introduced into H_1 which is phase-shifted by $\pi/2$ relative to the initial pulse. The second pulse is applied for a period t_p and when it is turned off it is followed by an induction decay signal. The first pulse has the effect of rotating M_0 into the xy-plane; the phase-shifted pulse makes H_1 and M_x co-linear (places the vector in the laboratory frame); H_1 remains constant during the phase shift. The magnetization M_x decays along H_1 in a similar manner to the relaxation of M_0 along H_0. The decay of M_x with time is often exponential and has a similar form to eqn (9.20)

$$M(t_p) = M_x \exp(-t_p/T_{1_\rho}).   \tag{9.21}$$

The two relaxation times T_1 and T_{1_ρ} are, respectively, measures of the rate of loss of memory of the magnetization along the H_0 and H_1 fields. The T_{1_ρ} measurements have been found to be useful in resolving overlapping relaxation processes and correspond to a different part of the relaxation spectrum from that observed with T_1.

9.3.3. Fourier transform experiments

The free induction decay which follows the switching of the applied radio frequency field contains all the information usually found in the absorption spectrum. The free induction decay observed at a frequency ω_1 is modulated by all frequencies different from this, so that constructive and destructive interference result. Fourier transformation of this decay pattern provides an absorption spectrum identical to that obtained from conventional spectrometers.[7,8] Use of multiple pulse sequences similar to those described for the Carr–Purcell experiment enables the spectrum to be observed as a function of time. The Fourier transform technique has been extensively used to study ^{13}C in its natural abundance, (ca. 0·5 per cent) or other nuclei of low concentrations when normal continuous-wave techniques lose sensitivity due to saturation effects.[9]

9.4. Theoretical consideration of n.m.r. relaxation in polymers

The Bloch approach outlined above describes the relationships between the macroscopic relaxation times and the observed magnetic fluxes, but does not provide a detailed understanding of the manner in which molecular motions produce the perturbing flux density.

The motion of the polymer will give rise not only to a change in spatial arrangement, but also to a fluctuating magnetic environment capable of promoting relaxation of the nuclear spin states. Molecular correlation times are obtained from n.m.r. and e.s.r. experiments by assuming the form of the magnetic interaction and deriving sets of equations relating the macroscopic fluxes to the detailed spatial arrangement. In this way models have been derived for liquids and for both isotropic and orientated mobile solids.[10-22] No attempt will be made to derive these equations as details are described adequately elsewhere.[4-9] Here we follow closely the treatment of McBrierty.[8]

Relaxation in polymers may be described by three equations[8,18]

$$1/T_1 = 3\gamma^4\hbar^2 I(I+1)/2(\tau_c/(1+\omega_0^2\tau_c^2)\sum_k \langle|f_{k1}|^2\rangle$$

$$+2\tau_c/(1+4\omega_0^2\tau_c^2)\sum_k \langle|f_{k2}|^2\rangle \tag{9.22}$$

$$1/T_{1_\rho} = 3\gamma^4\hbar^2 I(I+1)/8\left\{2\tau_c/(1+4\omega_e^2\tau_c^2)\sum_k \langle|f_{k0}|^2\rangle\right\} \tag{9.23}$$

$$1/T_2^2 = 2\gamma^4\hbar^2 I(I+1)/8\left\{\sum_k \langle|f'_{k0}|^2\rangle\right\} = \gamma^2 M_z/2 \tag{9.24}$$

where γ, $\hbar$, and I have their usual meanings, ω_0 and ω_e are respectively the resonance frequency and the precession frequency of the rotating frame, τ_c is the correlation time, $1/2\pi\nu_c$, which defines the exponential decay of the correlation function for the molecular motion. The positions of the nuclei are defined by

$$\langle|f'_{k0}|^2\rangle = 16\pi/5\langle[\![Y_2, (\Theta_k, \Phi_k)/r_k^3]\!]^2\rangle \tag{9.25}$$

$$\langle|f_{kn}|^2\rangle = C_n\{\langle[\![Y_{2,n}(\Theta_k, \Phi_k)/r_k^3]\!]_{lt}^2\rangle - \langle[\![Y_{2,n}(\Theta_k, \Phi_k)/r_k^3]\!]_{ht}^2\rangle\} \tag{9.26}$$

with subscripts $n = 0, 1, 2$, and $C_0 = 16\pi/5$, $C_1 = 8\pi/15$, $C_2 = 32\pi/15$. Θ_k and Φ_k are the polar angles of the internuclear vector, r_k relative to the laboratory field direction. The internuclear vector extends from a reference nucleus and the sum over k takes all other nuclei into account. The bracket $[\![-]\!]$ denotes the motional average over those molecular motions which contribute to relaxation and the subscripts lt and ht indicate averages over motions operative at temperature below and above the process of interest.

The equations for a fluid are essentially the same except that this latter distinction is dropped. T_2 derives from the static part of the local field at the reference nucleus ($T_2 \ll T_1$ for solid polymers) whereas T_1 and T_{1_ρ} depend on the time-dependent fluctuations of the local field.

Molecular motions only become apparent when $\tau_c \geqslant T_2$. Eqn (9.24) is analogous to the Van Vleck expression for the second moment.[13]

The expressions for T_1 and T_{1_ρ} are based upon the Bloembergen–Purcell–Pound (BPP) treatment of relaxation[10] for which $\tau_c \ll T_{2RL}$. T_{2RL} denotes the rigid lattice value of T_2. Eqn (9.22) is valid for T_1 if $\mathbf{H}_1^2 \gg \mathbf{H}_L^2$,[11] where $\mathbf{H}_L^2$ is the square of the local field and equals one-third of the second moment,[23] and $\mathbf{H}_1$ is the amplitude of the r.f. field.

At low temperatures the conditions $\tau_c > T_2$ and $\mathbf{H}_1 \sim \mathbf{H}_L$, in which case the BPP analysis no longer applies. This is the Slichter–Allison region[12] for which

$$1/T_{1_\rho} = 2(1-p)/\tau_c\{\mathbf{H}_L^2/\mathbf{H}_1^2 + \mathbf{H}_L^2)\} \sim 1/\tau_c. \qquad (9.27)$$

If $H_1 \sim H_L$ the expression for T_1 is:

$$1/T_1 = 2(1-p)/\tau_c\{3\mathbf{H}_L^2/4\mathbf{H}_1^2\}. \qquad (9.28)$$

The parameter p is normally set equal to zero.

Minimum values of T_1 and T_{1_ρ}, for which $\omega_e\tau_c$ and $\omega_0\tau_c \sim 0.5$, may be expressed in terms of T_2 by the approximate formulae[21]

$$T_{1_{min}} = (2)^{\frac{1}{2}}\pi\nu_0 T_{2LT}^2 \qquad (9.29)$$

$$T_{1\rho_{min}} = 4\mathbf{H}_1 T_{2LT}^2 \qquad (9.30)$$

where

$$1/T_{2LT}^2 = 1/T_{2lt}^2 - 1/T_{2ht}^2. \qquad (9.31)$$

The subscripts lt and ht refer respectively to the low- and high-temperature extremes of the relaxation, $T_{2LT} \sim T_{2lt}$ if the change in T_2 across the transition is large.

It is easy to see that T_2 values can be estimated from the corresponding T_1 and T_{1_ρ} data (minima) or, alternatively, these relaxation times may be computed directly from the appropriate equations.

The various expressions required to translate n.m.r. data into the corresponding correlation frequencies are listed in Table 9.1. The ranges of ν_c and τ_c monitored by the various measurements are also included, and indicate the potential of the technique for the study of local motions in polymer systems.

Table 9.1

Correlation frequency relationships and the ranges of ν_c and τ_c appropriate to each measurement

N.m.r. observation	Expression for ν_c	Range of ν_c/Hz	Range of τ_c/s
T_1 minimum	$\nu_c = (2)^{\frac{1}{2}}\nu_0$	$4 \times 10^6 - 4 \times 10^8$	$4 \times 10^{-8} - 4 \times 10^{-10}$
$T_{1\rho}$ minimum	$\nu_c = \gamma \mathbf{H}_1/2\pi$	$10^4 - 5 \times 10^5$	$2 \times 10^{-5} - 3 \times 10^{-7}$
S–A region of T_1	$\nu_c = 1/2\pi T_{1\rho}$	$10 - 10^3$	$2 \times 10^{-2} - 2 \times 10^{-4}$
T_2 transition	$\nu = 1/2T_{2LT}$	$10^4 - 10^5$	$2 \times 10^{-5} - 2 \times 10^{-6}$
Linewidth transitions	$\nu = \gamma \delta \mathbf{H}_{LT}/2$	$10^4 - 10^5$	$2 \times 10^{-5} - 2 \times 10^{-6}$

In the study of polymers there are two additional criteria which require consideration at this point: these are spin diffusion and chain orientation.

9.4.1. Spin diffusion

Since $T_1 \gg T_2$ in many polymers, excess energy can remain in the spin system for a time long compared with T_2 before being transferred to the lattice.[24] However the energy transfer is particularly efficient near paramagnetic impurities, lattice imperfections, or near molecular segments which are undergoing rapid motion.[4,10,25,26] Even minute concentrations of such relaxation sites can produce noticeable effects on the over-all relaxation behaviour.[26,27] The term 'spin-diffusion' relates to the transfer of excess spin energy to these preferred relaxation sites. The coefficient of spin diffusion is typically of the order of 10^{-14} m^2 s^{-1}.

For the case of one-dimensional diffusion, two situations can be identified:

$$\text{Fast diffusion limit} \qquad T_1 \propto L . T_1^r \propto X_m^{-1} T_1^r \qquad (9.32)$$

$$\text{Slow diffusion limit} \qquad T_1 \propto L^2/\pi^2 D \qquad (9.33)$$

where L is the length of the molecular chain and T_1^r is the characteristic relaxation time of the active site. X_m is the fraction of sites in the whole polymer. Analogous expressions can be written down for $T_{1\rho}$. In a linear polymer the active sites would be end groups. In the fast diffusion limit, the end groups act as a bottle-neck to the transfer of excess spin energy into thermal motions of the lattice. In the slow diffusion limit, the diffusion rate is slow compared with the rate of relaxation of the end groups.[28–31]

9.4.2. Orientation with respect to a magnetic axis

In bulk polymer the angular terms which define the orientation of the internuclear vectors with respect to $\mathbf{H}_0$, in the relaxation expressions (9.21)–(9.23), are averaged uniformly over all space. The n.m.r. signal is thus independent of the orientation of the sample. However if a single crystal, or a sample with aligned chains is observed, the n.m.r. response becomes a sensitive function of the relative orientation of the chain axis and the applied field.[15-20,22-45] The way in which anisotropy in the second moment of the spectrum, M_2, depends on angle provides data from which molecular orientation within the polymer can be deduced (Fig. 9.3).

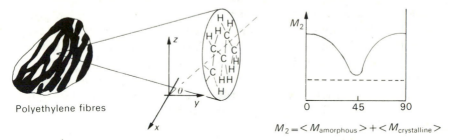

Fig. 9.3. Variation of the second moment for an orientated polyethylene mat. θ is the angle between the direction of orientation and the applied magnetic field.

In the study of orientational distribution in polymers, measurements are made at low temperatures to avoid the complications associated with the presence of molecular motion. The observed moment may be expressed as

$$M = x\langle M_{\mathrm{cryst}}\rangle + (1 - x)\langle M_{\mathrm{amorphous}}\rangle \tag{9.34}$$

where x is the crystalline mass fraction.

The orientation of a typical structural unit (with reference to a macroscopic set of axes) requires three Euler angles to be specified. The distribution of units can be described mathematically by $\mathscr{P}(\alpha_1, \Delta, \alpha_2)$, the functional forms of which may be expressed as a series of generalized spherical harmonics.[19,20,45]

$$\mathscr{P}(\alpha_1, \Delta, \alpha_2) = \sum_{l=0}^{\infty} \sum_{n,m=l}^{+l} P_{lmn}\mathscr{D}_m(l)\, n(\alpha_1, \Delta, \alpha_2). \tag{9.35}$$

The coefficients of P_{lmn} in the summation refer to the 'moments' of the distribution. The average of any function F, denoted by angled brackets, has the form

$$\langle F\rangle = \int_0^{2\pi} \int_0^{\pi} \int_0^{2\pi} \mathscr{P}(\alpha_1, \Delta, \alpha_2)F\, d\alpha_1 \sin\Delta\, d\Delta\, d\alpha_2. \tag{9.36}$$

A general expression can be derived for the crystalline contributions to M_2 in terms of the distribution moments P_{lmn}

$$\langle M_2\rangle_{\mathrm{cryst}} = \sum_{l=0,2,4}^{+l} \sum_{m,n=-1}^{+l} D_l P_{lmn} Y_{l,m}^* (\gamma, \phi_\gamma) S_{ln}) \tag{9.37}$$

where coefficients $D_l = 192\gamma^2\hbar^2 I(I+1)\pi^3 a_l(2l+1)^2$ in which $a_0 = \frac{1}{5}$, $a_2 = \frac{2}{7}$, and $a_4 = \frac{13}{35}$. S_{ln} are lattice sums defined as

$$S_{ln} = \frac{1}{N}\sum_k Y_{l,n}(\theta_k\phi_k)/r_k^6. \qquad (9.38)$$

The lattice sums are computed in pairs to avoid the complex terms. θ_k, ϕ_k, and γ, ϕ are respectively the polar and azimuthal angles of $\mathbf{r}_k$ with respect to the symmetry axis of the structural unit and of $\mathbf{H}_0$ with respect to the axis of the polymer.[46] The observed variation of the second moment is a function of the nature of the nuclei involved, lattice sums appropriate to a particular polymer, the relative direction of $\mathbf{H}_0$, and the even moments of the distribution up to order 4. The application of this approach to the study of drawn polymers will be described later in this chapter.

9.5. Resumé of the scope of n.m.r. in structural analysis

Although this text is concerned principally with the dynamics of polymer motion, the assignment of group relaxations requires prior understanding of the resonance spectrum. It is important to appreciate the way in which the technique is used in the study of macromolecular structures.

The ^{1}H spectra of polymers are often complicated by the complex array of interactions and distributions of neighbouring nuclei. Two factors may be present to complicate assignment of the resonances to particular structural units: the first is the presence of both geminal and vicinal coupling constants and secondly the polymer itself contains a distribution of structural units. In polymers such as poly(methylmethacrylate) the problem is simplified by the presence of the methyl group in the backbone. This removes one of the coupling parameters.[47–54]

In certain situations it is possible to remove the coupling between nuclei artificially by use of decoupling techniques. The experiment involves irradiating the sample with a very strong r.f. signal at the resonance of the nucleus to be decoupled and then observing the resulting simplified spectrum.

A totally new dimension in n.m.r studies appeared with the advent of Fourier transform techniques.[55,56] These have been applied to the study of sequence structure in polyethylene[57] and block copolymers.[58]

9.6. N.m.r. analysis of molecular motion in specific polymer systems

In the previous sections the theoretical relationships connecting the bulk magnetic response and the motion of individual nuclear magnetic moments have been briefly outlined. Since coupling can exist over relatively large distances in protonated materials it is useful to discuss the relaxation behaviour in terms of the degree of complexity of the molecular motions possibly with a particular polymer type. The simplest structure will obviously be the alkane chain, since this does not contain mobile-side chains which can (through spin diffusion) significantly modify the relaxation of the main chain. Polymers with side chains can be subdivided according to the degree of motion in the side chain. This approach will be adopted in this section.

9.6.1. Polymers without pendant groups—normal alkane chains

(a) *Molecular weight effects.* The spectra of the normal alkanes increase in complexity as the molecular weight is increased.[59-63] A number of the features detected in the lower molecular-weight polyethylenes are attributable to small concentrations of vinyl, vinylene, vinylidene, and methyl groups. Studies in chloronaphthalene and other aromatic solvents of low molecular weight have indicated that the methylene line undergoes splitting not observed in non-aromatic solvents. This splitting has been attributed to chain folding in solution and appears to indicate that short chains (C–6 to C–28) may not be completely random in aromatic solvents. Relaxation studies of the lower alkanes C–5 to C–14 in carbon tetrachloride indicate that T_1 varies only slightly with concentration. An increase in T_1 with concentration is observed with pentane, and the opposite effect is observed in hexadecane.[64] These observations have been used to suggest that n-pentane relaxes by rapid over-all rotation, while hexadecane behaves more like a polymer. T_1 then reflects the rate of intramolecular isomeric change. The reader will recall that a similar trend has been discussed in Chapter 5 in explaining the molecular-weight variation of the dielectric relaxation.

(b) *Relaxation in solid hydrocarbons.* The complexity of the relaxation behaviour of the stereochemically 'simple' hydrocarbons indicates the problems to be faced with stereochemically more complex molecules. Polyethylenes, as indicated above, do not conform to the ideal $(-CH_2-)_n$ structure and it is the existence of a small number of defects in the ideal structure which lead to very significant differences in relaxation behaviour.[65] Chain branching has two effects: first, it introduces into the system a number of relatively mobile end-groups which can act as sites for spin diffusion, and, secondly, it influences the degree of crystallinity of the polymer. The ^1H resonance in partially crystalline polyethylene at 333 K exhibits a narrow Lorentzian line which is split above the polymer crystallite melting point.[66] Dilute solutions of the polymer in tetrachloroethane give a single line at 393 K with a line width which is one-fifth of that of the narrow component in the solid polymer.

It appears that four possible contributions to the relaxation behaviour from different structural entities can be identified, not all of which may be present in every sample.[67-9] In order of decreasing linewidth these are crystalline phase or 'tight' tie molecules (α), grain boundaries or relaxed tie molecules (β_1), loose loops (β_2), and chain ends or branches (γ). In short-chain hydrocarbons up to C_{30} the relaxation is dominated by rigid chains (95 per cent). Chains of intermediate mobility and mobile end-groups make a very small contribution (4 and 1 per cent respectively) to the relaxation behaviour.[68] These materials are known from X-ray measurements to exhibit highly ordered structures and may indeed behave as almost ideal single crystals.

In solid polyethylene, spin diffusion has a dominant effect on the relaxation. Branched polymers containing a few per cent of methyl groups exhibit a well-defined low-temperature T_1 minimum with no corresponding T_2 transition. The small number of rotating methyl groups are able to relax the rigid protons by spin diffusion. In linear polyethylene, movement in the amorphous region produces a T_1 minimum at 253 K, with an activation energy of 8 kJ mol^{-1}.[69] The spins in the crystalline regions of the polymer are coupled (by spin diffusion) to the relaxed spins of the amorphous regions.[70-3] The magnitude of T_1 is given by the equation

$$1/T_{1_{\min}} = (1-X) T_1 \tag{9.39}$$

where X is the crystallinity and T_1 the relaxation time of the amorphous domain.[68]

The angular dependence of the second moment (M_2) spectra of polyethylene supports the suggestion that the over-all relaxation is composed of a number of components. Theoretical calculations of second moments have been attempted using a variety of models,[16,75,77] all of which use a common basis of three components (rigid, hindered,[76,77] and mobile or liquid-like) attributed to the different phases present. Using the formalism of § 9.5, it is possible to show that in a polymer with certain symmetry properties eqns (9.35) and (9.37) simplify considerably.[74] For example, if S is defined as the draw axis and there is a fibre symmetry, then $\mathscr{P}$ becomes independent of the angle $\alpha_1 (m = 0)$. If, in addition, the structural unit is transversely isotropic, that is, rotation of the unit about its symmetry axis leaves the value of $\langle M_2 \rangle_{\text{cryst}}$ unchanged, then the dependence on the angle α_2 also vanishes ($n = 0$). Eqns (9.35) and (9.37) then become[22,75]

$$\mathscr{P}(\Delta) = \sum_{l=0}^{\infty} (l + \tfrac{1}{2}) \overline{P_l (\cos \Delta)} \, P_l (\cos \Delta) \tag{9.40}$$

and

$$\langle M_2 \rangle_{\text{cryst}} = \sum_{l=0,2,4} D_l \overline{P_l (\cos \Delta)} \, P_l (\cos \gamma) \, S_l$$

$$= \sum_{l=0,2,4} C'(\gamma, l) \, \overline{P_l (\cos \Delta)}. \tag{9.41}$$

The lattice sums in this expression, S_l, have the form

$$1/N \sum_k P_l (\cos \theta_k)/r_k^6.$$

The moments P_{l00} have been expressed in their explicit form in terms of Legendre polynomials, $P_l(\cos \Delta)$. The coefficient $D_l' = 6\gamma^2 \hbar^2 I(I+1)a_l$.

There is a fourth moment analogue of (9.41)

$$\langle M_4 \rangle_{\text{cryst}} = \sum_{0,2,4,8} C(\gamma, l) \overline{P_l (\cos \Delta)} + \langle M_4 \rangle_{\text{cryst}}. \tag{9.42}$$

In general, when a material has crystallographic or statistical symmetry, not all the lattice sums are mutually independent, and the problem becomes experimentally resolvable. For example, in a fibre material consisting of orthorhombic crystal units, there are six non-zero independent P_{lmn} coefficients, namely P_{000}, P_{200}, P_{202}, P_{400}, P_{402}, and P_{404}.

Studies of the temperature and angular dependence of the second moments of the spectra in polyethylene in drawn and undrawn samples support the hypothesis of various mechanisms contributing to the over-all relaxation spectrum.[71,77] The orientational information on the drawn samples is in good agreement with that obtained from X-ray and mechanical studies. Recently the plasticization of the motion of the rigid phase by carbon tetrachloride has been demonstrated by the appearance of line splitting in the spectrum.

(c) *Poly(ethyleneoxide)*: *solution phase*. The presence of oxygen in the backbone partly decouples the long-range proton interactions. High-resolution studies in the pure liquid, and solutions of oligomers (trimer to octamer) indicate that in the liquid terminal 'alcoholic' CH_2 protons can be distinguished from the internal ether-like CH_2 protons.[78-83] Dilution with benzene or chloronaphthalene produces additional complexities attributed to positional isomerism of the CH_2 protons in the heptamer and hexamer but not in the pentamer. No such splitting is observed in chloroform or aqueous solutions. Comparison of the spectra from the oligomers with those of the higher polymers has led to the suggestion that at least in some solvents, there is a change at $n = 6$ to a conformation closely resembling the 7_2 helical structure identified from X-ray and infrared studies of the solid state. This structure requires a preference for *gauche* over the *trans* conformation, despite the higher coulombic repulsion of the former.[84-7]

Studies of T_1 and T_2 over a wide range of concentrations,[88] at several temperatures, for a variety of molecular weights and in solvents of varying viscosity and chemical structure support the existence of solvent-induced conformational changes in this polymer. The spin lattice relaxation arises almost entirely from intramolecular interactions, provided that the molecular weight is high and the concentration does not exceed 20 per cent. With the exception of oligomers in water, the magnetic dipole interaction between solvent and polymer appears to have negligible influence on the relaxation process. This observation is attributed to the fact that solvent molecules move much more rapidly than the polymer segments and make a negligible contribution to the density of states at the Larmor frequency. Support for the coil geometry hypothesis is obtained from studies of the variation of T_1 with concentration. If polymer–polymer magnetic dipole interactions were to originate from protons on different molecules then one would predict a rapid change in T_1 with concentration of polymer. This is not observed in practice, and suggests that the molecules occur in a coiled form. Further support for this hypothesis is obtained from the observation that the value of T_1 is insensitive to molecular-weight change once the degree of polymerization is greater than 30. The relaxation is assumed to occur via a mechanism which involves intramolecular segmental motion rather than a rotational diffusion of the entire molecule.

In highly dilute solutions T_2 is molecular-weight-independent. However as the concentration is increased T_2 decreases. The difference between the behaviour of T_1 and T_2 is attributed to the presence in the system of slow motions which contribute to $|f'_{ko}|^2$ but not to $|f_{kn}|^2$. Physically these motions are considered to arise from 'molecular entanglements', which describe the restriction of the motion of one polymer molecule by the interlacing of it with a different molecule.

The ^{13}C relaxation of polyethylene melts indicates that T_1 decreases with increasing molecular weight but rapidly attains a limiting value at a molecular weight of about 7000.[89] It is interesting to note that the apparent activation energy is independent of molecular weight.

(d) *Poly(ethyleneoxide)*: *relaxation in the solid state*. Just as in the case of polyethylene, the relaxation behaviour of poly(ethyleneoxide) is complicated by the partial crystallinity of the material. The crystallinity of poly(ethyleneoxide) has been studied by X-ray diffraction as a function of molecular weight[90] and appears to go through a maximum at 10^4. The lowest molecular weight (M-550) samples show fairly simple behaviour (Fig. 9.4) exhibiting two minima in T_1 and T_{1_ρ} measurements. The magnetization appears to decay exponentially over the whole temperature range and has an apparent activation energy of 74 kJ mol^{-1}. A lower temperature minimum associated with the motion of the $-OCH_3$ group, which forms one end of the polymer, exhibits an activation energy of 12·8 kJ mol^{-1}. It has been suggested that in this process spin diffusion acts as a contributory relaxation mechanism.[91,92]

Observation of T_1 and T_{1_ρ} for higher molecular-weight samples (*ca.* 6000) as a function of temperature (Fig. 9.5) show a discontinuity at 235 K. This abrupt change is attributed to the onset of rotational motion associated

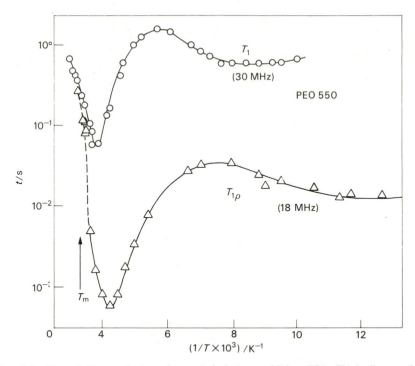

Fig. 9.4. T_1 and T_{1_ρ} variations for poly(ethyleneoxide) -550. T_m indicates the approximate melt temperature.

with the crystallite melting process. Below this temperature the decay is found to be non-exponential and may be attributed to the large molecular-weight distribution in this material. No end-group motion has been observed in this system.

The very highest molecular-weight polymer (2.8×10^6), shows a more complex relaxation behaviour than either of the two other systems. The T_{1_ρ} decays are non-exponential over the whole temperature range studied. There appears to be a discontinuity at the softening point at 336 K. At lower temperatures crystalline and amorphous regions undoubtedly contribute to two-phase behaviour and the observation of associated minima in the temperature dependence of T_1 and T_{1_ρ}. Similar observations have been reported for poly(methyleneoxide).[93–6]

(e) *Poly(tetrafluoroethylene)*. This polymer is an excellent example of one which exhibits specialized motions in the crystalline phase. Features have been recognized which may be attributed to both the amorphous and the crystalline phases.[97–102]

X-ray scattering studies of the crystalline solid indicate that the polymer exhibits both a triclinic and pseudo-hexagonal phase.[98] The change in

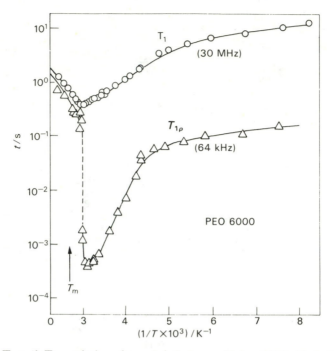

Fig. 9.5. T_1 and $T_{1\rho}$ variations for poly(ethyleneoxide) -6000. T_m indicates the approximate melt temperature.

structure between these two forms occurs at a temperature of 293 K. Studies of the ^{19}F relaxation times indicate the presence of two T_2s over a wide temperature range. The shorter T_2 measurements show a sharp change at 293 K and may be associated with the effects of the structural changes observed by X-ray.[95] A more subtle change in structure of the crystalline region above 308 K has also been identified from T_2 measurements.[102]

Studies of the effects of orientation on PTFE fibres have shown the presence of two components in the relaxation. There is a short decay due to the crystalline regions and a long decay due to the amorphous regions.[101] Changes with temperature of T_2 for chains in the crystalline region are consistent with the onset of rotational motion at 293 K.

9.6.2. Molecules with pendant groups—single substituents

The interpretation of n.m.r. data is difficult in these systems because the introduction of a single substituent may either simplify or complicate coupling between neighbouring nuclei.

(a) *Methyl substitution.* The introduction of a methyl group into the backbone, as in polypropylene, introduces complications both from the point of view of additional nuclear spin couplings and also because it creates sites for spin diffusion. This latter effect produces a high degree of uncertainty in the relaxation studies of the chain motion.[103]

Studies on solutions and melts have been performed and ν_c and E_a for segmental and methyl group motion in polypropylene estimated. Correlation times for the backbone calculated using these data agree well with those obtained from mechanical studies. The principal cause of uncertainty is the magnitude of spin diffusion effects. It has been estimated that methyl groups relax about 50 per cent of the backbone protons by spin diffusion.[104] Whether or not this is correlated with the sequence structure of these polymers is not clear at present.

Poly(propyleneoxide) behaves very similarly to poly(ethyleneoxide). There is a low temperature minimum in T_1. This is more pronounced than for the analogous poly(ethyleneoxide), suggesting that the feature is due to methyl group relaxation.[105-7] The activation energy for methyl group rotation is found to be $16 \cdot 8 \text{ kJ mol}^{-1}$ and this agrees well with a value derived from neutron scattering data. The high temperature minimum has a mean activation energy of 134 kJ mol^{-1} and lies in the region of a correlation diagram which agrees well with that for main backbone motions as observed in dielectric studies.

(b) *Phenyl substitution.* The phenyl group simplifies the coupling between neighbouring protons in the chain backbone and it cannot act as a site for spin diffusion relaxation. Polystyrene is perhaps the most widely studied of all polymers,[107-12] and can be obtained in atactic and predominantly isotactic forms (Fig. 9.6). Both ^1H and ^{13}C studies of solids, melts, and solutions have been reported.

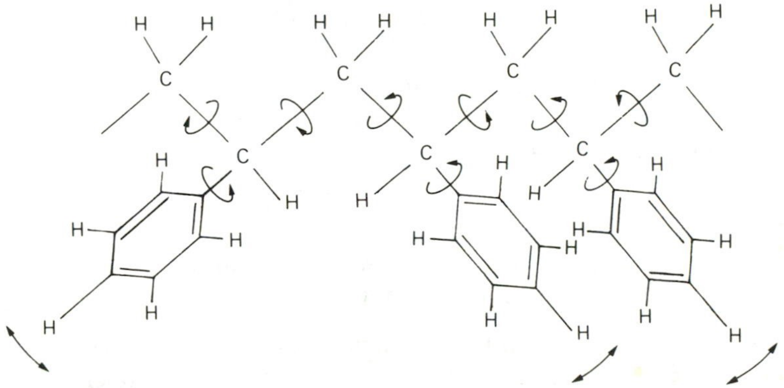

Fig. 9.6. A representation of some of the possible motion which lead to relaxation in polystyrene.

Perhaps the most definitive study of the relaxation behaviour is that of Allerhand and Hailstone[113] using proton-decoupled Fourier transform spectra. The ^{13}C lattice relaxation times (T_1) of protonated carbons in large molecules are relatively simple to interpret, the relaxation mechanism being dominated by the dipolar interaction with the attached protons.[114] Studies of 'atactic' polystyrene in tetrachloroethylene show a molecular-weight independence (above 10^4) of the relaxation.[115,116] This is because the 'effective' rotational correlation time is determined by segmental motion and not by over-all reorientation of the whole molecule. Below molecular weights of 10^4 the relaxation is molecular weight-dependent, reflecting the comparability of segmental and over-all reorientational correlation times.

Molecular weight effects have been reported in the variation with temperature (90–500 K) of T_1 and T_{1_ρ} for ^1H relaxation.[117] The higher-temperature minimum observed in the solid polymer is molecular-weight-dependent and is associated with the glass-transition process, the minimum moving to higher temperatures as the chain length increases. The position of the lower temperature minimum, on the other hand, is independent of temperature, but its intensity decreases as the chain length increases. This relaxation is attributed to motion of an n-butyl group, introduced into the polymer during the initiation of polymerization. Since the minimum is observed in very high molecular-weight materials it seems likely that a spin diffusion mechanism operates in this situation.

T_1 measurements on low molecular-weight samples indicate a minimum at 130 K. This minimum is not as pronounced as that in high molecular-weight polymers, and is ascribed to spin diffusion from backbone protons to rotating terminal methyl groups. A direct relationship between T_1^{-1} and M_n^{-1} has been observed, implying correlation in the motion of the segments of the polymer.[118]

It is apparent that, in the case of polystyrene, although spin diffusion is less dominant than in polypropylene or poly(propyleneoxide), it is still significant enough to influence the relaxation behaviour.

Studies of the isotactic form in toluene over a temperature range 283–383 K suggest that a form of second-order transition associated with loss of a helical conformation occurs at 338 K.[119,120]

9.6.3. Polymers with two or more substituents

The possible simplification in coupling between spin states has been mentioned already, and this is most marked in poly(methylmethacrylate).

(a) *Effects of tacticity.* Although tacticity effects are known to exist in these polymers, the way in which they influence the nuclear spin relaxation is not as obvious as in dynamic mechanical relaxation. T_1 measurements made on poly(methylmethacrylate)[121–3] exhibit minima associated with the motions of the ester methyl, main-chain methyl, and main-chain reorientation at 100, 250, and 490 K, respectively. The whole ester group rotation is not resolvable from the main chain motion.

The T_{1_ρ} data, unlike the T_1 data, show only two minima, although it has been suggested that the origin of this observation is an inability to resolve two of these processes. The relaxation of the ester methyl group (the higher temperature minimum) is connected with the glass transition temperature and syndiotactic and isotactic forms show appropriately spaced minima. The lower temperature minimum is associated with the rotation of the CH_3 group. The activation energies obtained for the syndiotactic polymer are respectively 208 and 21·6 kJ mol^{-1} and those for the isotactic material are 126 and 14·8 kJ mol^{-1}.

Three types of segments of different mobility have been identified from T_2 measurements[124–6] made on 10–15 per cent solutions of syndiotactic and isotactic polymers spinning at the magic angle, a technique which removes dipolar coupling and usually leads to line narrowing (§ 9.8.3).

9.6.4. Use of deutero-substitution

In certain instances it is convenient to simplify the relaxation spectrum by chemically substituting one or more of the protons by deuterium.[127–30] An example of this type of approach is a study of the T_1, T_{1_ρ}, and T_2 relaxations in poly(vinylacetate) and poly(vinyldeuteroacetate).[124] The glass transition for this polymer is in the region of 370 K and T_1 and T_{1_ρ} both exhibit minima in this region, in agreement with dielectric and mechanical observations. Application of an hydrostatic pressure of 680 atm leads to an increase in the temperature of the minima equivalent to 27 K/1000 atm. It was deduced from this observation that the activation volume of the molecular process is 160 cm^3 mol^{-1}, the process having an activation energy of 21 kJ mol^{-1}.

The methyl rotation exhibits a minimum at between 20–60 K. T_1 measurements on the 90 per cent deuterated polymer exhibit a much shallower minimum at very low temperatures than is observed in the protonated material. This is consistent with a quantum mechanical mechanism for relaxation of the methyl group involving tunnelling.

Reduced linewidths and changes in the relaxation mechanism have also been reported for deuterated poly(propylenesulphide) where deuteron decoupling removes complexities due to the diad structure.[126]

9.6.5. Fillers and plasticizers

In polymers such as poly(dimethylsiloxane) the relaxation behaviour is sufficiently simple for the effects of plasticization to be observed in T_1 and T_2 measurements.[131–132] A study of this material plasticized by argon or nitrogen has indicated that the glass transition is lowered in a way similar to that observed with normal solvents.

Carbon-13 studies have been reported on *cis*- and *trans*-polyisoprene, on *cis*-poly (3-^{2}H) isoprene and on carbon black-filled *cis*-polyisoprene.[130–4] The presence of filler broadens the ^{13}C n.m.r. lines by a factor of about 5–10, but has little effect on the T_1 values. This is attributed to restriction of the polymer to certain conformations and to the influence of spin diffusion on the over-all relaxation. Proton T_1 and T_2 measurements on such systems indicate that the bulk of the rubber acts normally[133–5] although a number of the rubber chains are immobilized adjacent to the carbon particles.

9.6.6. *Effect of solvent*

A number of studies have been performed on polymer solutions.[136–9] Activation energies have been obtained for methyl and methylene protons in poly(isobutylene) in a variety of solvents,[137] poly(vinylpyridine) in CDCl$_3$,[128] and polystyrene in various solvents.[116–19]

From a comparison of ^{1}H and ^{13}C n.m.r. relaxation data it has been possible to separate contributions arising from intra- and intermolecular interactions in poly(methylmethacrylate) solutions.[140] It was found that solvent motion has little effect on the polymer relaxation up to a concentration of 20 per cent, despite the large concentration dependence of the bulk viscosity.[141]

9.6.7. *Polymer melts*

Measurements of T_1 and T_2 in fractionated samples (with different molecular weights) of poly(dimethylsiloxane) have been reported.[131] The low density and the large Si—O bond distance means that the most important relaxation mechanism is between protons on the same methyl group.[137] Rotation of the methyl is so fast that it does not provide an effective relaxation mechanism and it is the slower positional changes relative to the main chain which are observed. Approximate calculations of the rotational correlation time of the chain segments, obtained from the non-exponential relaxation decays show that these are roughly proportional to the square root of the viscosity of the liquid.

9.6.8. *Fibres*

It is possible to induce orientation in polymer chains by drawing and hence to obtain an orientation dependence of the second moment. The use of this technique in studies of polyethylene has been discussed already. Poly(vinylchloride),[138] poly(methylmethacrylate),[139] poly(ethylene-terephthalate),[142] and polyoxymethylene[143] have all been studied as a function of draw ratio.

Considerable anisotropy has been reported in nylon[144,145] and the experimental observations have been compared with various models. Over the temperature range 77–473 K measurements on drawn fibres made at 0, 45, and 90° to the fibre axis yield degrees of orientation which agree well with those obtained by X-ray and birefringence analysis. The second moment falls with increasing temperature in two steps, a low-temperature transition at 173 K and two high-temperature transitions beginning at 313 K. A minor contribution to the high-temperature step comes from a transition in the amorphous polymer chain often associated with the glass transition. The major part of the high-temperature transition (α-process) has been ascribed to changes in the crystalline parts of the polymer. The low temperature (γ) process has been assigned to amorphous parts of the polymer. This process is consistent with a model in which non-crystalline chains rotate about axes fixed in space.

The decrease on alignment of the second moment due to the process can be ascribed to orientation of tie molecules. The high-temperature decrease in second moment is consistent with models which involve $-CH_2-$ segmental oscillations and rotational jumps of the crystalline polymer chains.

Similar measurements have been reported on drawn poly(ethylene-terephthalate). The dipolar effect of the protons in CH_2 groups dominates the band shapes and demonstrates that CH_2 segmental motion begins at 173 K. At 398 K an abrupt decrease in the second moment occurs, indicating the onset of rotation of the aromatic ring.

9.6.9. Phase separation

A number of studies have been reported of the effects of phase separation on molecular motion.[146–51] One example is the network copolymers formed from poly(vinyltrichloroacetate) and poly(methylmethacrylate).[148] It is found that T_1 varies with temperature in a similar manner to that observed for poly(methylmethacrylate) except that the relaxation curves are displaced on the temperature axis. Correlation with electron microscopy suggests that the onset of motion of the α-methyl group occurs at different temperatures depending upon whether there is significant microphase separation. In samples with a second phase of μm-size the relaxation was essentially identical to that of poly(methylmethacrylate), whereas if no phase separation occurs relaxations are shifted to lower temperatures.

Similar measurements on styrene–butadiene–styrene copolymers[151] have indicated that the relaxation behaviour resembles that of the component homo polymers. The lowest-temperature feature is associated with the relaxation of polybutadiene and is strongly spin-coupled to the polystyrene domain. The effects of spin diffusion are so strong that the T_1 data are dominated by the butadiene motion and the styrene relaxation is almost insignificant in comparison.

9.7. Models of polymer motion

9.7.1. A comment on the glass transition

The large-scale motion of the polymer backbone is usually identified by studying the shape of the T_1 or T_{1_ρ} dependence on temperature. Extensions of this type of observation on rubbers using pressure as well as temperature[152] indicate that configurational entropy, rather than volume, plays a dominant role in determining the glass transition. Whether this observation is generally applicable, and how it can be reconciled with the free-volume dependence observed by other methods, is open to further experimental investigation.

9.7.2. Segmental motion and whole-molecule tumbling

In the first place it is necessary to calculate how the non-zero average magnetic coupling of the lattice affects the linewidth (both through the magnetic screening of the environment and through the relaxation of magnetization). The relationship used in the calculation of the line shape as a function of concentration and temperature has been justified[152-4] in two ways, one using a theory based on the dynamics of entanglements in polymerized solids[155] and the other using the lattice theory approach.[156]

For dilute polymer solutions the theoretical problem simplifies to one of determining the rates and relative importance of local segmental motions (conformational jumps or torsional oscillations) on the one hand and whole-molecule tumbling on the other, and an attempt has been made to model such motions on the basis of a diamond lattice.[157]

A study of poly(dimethyl- and diphenylphenylene oxides) in solution indicates that the dominant relaxation mechanism is one of torsional oscillation about the ether linkage. However similar studies of polystyrene and monosubstituted polystyrenes[157-9] in solution indicate a considerably more complex range of interactions involved in the relaxation process. These may be separated artificially into whole-molecule tumbling, torsional oscillation of the backbone carbons, rotation of the phenyl group relative to the backbone, and a librational motion of the phenyl group (Fig. 9.7). It appears that around ambient temperature the motions of the side-group and backbone are coupled. However at elevated temperatures the side-group motion is less strongly coupled to the backbone motion and apparently has a lower activation energy.

A related study of alkane copolymers with styrene and α-methylstyrene (Fig. 9.7) shows a similar decoupling of the side-group and backbone motions with increasing size of the alkane block.[160]

Attempts to probe the mechanism of relaxation in polyelectrolytes such as poly(vinylpyridine) and related quaternized polymers have been reported.[161,162] Quaternization markedly influences the dynamics of the chain

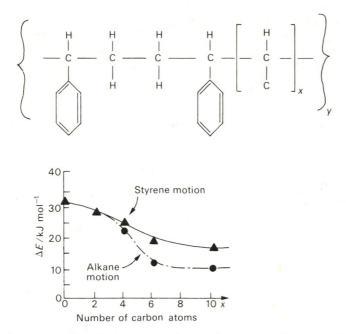

Fig. 9.7. Structure and activation energy changes for styrene–alkane copolymers. x, number of carbon atoms; y, number of blocks—typically 50 units.

and is thought to modify the value of the activation energy to segmental motion. Similarly it appears that the backbone motions are dominant only at high pH. At lower pH-values, exchange and over-all tumbling processes make a major contribution to the relaxation behaviour.[163]

One of the most serious and continuing problems with the interpretation of n.m.r. data is that the usual experiment samples the density of states at only one point. Change in frequency is sometimes possible and this can lead to a modification of the contributions arising from the possible relaxation mechanisms. Such a study[164] has been attempted using poly(ethyleneoxide) in the melt phase. The non-exponential correlation function obtained from this study is ascribed to defect diffusion associated with the process of rotational isomerism. The occurrence of both segmental motion and over-all tumbling generally means, too, that a distribution of correlation times,[165] or a non-exponential correlation function[166] is observed experimentally.

9.7.3. Energetics of conformational change

Studies of the time-dependent distributions of conformations in polymer molecules are usually complicated by the complexity of the spectra and as such have met with little success. A temperature-dependent study of the CH_3—CH_2 chemical shift in n-$C_{44}H_{90}$ indicates that the *gauche–trans* energy difference has a value of approximately $2 \cdot 1 \pm 0 \cdot 2$ kJ mol^{-1}.[167] Similar studies have recently been reported on polypropylene,[168] poly-(isopropylethyleneoxide),[169] poly(methylmethacrylate),[170,171] and poly(ethyleneoxide).[172] In general the complexity of the spectrum leads to considerable uncertainty in the energy parameters.

9.8. High-resolution ^{13}C linewidth studies of solids

9.8.1. General comments

The normal n.m.r. spectrum of a solid is made up of a series of broad features and peak widths of several kHz are not abnormal. Such broad line studies have been used by a number of workers to probe the microscopic motions of polymers. However such spectra are not very useful for structural assignment, nor for studies of the detailed molecular dynamics. Fourier transform spectrometry has introduced the possibility of artificially destroying coupling, and so made it possible to obtain high-resolution spectra of even glassy solids. Early experiments achieved spectacular narrowing and splitting of the resonances by spinning the sample at the 'magic' angle.[173-80]

Narrowing occurs when the frequency of macroscopic rotation Ω_r becomes comparable with the static line width $\delta\omega$. The internal restricted motion of the polymer reduces the rigid lattice second moment by an amount ω_{1a}^2 to a residual value of ω_{1b}^2. The motion responsible for the internal narrowing is characterized by the correlation time $\tau_a(\omega_{1a}\tau_a < 1)$. The limiting value for the narrowing is given by

$$\delta\omega_{min} = \omega_{1a}\tau_a + \left(\frac{\omega_{1b}}{\Omega_r}\right)^2 \tau^{-1}. \qquad (9.43)$$

Other motions which may be present are usually too slow to produce line narrowing, and are characterized by the correlation time, $\tau_b(\omega_{1b}\tau_b < 1)$. The linewidth will be narrowest under macroscopic rotation when τ_a is very short and τ_b is very long. For viscous liquids $\omega_{1b} \sim 0$ and therefore $\delta\omega = \omega_{1a}^2\tau_a$ whether at rest or rotated. Generally speaking there is no further narrowing possible when there is extensive internal isotropic motion. It is therefore not surprising that narrowing cannot be achieved in a polymer which is well above its T_g at room temperature, such as polyisobutene.

The above expression for the limiting line width also applies to multiple-pulse experiments. Here Ω_r is the cycle frequency of the pulses instead of the specimen rotation frequency.

This line narrowing by spinning can affect different peaks to different extents. Thus in partially crystalline polyethylene it is possible to resolve a peak associated with amorphous regions into a central line with two side bands, while leaving a band associated with the crystalline region unaffected.

9.8.2. Decoupling ^1H dipolar broadening

Dipolar broadening of the ^{13}C lines by protons is removed by strong resonant decoupling (referred to as dipolar decoupling) using ^1H decoupling radio frequency fields of about 10 G. These decoupling fields are comparable to the proton linewidths. In most cases the resulting ^{13}C n.m.r. spectra are still severely complicated by overlapping chemical shift anisotropy, so that in general only a few lines are observed. A dramatic improvement can be achieved by fashioning rotors from the solid polymers and obtaining dipolar-decoupled spectra while mechanically spinning these rotors at the 'magic angle'.

9.8.3. Overcoming the long ^{13}C relaxation time

The quality of the spectra obtained under the above conditions suffer from the long ^{13}C spin-lattice relaxation times, which are of the order of tens of seconds. An excessive sampling time can be avoided by performing a matched spin-lock (or Hartmann–Hahn) cross-polarization experiment.[181,182] Polarization of the carbon in this type of experiment is achieved by a transfer of polarization from nearby protons, spin-locked in their own r.f. field, via static dipolar interactions in a time T_{CH}.[183] This spin–spin transfer takes less than 100 μs, and more importantly can be repeated so that more data can be accumulated after allowing the protons to repolarize in the static field. Spectra obtained using this technique have theoretical line widths of the order of Hz and are comparable to those obtained in solutions (Fig. 9.8).

In practice, the line widths in glassy solids reflect the distribution of environments to be found in the solid. An initial survey of the correlation times for glassy polymers ranging from poly(methylmethacrylate) to polycarbonate has been reported.[184,185] When compared with the results of impact testing, these data indicated an apparent correlation between the two properties. This correlation is not as surprising as it might seem, since the n.m.r. correlation time is dominated by motions which have time constants of the order of the impact contact time, typically 10^{-4} s.[186-8]

9.8.4. Isotope dilution

As indicated in earlier sections, the principal problem in the study of bulk polymer systems is the separation of the contributions from intra- and intermolecular interactions. The isotopic dilution method,[189-94] by the dilution of a protonated polymer by its deuterated analogue, allows the quantitative estimation of the effects of proton–proton dipolar interactions between nuclei on neighbouring chains. In the case of polyethylene[189] at 453 K the intra- and intermolecular T_1s have values of respectively 10^{-12}–10^{-13} s and 10^{-4} s. The former relaxation corresponds to segmental motion

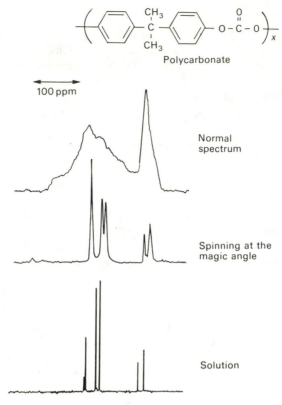

Fig. 9.8. ^{13}C n.m.r. spectra of solution and solid polycarbonate using various spinning conditions.

and has an activation energy of $18\cdot8\,\mathrm{kJ\,mol^{-1}}$ whereas the latter has an activation energy of $59\cdot3\,\mathrm{kJ\,mol^{-1}}$ and corresponds to translational diffusion. It has been proposed (§ 7.5.3) that the over-all relaxation spectrum of the polymer may be described in terms of three regions; segmental motion of the chain backbone, diffusion of the polymer in a 'tube', and conformational changes of the 'tube'. Using this model and the theory of de Gennes,[195] a correlation time of $10^{-5}\,\mathrm{s}$ was predicted for the reptation motion. This prediction is in qualitative agreement with the observed value of T_1 (inter).

9.8.5. Models of molecular motion in solution

The lattice model described in § 4.4.1. has been used to generate a description of the segmental motion of flexible polymer molecules in solution which appears to fit both ^{1}H- and ^{13}C-observations. The nature of the exact equations used will obviously depend on the spin system chosen.

However, the final form of the spectral density usually contains one, or at the most two, adjustable parameters in the theory. For backbone motions, the autocorrelation function has the form:[158,166,196,197]

$$G(\tau) = \exp(\tau/\tau_0) \exp(\tau/\tau_D) \, \text{erfc}(\tau/\tau_D) \tag{9.44}$$

which is directly derived from the theory outlined in § 4.4.1. This is based on the probability of three- and four-bond conformational transitions occurring in an infinitely long polymer chain on a diamond lattice. The correlation time, τ_D, characterizes three-bond transitions and the correlation time τ_0 characterizes four-bond motions and other processes. The spectral density $J(\omega)$ resulting from (9.44) is given by

$$J(\omega) = \tfrac{1}{2} \int_{-\infty}^{+\infty} G(\tau) \, e^{i\omega\tau} \, d\tau$$

$$= \frac{\tau_0 \tau_D (\tau_0 - \tau_D)}{(\tau_0 - \tau_D)^2 + \omega^2 \tau_D^2 \tau_0^2} \left\{ \left(\frac{\tau_0}{2\tau_D}\right)^{\frac{1}{2}} \left[\frac{(1+\omega^2 \tau_0^2)^{\frac{1}{2}} + 1}{1+\omega^2 \tau_0^2} \right]^{\frac{1}{2}} \right.$$

$$\left. + \left(\frac{\tau_0}{2\tau_D}\right)^{\frac{1}{2}} \frac{\omega \tau_0 \tau_D}{(\tau_0 - \tau_D)} \left[\frac{(1+\omega^2 \tau_0^2)^{\frac{1}{2}} - 1}{1+\omega^2 \tau_0^2} \right]^{\frac{1}{2}} - 1 \right\}. \tag{9.45}$$

This equation can be rearranged so that the adjustable parameter becomes the ratio of τ_0 to τ_D. It has been used to interpret the temperature and frequency dependence of ^{1}H-relaxation in poly(vinylacetate),[189,199] poly(methylmethacrylate) (syndiotactic),[200] and polystyrene[201,202] and ^{13}C-relaxation in polystyrene. In the study of syndiotactic poly(methylmethacrylate) it was observed that the rate of backbone motion was influenced by the viscosity and thermodynamic quality of the solvent, whereas the rate of internal rotation of the α-methyl group was essentially independent of solvent.

9.9. Electron spin resonance studies

9.9.1. Fundamental concepts

In the last thirty years electron spin resonance (e.s.r.) has been used for the study of free radicals generated chemically during polymerization reactions, by the action of high-energy irradiation (γ-rays, and fast electrons), by photolysis or mechanical degradation and specifically inserted as spin labels.[203,204] Information relating to the dynamic properties of polymers can be derived from studies of the temperature dependence of the spectral line widths.

The process of electron spin resonance is associated with the absorption of radio frequency radiation to change the spin state of unpaired electrons for which the degeneracy has been destroyed by a strong magnetic field. The interaction energy between the electron magnetic moment and the applied field **H** in the z-direction may be represented by the Hamiltonian

$$\mathcal{H} = g\beta \mathbf{H} S_z, \tag{9.46}$$

where S is the electron spin state, β is the electronic Bohr magneton = $e/2mc$, and g is the spectroscopic splitting constant. The absorption of energy with value $h\nu$, where ν is the exciting frequency (typically 9·5 GHz) occurs when $h\nu = g\beta \mathbf{H}$ (Fig. 9.9).

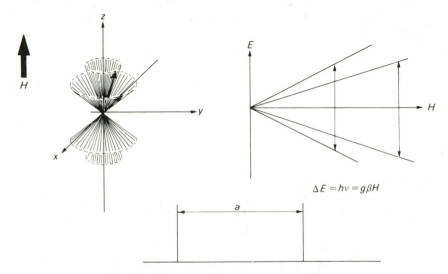

Fig. 9.9. Electron spin resonance experiments.

Absorption of energy assumes a net excess of electrons in the lower energy state (n_1) over that in the upper state (n_u). The population of the levels is determined by the Boltzmann law $n_u/n_1 = \exp(-\Delta E/kT)$. At ordinary temperatures $E \ll kT$ and therefore the upper state can easily be saturated.

Most organic free radicals exhibit complex spectra which arise from splitting of the electron energy level by neighbouring magnetic nuclei with spins greater than zero. The most common situation is interaction with the hydrogen nucleus ($I = \frac{1}{2}$) which leads to the hyperfine splitting shown in Fig. 9.9. The allowed electronic transitions are given by the selection rules $\Delta M_I = 0$, $\Delta M_S = 1$. As a result interaction of the electron spin with a single proton gives rise to a doublet spectrum with hyperfine splitting constant a. Interaction with other nuclei causes further splitting of the energy levels in a similar manner.

The spin hamiltonian which describes the interaction between a nuclear magnetic spin $\mathbf{I}$ and an electron spin $\mathbf{S}$ in a magnetic field $\mathbf{H}$, ignoring the nuclear Zeeman terms, is

$$\mathcal{H} = \beta g \mathbf{SH} + \mathbf{STI} \tag{9.47}$$

where the term $\mathbf{STI}$ represents the hyperfine interaction and $\mathbf{T}$ is the hyperfine tensor. This tensor is composed of two parts, isotropic and anisotropic, $a\mathbf{SI} + \mathbf{ST'I}$, where the quantity a is given by

$$a = \frac{8\pi g \beta g_N \beta_N}{3} |\psi(0)|^2 \tag{9.48}$$

where g_N and β_N are respectively the nuclear g factor and the nuclear magneton. The above equation implies hyperfine splitting if there is a finite probability of finding the unpaired electron at the nucleus, as indicated by the terms $|\psi(0)|^2$.

The anisotropic part of the hyperfine interaction is the angular dependent electron spin–nuclear spin, dipole–dipole interaction. The Hamiltonian of the dipolar interaction energy is given by

$$\mathcal{H}_{aniso} = \frac{-g\beta g_N \beta_N}{\mathbf{r}^3} \left[\mathbf{IS} \frac{-3(\mathbf{I} . \mathbf{r})(\mathbf{S} . \mathbf{r})}{\mathbf{r}^2} \right] \tag{9.49}$$

where $\mathbf{r}$ is the radius from the nucleus to the electron. The resulting equations can now be expressed in the same form as that used in the n.m.r. experiment.[204-6]

In relaxation studies it is usual to employ concentrations of radicals sufficiently low (10^{-3} M) that the intermolecular dipole–dipole interaction and electron exchange contribution are negligible.

The g and a parameters are the isotropic g-value and the isotropic hyperfine coupling constant respectively. They are defined by

$$g = \tfrac{1}{3}(g_x + g_y + g_z) = \tfrac{1}{3}\mathbf{T}, \bar{\mathbf{g}}'$$
$$a = \tfrac{1}{3}(T_x + T_y + T_z) = \tfrac{1}{3}\mathbf{T}, \bar{\mathbf{T}}' \tag{9.50}$$

where $\bar{\mathbf{g}}'$ and $\bar{\mathbf{T}}'$ are the traceless tensors. Their components are related to the total $\bar{\mathbf{g}}$ and $\bar{\mathbf{T}}$ tensors through the relations

$$\mathbf{g}_i = \mathbf{g}_i' + g$$
$$\mathbf{T}_i = \mathbf{T}_i' + a \quad (i = x, y, z). \tag{9.51}$$

9.9.2. Spin probes and labels

Two types of approach have been widely adopted for dynamic studies of polymers; the first involves the introduction of a spin-probe molecule into a polymer matrix, in which case the relaxation behaviour reflects the inter-action between the probe and the surrounding environment. The second involves chemical combination of the probe with the polymer, in which case the relaxation reflects the motion of the polymer and of its environment.

One of the most popular spin labels is the nitroxide radical.

The side-groups R and R' may be aliphatic or aromatic.[207,208] Very stable radicals are achieved if tertiary carbons are bonded to the nitrogen so that disproportion reactions are impossible. Such radicals are di(tert-butyl)nitroxide(I),[209] 2,2-dimethyloxazoline derivatives (II),[210] and pyrol-ine(III)[211] and piperidine(IV) nitroxides[212] (Fig. 9.10). These radicals are most conveniently prepared by oxidation of the corresponding secondary amines.

If the nitroxide radicals are in solution of low viscosity (e.g. benzene) at room temperature, relaxation occurs via rapid Brownian motion. The resultant spectrum leads to an averaged $\bar{g}$ and $\bar{T}$ tensor from which a rotational correlation time can be computed. In a viscous liquid the tumbling rates are slower and experimentally the longer relaxation time is reflected in broad peaks. Using an approach similar to that used in deriving the n.m.r. relaxation times it can be shown that linewidth is defined by[213-17]

$$\frac{1}{T_2(\mathbf{M})} = [\tfrac{1}{20} b^2(3+7u) + \tfrac{1}{15} (\Delta\gamma\mathbf{H})^2(4/3+u)$$

$$+ \tfrac{1}{8}b^2N^2(1-u/5) + \tfrac{1}{5}b\Delta\,\mathbf{H}M(4/3+u)]\,.\,\tau \qquad (9.52)$$

where

$$\Delta = |\beta|/[g_z - \tfrac{1}{2}(g_x + g_y)]$$

$$b = \tfrac{2}{3}(T_z - \tfrac{1}{2}(T_x + T_y))$$

$$u = (1 + \omega_0^2\gamma^2)^{-1} \qquad (9.53)$$

where ω_0 is the Larmor frequency of the electron in the applied magnetic field $\mathbf{H}$. Eqn (9.53) is valid only when the following conditions are satisfied.[213-20]

(a) $(T_x + T_y + T_z)/3\omega_0 \ll 1$;
(b) $T_x \sim T_y$ and $g_x \sim g_y$ (e.g. the tensors are axially symmetrical);
(c) The rotation is isotropic;
(d) $T_2(\mathbf{M}) \gg 1$.

t − butylnitroxide

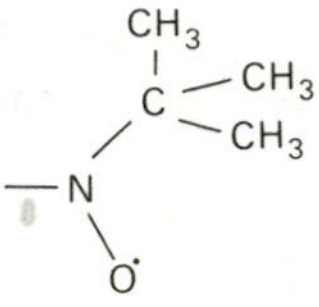

CH_3
C—CH_3
CH_3
—N
O·

2,2 − dimethyloxazolinenitroxides

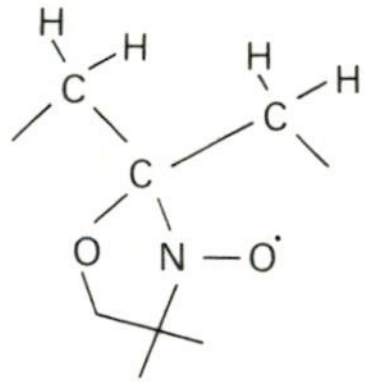

Pyrrolinenitroxides

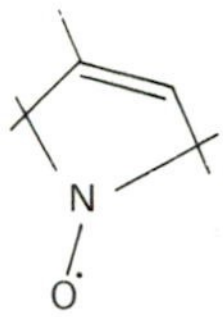

Piperidinenitroxides

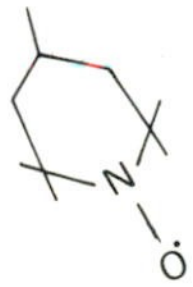

Fig. 9.10. Structures of some stable probe molecules.

Combining the equations for different hyperfine components ($M = 0, \pm 1$), and assuming sufficiently slow rotations to make $u = 0$, the following equations for calculation of τ are obtained

$$\tau(1) = \left(\frac{\Delta H_1}{\Delta H_0} - \frac{\Delta H_{-1}}{\Delta H_0}\right)\left(\frac{15\pi\sqrt{3}\Delta H_0}{8Hb\Delta\gamma}\right) \tag{9.54}$$

$$\tau(2) = \left(\frac{\Delta H_1}{\Delta H_0} - \frac{\Delta H_1}{\Delta H_0} - 2\right)\left(\frac{4\pi\sqrt{3}h}{g\beta b^2}\right) \tag{9.55}$$

where $\Delta H_1 =$ the peak-to-peak width of the hyperfine component ($m = i$) in Gauss and the unit of b is Gauss. The equations are valid in the range from 5×10^{-11} to 5×10^{-9} s and in principle they should yield equal values for τ.

The correlation time τ is related to the radius (r) of the tumbling molecule and to the viscosity (η) of the solvent by the simple Stokes expression

$$\tau = 4\pi\eta r^3/3kT. \tag{9.56}$$

It has been found that this approach is valid for motions with correlation times $5 \times 10^{-9}\,\text{s} < \tau < 10^{-6}\,\text{s}$.[220-4] Various modifications and simplifications of the above approach have been reported but all support the general validity of the basic theory. It has also been demonstrated[224] that it is possible to express τ as a function of S $(= A'_z/A_z)$, where A'_z is one-half of the separation of the outer hyperfine extreme of the e.s.r. spectrum and A_z is the rigid limit value for the quantity. Thus

$$\tau = a(1-S)^b \tag{9.57}$$

where a and b are dependent upon the diffusion model adopted and the intrinsic linewidth characteristics of the spectrum. This relation is valid for τ in the range 10^{-8}–$10^{-6}\,\text{s}$.[227,228]

9.9.3. Spin probe studies of linear polymers

Some of the earliest observations of the dynamics of radicals were conducted on nitroxide radicals dispersed in polymer matrices.[225,226] At temperatures above T_g the e.s.r spectrum consists of a well-resolved triplet indicative of a high degree of rotational freedom in partially amorphous polymers. The fact that an irreversible deformation which reduces the number of degrees of freedom of the polymer molecule also retards the rotational movements of the probe radicals shows the close connection between the two.

Numerous studies have demonstrated that in linear polymers and elastomers the rotational correlation times obey an Arrhenius type of relationship.[229,230] Thus

$$\tau = \tau_0 \exp(\Delta E_a/RT) \tag{9.58}$$

when $T > T_g$. Values of the Arrhenius coefficients for a range of polymers are listed in Table 9.2. It can be seen that the nature of the radical does influence slightly the observed magnitude of the rotational activation energy. Comparison of the values with the activation energies observed by other techniques (n.m.r.,[229,230] dielectric relaxation,[234] neutron quasi-electric scattering[235]) shows a good agreement for small spin probes.[232]

9.9.4. Spin-labelling studies of linear polymers

(1) *Dilute solutions.* The first spin-labelled synthetic polymers to be studied were poly(ethyleneoxide)[236] polystyrene,[237] polyamides,[237] and polyesters.[237] Studies of the relaxation behaviour in polystyrene indicated a

Table 9.2
Activation energies for segmental motion using spin probes

Polymer	Spin probe	Solvent	Viscosity (cP)	Activation energy (kJ mol^{-1})	Ref.
Polystyrene	p-t-butylnitroxide	toluene	0·465	9·0	245
	p-t-butylnitroxide	cyclohexane	0·704	12·55	245
	p-t-butylnitroxide	chloronaphthalene	2·09	17·78	245
	phenylnitroxide	toluene	0·465	9·00	239
Poly(methyl-methacrylate)	t-butylnitroxide	ethylacetate		20	250
Poly(methylacrylate)	t-butylnitroxide	ethylacetate		19	242
Poly(n-butyl-methacrylate)	t-butylnitroxide	ethylacetate		14·6	242

1 cP = 10^{-3} kg m^{-1} s^{-1}.

close correspondence between the e.s.r correlation times and those reported for n.m.r. studies.[238–42] An Arrhenius plot of the correlation times produced an activation energy of 18·0 kJ mol^{-1} which is in reasonable agreement with a value of 20·0 kJ mol^{-1} obtained from dielectric studies.[243]

Using narrow molecular-weight fractions it was observed that the rotational correlation time was sensitive to $\bar{M}_n$ in the range 1950 to 196 000 (Fig. 9.11). The molecular weight-independent region is assumed to reflect a

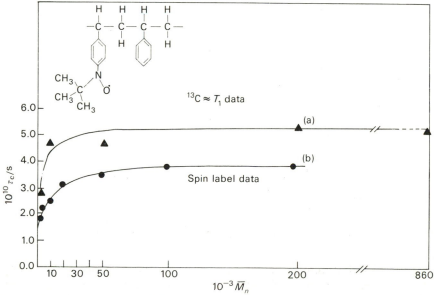

Fig. 9.11. A comparison of ^{13}C and spin probe studies on narrow molecular-weight distribution polystyrenes in toluene.

predominantly local-mode or segmental relaxation mechanism associated with a correlation time τ_{lm}. At lower molecular weights other modes contribute to the observed correlation time τ_c, in particular the first normal mode which represents end over end rotation of the macromolecule.[244] The total correlation time will be

$$1/\tau_c = (1/\tau_{eoe}) + (1/\tau_{lm}). \tag{9.59}$$

Correlation times in magnetic resonance normally involve the time evolution of certain second-order spherical harmonics of the position coordinates. On the other hand, τ_1, the relaxation time of the first normal mode of a polymer chain, describes the time dependence of the first-order spherical harmonic, $\cos \theta$, where θ is the angle between the vector joining the end of the chain and some vector fixed in an external frame of reference. Nevertheless, τ_{eoe} can be obtained from estimates of τ_1. Using this approach, the temperature dependence of τ_{eoe} was found to be close to that for the activation energy for viscous flow of the solvent. Recent extensions of these studies to poly(α-methylstyrene) and poly(methylmethacrylate) suggests that the calculation of the relaxation time attributed to over-all rotation should also include contributions from higher modes of motion.[245-7]

(2) *Concentrated polymer solutions.* As the concentration of polymer is increased so the rotational correlation time changes. A plot of the reduced correlation time τ_c/τ_0 (where τ_0 is the limiting value as concentration tends

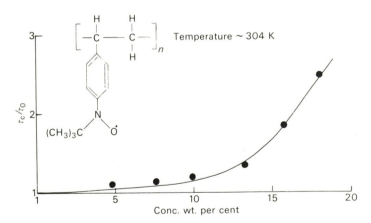

Fig. 9.12. Concentration effects of the ratio τ_c/τ_0.

to zero) of polystyrene in toluene (Fig. 9.12) indicates a marked concentration dependence above 10 per cent by weight. The effects observed in the e.s.r. spectrum reflect local segmental motions, and it is the correlation times that deviate at the point when an entanglement network is formed.

9.9.5. Relaxation in solid polymers

It is possible, using eqn (9.57), to obtain values of the parameters a and b which relate to a diffusion model of the relaxation behaviour. Examination of narrow fractions of polystyrene indicate that these parameters are insensitive to molecular weight.[240] Three models have been considered,[248,240] Brownian diffusion, 'moderate jump' diffusion, and 'large jump' diffusion. There is no *a priori* reason for favouring one model over another. Nevertheless the activation energy obtained from the moderate jump model ($E_a \sim 6 \cdot 3$ kJ mol^{-1}) is close to that reported from mechanical studies.[249]

A study of the relaxation of polyethylene in which the nitroso radical is incorporated on to the backbone[250] (Fig. 9.13) provides an interesting

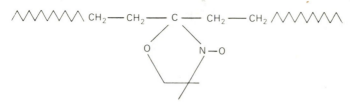

Fig. 9.13. Structure of the polyethylene probe.

example of the use of e.s.r. to determine a T_g. This is particularly significant since confusion has arisen from mechanical and calorimetric studies of this transition. The e.s.r. values are 205 K for amorphous and 220 K for annealed polymer. The activation energy for the α relaxation is 48 kJ mol^{-1} and that for the β relaxation is 28 kJ mol^{-1}. The α relaxation is ascribed to the large-scale reorientation of the polymer backbone while the β relaxation being associated with more local motions. The corresponding segmental reorientation processes of polyethylene have been shown to have activation energies of $15 \cdot 2$ and $22 \cdot 9$ kJ mol^{-1} in the melt and solution phases, respectively. Allowing for the effects of viscous drag of the solvent, the two activation energies are similar, suggesting that in the fluid state relaxation in the bulk and of the isolated molecule occur by the same mechanism.

References

1. Slichter, C. P. *Principles of magnetic resonance*. Harper and Row, New York (1978).
2. Carrington, A. and McLauchlen, A. D. *Introduction to magnetic resonance*. Harper and Row, New York (1967).
3. Bovey, F. A. *Nuclear magnetic resonance spectroscopy*. Academic Press, New York (1969).

4. Abragam, A. *The principles of nuclear magnetism*. Clarendon Press, Oxford (1961).
5. Bloembergen, N. *Nuclear magnetic relaxation*. Benjamin, New York (1961).
6. Slonim, I. Ya and Lyubimov, A. N. *The NMR of polymers*. Plenum Press, New York (1970).
7. Connor, T. M. *NMR basic principles and progress*, Vol. 4, p. 247. Springer Verlag, Berlin (1971).
8. McBrierty, V. J. *Polymer* **15**, 503 (1974).
9. Levy, G. C. *Topics in Carbon-*13 *NMR Spectroscopy* **1**, 79 (1974).
10. Bloembergen, N., Purcell, E. M. and Pound, R. V. *Phys. Rev.* **73**, 679 (1948).
11. Jones, G. P. *Phys. Rev.* **148**, 332 (1966).
12. Slichter, C. P. and Ailion, D. *Phys. Rev.* **A135**, 1099 (1964).
13. Van Vleck, J. H. *Phys. Rev.* **74**, 1168 (1948).
14. Yamagata, K. and Hirota, S. *Rep. Progr. polym. Phys., Jpn* **5**, 236 (1962).
15. McBrierty, V. J. and Ward, I. M. *J. Phys. D* **1**, 1529 (1968).
16. Olf, H. G. and Peterlin, A. *J. polym. Sci. A-2* **8**, 753 (1970).
17. Olf, H. G. *J. polym. Sci. A-2* **9**, 1851 (1971).
18. McBrierty, V. J. and Douglass, D. C. *J. magnet. Reson.* **2**, 352 (1972).
19. Roe, R. J. *J. polym. Sci. A-2*, **8**, 1187 (1970).
20. Kashiwagi, M., Cunningham, A., Manuel, A. J. and Ward, I. M. *Polymer* **14**, 111 (1973).
21. McCall, D. W. *Acc. chem. Res.* **4**, 233 (1971).
22. McBrierty, V. J., McDonald, I. R. and Ward, I. M. *J. Phys. D* **4**, 88 (1971).
23. Douglass, D. C. and Jones, G. P. *J. chem. Phys.* **45**, 956 (1966).
24. McCall, D. W. and Douglass, D. C. *Polymer* **4**, 433 (1963).
25. Bloembergen, N. *Physica* **15**, 386 (1949).
26. Trappeniers, N. J., Gerritsma, C. J. and Oosting, P. H. *Physica* **30**, 997 (1964).
27. Powles, J. G. and Hunt, B. J. *Phys. Lett.* **14**, 202 (1965).
28. McCall, D. W. and Falcone, D. R. *Trans. Faraday Soc.* **66**, 262 (1970).
29. Crist, B. and Peterlin, A. *J. polym. Sci. A-2* **7**, 1165 (1969).
30. Crist, B. and Peterlin, A. *Macromol. Sci. B* **4**, 791 (1970).
31. Crist, B. *J. polym. Sci. A-2* **9**, 1719 (1971).
32. Slichter, W. P. *J. polym. Sci.* **24**, 173 (1957).
33. McCall, D. W. and Slichter, W. P. *J. polym. Sci.* **26**, 171 (1957).
34. Hyndmann, D. and Origlio, G. E. *J. polym. Sci.* **39**, 556 (1959).
35. Hyndmann, D. and Origlio, G. F. *J. appl. Phys.* **31**, 1849 (1960).
36. Hyndmann, D. and Origlio, G. F. *J. polym. Sci.* **46**, 259 (1960).
37. Yamagata, K. and Hirota, S. *Rep. Progr. polym. Sci., Jpn* **5**, 236 (1962).
38. Peterlin, A. and Olf, H. G. *J. polym. Sci. B* **2**, 409 (1964).
39. Peterlin, A. and Olf, H. G. *J. polym. Sci. B* **2**, 769 (1964).
40. Olf, H. G. and Peterlin, A. *J. appl. Phys.* **35**, 3108 (1964).
41. Lander, J. B., Olf, H. G. and Peterlin, A. *J. polym. Sci. A-1* **4**, 941 (1966).
42. McMahon, J. *J. polym. Sci. A-2* **4**, 639 (1966).
43. McBrierty, V. J., Douglass, D. C. and Falcone, D. R. *J. Chem. Soc. Faraday Trans. II* **68**, 1051 (1972).
44. McBrierty, V. J., McCall, D. W., Douglass, D. C. and Falcone, D. R. *Macromolecules* **4**, 548 (1971).
45. Fuschillo, N. and Sauer, J. A. *J. appl. Phys.* **28**, 1073 (1957).
46. Roe, R. J. *J. appl. Phys.* **36**, 2024 (1963).
47. Emsley, J. W., Feeney, J. and Sutcliffe, L. H. *High resolution nuclear magnetic resonance*, Vols. 1 and 2. Pergamon Press, Oxford (1966).

48. Goodman, M. In *Topics in stereochemistry*, Vol. 2 (ed. N. L. Allinger and E. E. Eliel). Wiley, New York (1967).
49. Bovey, F. A. *High resolution NMR of macromolecules* Academic Press, New York and London (1972).
50. Natta, G., Corradini, P. and Ganis, P. *J. polymer Sci.* **58**, 1159 (1962).
51. Powles, J. G., Hunt, B. I. and Sandiford, D. J. *Polymer* **5**, 505 (1964).
52. Powles, J. G. and Mansfield, P. *Polymer*, **3**, 340 (1962).
53. McCall, D. W. and Falcone, D. R. *Trans. Faraday Soc.* **66**, 262 (1970).
54. Lyerla, J. R. and Grant, D. M. In *MTP International review of science, physical chemistry*, Series One, Vol. 4 (ed. A. D. Buckingham), p. 155. Butterworth, London (1972).
55. Vold, R. L., Waugh, J. S., Klein, M. P. and Phelps, D. E. *J. chem. Phys.* **48**, 3831 (1968).
56. Freeman, R. and Hill, H. D. W., *J. chem. Phys.* **53**, 4103 (1970).
57. Mooney, E. F. and Wilson, P. H. In *Annual review of NMR Spectroscopy* (ed. E. F. Mooney), p. 153. Academic Press, London, New York (1969).
58. Cudby, M. E. A. and Willis, H. A. In *Annual Review of NMR Spectroscopy* (ed. E. F. Mooney), p. 353. Academic Press, London, New York (1969).
59. Ferguson, R. C. and Phillips, W. D. *Science* **157**, 257 (1967).
60. Lui, K. J. *J. polym. Sci. A-2* **5**, 1209 (1967).
61. Lui, K. J. *J. polym. Sci. A-2* **6**, 947 (1968).
62. Kato, Y. and Nishioka, A. *Bull. chem. Soc. Jpn* **37**, 1614 (1964).
63. Lui, K. J. and Ullman, R. *J. polym. Sci. A-2* **6**, 451 (1968).
64. Woessner, D. E. *J. chem. Phys.* **41**, 84 (1964).
65. McBrierty, V. J. and McDonald, I. R. *Polymer*, **16**, 125 (1975).
66. Golz, W. L. F. and Zacmann, H. G. *Kolloid Z.* **247**, 814 (1971).
67. Phaovibul, O., Loboda-Cackovic, J., Cakovic, H. and Hosemann, R. *Makromol. Chem.* **175**, 2991 (1974).
68. Phaovibul, O., Cackovic, H., Laboda-Cackovic, J. and Hosemann, R. *J. polym. Sci. A-2* **11**, 2377 (1973).
69. Ahamad, S. R. and Charlesby, A. *Eur. polym. J.* **11**, 91 (1975).
70. Bergmann, K. and Nowotki, K. *Kolloid Z.* **250**, 1094 (1972).
71. Smith, J. B., Manuel, A. J. and Ward, I. M. *Polymer* **16**, 57 (1976).
72. Ando, I. and Nishioka, A. *Makromol. Chem.* **160**, 145 (1972).
73. Zeta, L. and Gatti, G. *Macromolecules.* **5**, 535 (1972).
74. Roe, R. J. *J. appl. Phys.* **36**, 2024 (1963).
75. McBrierty, V. J. and Ward, I. M. *J. Phys. D* **1**, 1529 (1968).
76. Roe, R. J. *J. polym. Sci. A-2* **8**, 1187 (1970).
77. Schmedding, P. and Zachmann, H. G. *Kolloid Z.* **250**, 1105 (1972).
78. Connor, T. M. and McLauchlen, K. A. *J. phys. Chem.* **69**, 1888 (1956).
79. Lui, K. J., International Symposium Macromolecular Chemistry, Tokyo, Japan, (1966); *Polym. Prepr., Amer. chem. Soc., Div. polym. Chem.* **7**, 48 (1966).
80. Lui, K. J. *Macromolecules* **1**, 213 (1968).
81. Lui, K. J. *Makromol. Chem.* **116**, 146 (1968).
82. Lui, K. J. and Anderson, J. E. *Macromolecules* **2**, 235 (1969).
83. Lui, K. J. and Parson, J. P. *Macromolecules* **2**, 529 (1969).
84. Kuroda, Y. and Kubo, M. *J. polym. Sci.* **36**, 453 (1959).
85. Barnes, W. H. and Ross, S. *J. Amer. chem. Soc.* **58**, 1129 (1963).
86. Tadokoro, H., Chatani, Y., Yoshihara, T., Tahara, S. and Murahashi, S. *Makromol. Chem.* **74**, 109 (1964).

87. Tadokoro, H. *Macromol. Rev.* **1**, 119 (1967).
88. Lui, K. J. and Ullman, R. *J. chem. Phys.* **48**, 1158 (1968).
89. Lindberg, J. J., Sirenn, I., Rahkamanaa, E. and Tormala, T. *Angew. makromol. Chem.* **50**, 187 (1976).
90. Allen, G., Connor, T. M. and Pursey, H. *Trans. Faraday Soc.* **59**, 1525 (1963).
91. Haeberlin, U. *Polymer* **9**, 50 (1968).
92. Connor, T. M. *Polymer* **7**, 426 (1966).
93. Matsuzaki, K. and Ito, H. *J. polym. Sci.* (*B*) **12**, 2507 1974).
94. Fritzche, Ch. and Fisher, E. W. *Kolloid Z.* **251**, 721 (1973).
95. Tanaka, A. and Ishida, Y. *Rep. Progr. polym. Phys. Jpn* **15**, 457 (1972).
96. Passing, G. and Noak, F. *Kolloid Z.* **247**, 811 (1971).
97. Wilson, G. W. and Pake, G. E. *J. polym. Sci.* **10**, 503 (1953).
98. Bunn, C. W. and Howells, E. R. *Nature* (*Lond.*) **174**, 549 (1954).
99. McCall, D. W., Douglass, D. C. and Falcone, D. R. *J. Phys. Chem.* **71**, 998 (1967).
100. Slichter, W. P. *J. polym. Sci.* **24**, 173 (1957).
101. McBrierty, V. J., McCall, D. W., Douglass, D. C. and Falcone, D. R. *J. chem. Phys.* **52**, 512 (1970).
102. McBrierty, V. J., McCall, D. W., Douglass, D. C. and Falcone, D. R. *Macromolecules* **4**, 584 (1971).
103. Inoue, Y., Nishioka, A. and Chujo, R. *Makromol. Chem.* **168**, 163 (1973).
104. Ferguson, R. C. *Macromolecules*, **4**, 324 (1971).
105. Connors, R. M. and Blears, D. J. *Polymer*, **6**, 385 (1965).
106. Connor, T. M. and Hartland, A. *Polymer*, **9**, 592 (1968).
107. Segre, A. L., Ferruti, P., Toja, E. and Danusso, F. *Macromolecules* **2**, 35 (1969).
108. Bovey, F. A., Hood, F. P., Anderson, E. W. and Snyder, L. C. *J. chem. Phys.* **42**, 3900 (1965).
109. Heatley, F. and Bovey, F. A. *Macromolecules* **1**, 301 (1968).
110. Williams, A. D. and Flory, P. J. *J. Amer. Chem. Soc.* **91**, 3111 (1969).
111. Fujiwara, Y. and Flory, P. J. *Macromolecules* **3**, 43 (1970).
112. Johnson, L. F., Heatley, F. and Bovey, F. A. *Macromolecules* **3** 175 (1970).
113. Allerhand, A. and Hailstone, R. K. *J. chem. Phys.* **56**, 3718 (1972).
114. Vold, R. L., Waugh, J. S., Klein, M. P. and Phelps, D. E. *J. chem. Phys.* **48**, 3831 (1968).
115. Allerhand, A., Dodrell, D. and Komoroski, R. *J. chem. Phys.* **55**, 189 (1971).
116. Kuhlmann, K. F., Grant, D. M. and Harris, R. K. *J. chem. Phys.* **52**, 3439 (1970).
117. Jentoft, R. E. and Gouw, T. H. *J. polym. Sci. B* **7**, 811 (1969).
118. Crist, B. *J. polym. Sci.* A-*2* **9**, 1719 (1971).
119. Helms, J. B. and Calla, G. *J. polym. Sci. A-2* **10**, 1447 (1972).
120. Ikuto, Y., Ando, I. and Nishioka, A., *Rep. Progr. polym. Phys.* (*Jpn*) **15**, 461 (1972).
121. Powles, J. G., Hunt, B. I. and Sandiford, D. J. *Polymer* **5**, 505 (1964).
122. Powles, J. G. and Mansfield, P., *Polymer* **3**, 340 (1962).
123. McCall, D. W. and Falcone, D. R. *Trans. Faraday Soc.* **66**, 262 (1970).
124. Spevacek, J. and Schneider, B. *Makromol. Chem.* **176**, 729 (1975).
125. Kosfled, R. and Mylius, U. V. *Kolloid Z.* **250**, 1081 (1972).
126. Emsley, J. W., Ivin, K. J. and Little, E. D. *J. polym. Sci. B* **11**, 295 (1973).
127. Rhoch, M. J., Bovey, F. A., Davis, D. D., Douglass, D. C., Falcone, D. R., McCall, D. W. and Slichter, W. P. *Macromolecules* **4**, 712 (1971).

128. Chachaty, C., Forchioni, A. and Ronfard-Haret, J. C. *Makromol. Chem.* **173**, 213 (1973).
129. Levy, G. C. *J. Amer. chem. Soc.* **95**, 6117 (1973).
130. Fajt, B., Pumpernik, D., Jagodic, F., Penko, M. and Azman, A. *Colloid polymer Sci.* **252**, 997 (1974).
131. Assink, R. A. *J. polym. Sci.* **12**, 2281 (1974). Barrie, J. A., Freedrickson, M. J. and Sheppard, R. *Polymer* **13**, 431 (1972).
132. Zachmann, H. G. *J. polym. Sci., polym. Symp.* **43**, 111 (1973).
133. Nishi, T. *J. polym. Sci. A-2* **12**, 685 (1974).
134. Svoboda, J. and Dehnal, M. O. *Makromol. Chem.* **164**, 295 (1973).
135. Schaefer, J. *Macromolecules* **5**, 427 (1972).
136. Nishioka, A. and Shimazaki, K. *Kobunshi Ronbunshi* **31**, 571 (1964).
137. Cuniberti, C. *J. polym. Sci. A-2* **8**, 2051 (1970).
138. Kashiwagi, M. and Ward, I. M. *Polymer* **13**, 145 (1972).
139. Kashiwagi, M., Folkes, M. J. and Ward, I. M. *Polymer* **12**, 697 (1971).
140. Heatley, F. and Scrivens, J. H. *Polymer* **16**, 489 (1975).
141. Vogl, O. and Hatada, K. *J. polym. Sci., polym. Lett.* **13**, 603 (1975).
142. Kashiwagi, M., Cunningham, A., Manuel, A. J. and Ward, I. M. *Polymer* **14**, 111 (1973).
143. McBrierty, V. J. and McDonald, I. R. *J. Phys. D* **6**, 131 (1973).
144. Olf, H. G. and Peterlin, A. *J. polym. Sci. A-2* **9**, 1449 (1971).
145. Olf, H. G. *J. polym. Sci. A-2* **9**, 1851 (1971).
146. Ito, S. E., Okajima, S. and Kasa, T. *Kolloid Z.* **248**, 899 (1971).
147. Ashworth, J., Bamford, C. H. and Smith, E. G. *Pure appl. Chem.* **30**, 25 (1972).
148. Bamford, C. H., Eastmond, G. C. and Whittle, D. *Polymer* **16**, 377 (1975).
149. Preissing, V. G. and Noak, F. *Progr. colloid polym. Sci.* **57**, 216 (1975).
150. Lipatov, Yu. S. and Fabulyak, F. Y. *J. appl. polym. Sci.* **16**, 2131 (1972).
151. Wardell, G. E., McBrierty, V. J. and Douglass, D. C. *J. appl. Phys.* **45**, 3441 (1974).
152. Cohen-Addad, J. P. and Roby, C. *J. chem. Phys.* **63**, 3095, (1975).
153. Cohen-Addad, J. P. *J. chem. Phys.* **64**, 3438 (1976).
154. Cohen-Addad, J. P. *J. chem. Phys.* **63**, 4880 (1975).
155. Edwards, S. F. and Kerr, J. W. *J. Phys. C* **5**, 2889 (1972).
156. Guggenheim, E. A. *Proc. Roy. Soc.* **A187**, 1007 (1944).
157. Laupretre, F., Noel, C. and Monnerie, L. *J. polym. Sci. polym. Phys.* **15**, 2127 (1977).
158. Laupretre, F., Noel, C. and Monnerie, L. *J. polym. Sci., polym. Phys.* **15**, 2143 (1977).
159. Inoue, Y. and Konno, T. *Polym. J.* **5**, 457 (1976).
160. Cunliffe, A. V. and Pethrick, R. A. *Polymer* **21**, 1025 (1980).
161. Ghesquiere, D., Chachaty, C., Bunn, B. and Loucheux, C. *Makromol. Chem.* **177**, 1601 (1976).
162. Yasukawa, T., Ghesquiere, D. and Chachaty, C. *Chem. Phys. Lett.* **45**, 279 (1977).
163. Cutnell, J. D. and Glasel, J. A. *Macromolecules* **9**, 71 (1976).
164. Kimmich, R. *Polymer*, **16**, 851 (1975).
165. Heatley, F. and Begum, A. *Polymer* **17**, 399 (1976).
166. Valeur, B., Jarry, J. P., Geny, F. and Monnerie, L. *J. polym. Sci., polym. Phys.* **13**, 2251 (1975).
167. Ando, I. and Nishioka, A. *Makromol. Chem.* **176**, 3089 (1975).
168. Asakura, T., Ando, I. and Nishioka, A. *Makromol. Chem.* **177**, 523 (1976).

169. Tsuji, K., Hiano, T. and Tsuruta, T. *Makromol. Chem.* **176**, 55 (1975).
170. Day, P. J., Kelly, D. P., Milgate, G. I. and Treloar, F. E. *Makromol. Chem.* **177**, 885 (1976).
171. Spevacek, J. and Schneider, B. *Makromol. Chem.* **176**, 3409 (1975).
172. Zinchenko, V. D., Mank, V. V., Moieseev, V. A. and Qvcharenko, F. D. *Kolloid Z.* **38**, 44 (1976).
173. Schneider, B., Doskocilova, D. and Pivcova, H. *IUPAC Macromol. Symp. Prepr.* **12**, 869 (1970).
174. Schneider, B., Pivcova, H. and Doskocilova, D., *Macromolecules* **5**, 120 (1972).
175. Schnabel, B. and Taplick, T. *Phys. Lett. A* **27**, 310 (1968).
176. Moritz, P. and Schnabel, B. *Plaste Kautsch*, **19**, 281 (1972).
177. Andrew, E. R., Farnell, L. F., Firth, M., Gledhill, T. D. and Roberts, I. *J. magnetic Resonance* **1**, 27 (1969).
178. Andrew, E. R. and Jasinski, A. *J. Phys. C* **4**, 391 (1971).
179. Schaefer, J., Stejskal, E. O. and Buchdahl, R. *Macromolecules* **8**, 291 (1975).
180. Schaefer, J. and Stejskal, E. O. *J. Amer. chem. Soc.* **98**, 1031 (1976).
181. Hartmann, S. R. and Hahn, E. L. *Phys. Rev.* **128**, 2042 (1962).
182. Pines, A., Gibby, M. G. and Waugh, J. S. *J. chem. Phys.* **59**, 569 (1973).
183. McArthur, D. A., Hah, E. L. and Walstadt, R. E. *Phys. Rev.* **188**, 609 (1969).
184. Stejskal, E. O., Schaefer, J. and McKay, R. A. *J. magnetic Resonance*, **25**, 569 (1977).
185. Schaefer, J., Stejskal, E. O. and Buchdahl, R. *Macromolecules* **10**, 384 (1977).
186. Sternstein, S. S. *Polymeric materials*, p. 369. American Society of Metals, Metals Park, Ohio (1975).
187. Argon, A. S. *Polymeric materials*, p. 411. American Society of Metals, Metals Park, Ohio (1975).
188. Boyer, R. F. *Polym. Engng Sci.* **8**, 161 (1968).
189. Morita, M., Ando, I. and Nishioka, A., *Polymers Letters* **18**, 109 (1980).
190. Sillesco, H. *Macromol. Chem.* **178**, 2759 (1977).
191. Kroon, P. A., Kainosho, M. and Chan, S. I., *Biochem. Biophys. Acta* **433**, 282 (1976).
192. Kimmich, R., *Polymer* **18**, 233 (1977).
193. Kimmich, R. and Schmauder, K., *Polymer* **18**, 239 (1977).
194. Kimmich, R., *Polymer* **20**, 132 (1979).
195. De Gennes, P. G., *J. chem. Phys.* **55**, 572 (1971).
196. Jones, A. A., Lubianez, R. P., Hanzon, M. A. and Shostak, S. L., *J. polymer Sci., polymer Phys.* **16**, 1685 (1978).
197. Jones, A. A. and Stockmayer, W. J., *J. polymer Sci., polymer Phys.* **15**, 847 (1977).
198. Heatley, F. and Cox, M. K. *Polymer* **19**, 63 (1978).
199. Heatley, F., Begum, A. and Cox, M. K. *Polymer* **18**, 637 (1977).
200. Heatley, F. and Cox, M. K. *Polymer* **21**, 381 (1980).
201. Heatley, F. and Wood, B. *Polymer* **19**, 1405 (1978).
202. Heatley, F. and Wood, B. *Polymer* **20**, 1512 (1979).
203. Ingram, D. J. E. *Free radicals as studied by electron spin resonance*. Butterworths, London (1958).
204. Carrington, A. and McLachlen, K. V. *Introduction to magnetic resonance*. Harper and Ross, New York 1967.
205. Campbell, D. *Macromolecular reviews*, Vol. 4. Interscience, New York (1970).
206. Sohma, J. and Sakaguski, M. In *Advances in polymer science*, Vol. 20, p. 109. Springer Verlag, Berlin (1976).

207. Tsuji, K. *Advances in polymer science*, Vol. 12, p. 131. Springer Verlag, Berlin (1973).
208. Kneubuhl, F. K. *J. chem. Phys.* **33**, 1074 (1960).
209. Hoffman, A. K. *J. Amer. chem. Soc.* **83**, 4671 (1961).
210. Keana, J. F. W., Keana, S. B. and Beetham, D. *J. Amer. chem. Soc.* **89**, 3055 (1967).
211. Rozantsev, E. G. *Free nitroxyl radicals*. Plenum, New York (1970).
212. Briere, R., Lemaire, H. and Rassat, A. *Bull. Soc. chem. Fr.* 327 (1965).
213. Freed, J. H. and Fraenkal, G. K. *J. chem. Phys.* **39**, 326 (1963).
214. Freed, J. H. and Fraenkel, G. J. *J. chem. Phys.* **40**, 1815 (1964).
215. Kivelson, D. *J. chem. Phys.* **27**, 1087 (1957).
216. Kivelson, D. *J. chem. Phys.* **33**, 1094 (1960).
217. Poggi, G. and Johnson, S. *J. magnetic resonance* **3**, 436 (1970).
218. Stone, T. J., Buckman, T., Nordio, P. C. and McConnell, H. M. *Proc. Nat. Acad. Sci. US* **54**, 1010 (1965).
219. Waggoner, A. S., Griffith, O. H. and Christensen, C. R. *Proc. Nat. Acad. Sci. US* **57**, 1198 (1967).
220. Korst, N. N. and Lazarev. A. V. *Mol. Phys.* **17**, 481 (1969).
221. Alexandrov, I. V., Ivanov, A. N., Korst, N. N., Lazarev, A. V., Prikhozenko, A. J. and Stryuko, V. B. *Mol. Phys.* **18**, 681 (1970).
222. Freed, J. H., Bruno, G. U. and Polnasek, C. F. *J. phys. Chem.* **75**, 3385 (1971).
223. McCalley, R. C., Shimshick, E. J. and McConnell, H. M. *Chem. Phys. Lett.* **13**, 115 (1972).
224. Goldman, S. A., Bruno, G. V. and Freed, J. H. *J. phys. Chem.* **76**, 1858 (1972).
223. Styukov V. B., Karimov, Yu. S. and Rozantsev, E. G. *Vysokomol Soedinenya B* **7**, 493 (1967).
226. Stryukov, V. B. and Rozantsev, E. G. *Vysokomol Soedinenya A* **10**, 626 (1968).
227. Hyde, J. S. and Dalton, L. *Chem. Phys. Lett.* **16**, 568 (1972).
228. Brown, I. M. *Chem. Phys. Lett.* **17**, 404 (1972).
229. Wasserman, A. M., Buchachenko, A. L., Kovarskii, A. L. and Neiman, M. B. *Vysokomöl Soedinenya A* **10**, 1930 (1968).
230. Rousseau, A. and Lenk, R. *Mol. Phys.* **15**, 425 (1968).
231. Rabold, G. P. *J. polym. Sci. A-1* **7**, 1203 (1969).
232. Tormala, P., Lattila, H. and Lindberg, J. J. *Polymer* **14**, 481 (1973).
233. Wasserman, A. M., Buchachenko, A. L., Kovarskii, A. L. and Neiman, M. B. *Eur. polym. J.* **5**, 473 (1969).
234. Tormala, P. *Angew. makromol. Chem.* **37**, 135 (1974).
235. Allen, G., Higgins, J. S. and Wright, C. J. *J. Chem. Soc. Faraday Trans. II* **70**, 348 (1974).
236. Tormala, P., Martinmaa, J., Silvennoninen, K. and Vaahtera, K. *Acta Chem. Scand.* **24**, 3066 (1970).
237. Bullock, A. T., Butterworth, J. H. and Cameron, G. G. *Eur. polym. J.* **7**, 445 (1971).
238. Bullock, A. T., Cameron, G. G. and Smith, P. M. *Polymer* **13**, 89 (1972).
239. Bullock, A. T., Cameron, G. G. and Smith, P. M. *J. phys. Chem.* **77**, 1635 (1973).
240. Bullock, A. T., Cameron, G. G. and Smith, P. M. *J. polym. Sci., polym. Phys.* **11**, 1263 (1973).
241. Bullock, A. T., Cameron, G. G. and Smith, P. M. *Polymer* **14**, 525 (1973).
242. Bullock, A. T., Cameron, G. G. and Elsom, J. M. *Polymer* **15**, 74 (1974).

243. Stockmayer, W. H. *Pure appl. Chem.* **15**, 539 (1967).
244. Zimm, B. H. *J. chem. Phys.* **24**, 269 (1956).
245. Bullock, A. T., Cameron, G. G. and Smith, P. M. *J. Chem. Soc. Faraday Trans. II* **70**, 1202 (1974).
246. Porter, R. S. and Johnson, J. F. *Chem. Rev.* **66**, 4 (1966).
247. Onogi, S., Kobayashi, T., Konjuna, Y. and Taniguchi, Y. *J. Appl. polym. Sci.* **7**, 847 (1963).
248. Goldman, S. A., Bruno, G. V. and Freed, J. H. *J. phys. Chem.* **76**, 1858 (1972).
249. Yano, O. and Wada, Y. *J. polym. Sci. A-2* **9**, 669 (1971).
250. Bullock, A. T., Cameron, C. G. and Smith, P. M. *Eur. polym. J.* **11**, 617 (1975).

MECHANICAL RELAXATION STUDIES

10.1. Introduction

Of the techniques discussed in this book, mechanical methods are amongst the oldest and perhaps the most useful. The simplest way to characterize the mechanical properties of a polymer is to measure its elastic modulus as a function of temperature. Since polymers are viscoelastic, the modulus will depend on the time and method of measurement. For a typical polymer, five regions can be identified. Exemplified by polystyrene,[1-11] these are:

(i) Below 90 °C a glassy region with a modulus in the range $10^{9.5}$–10^9 N m^{-2} is observed (Fig. 10.1). In this region the polymer is glassy, hard, and brittle.

(ii) Above 90 °C the modulus drops to a value of $10^{5.8}$ N m^{-2} at approximately 120 °C. In this region of rapidly dropping modulus the material is best described as leathery.

(iii) From 120 to 150 °C the modulus remains fairly constant at a value of $10^{5.7}$ to $10^{5.4}$ N m^{-2}, the precise value and the temperature range depending on the molecular weight of the polymer. This is the rubbery plateau region.

(iv) Increasing the temperature further leads to a region characterized by a modulus range of $10^{5.4}$ to $10^{4.5}$ N m^{-2}. Some properties of this region are similar to those of the rubbery plateau although there is now a marked component of flow.

(v) The highest temperature interval is characterized by a modulus falling below $10^{4.5}$ N m^{-2}; the polymer exhibits very little elastic recovery and manifests an apparent state of flow.

The precise temperature at which these intervals are observed will depend on the time scale of the observation, although the general characteristics are similar for most common polymers. The dynamic properties of regions (iv) and (v) have been discussed in Chapter 7. In this chapter we shall concentrate on regions (i) to (iii), although to achieve a consistency in the discussion certain of the properties of region (iv) will be considered.

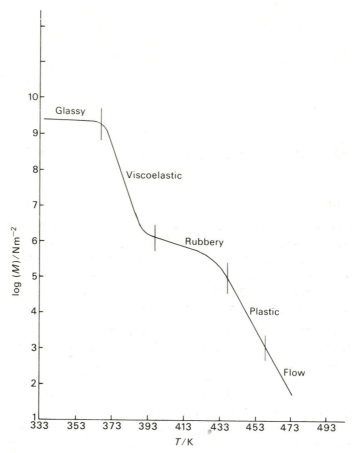

Fig. 10.1. Viscoelastic behaviour of polystyrene.[1-11]

Studies of a large number of polymer systems have indicated that proper-ties of the glassy and transition regions are independent of chain length for a polymer of reasonable molecular weight. The value of the modulus in the region of the rubbery plateau is likewise independent of chain length although the range of temperature which it spans is a function of molecular weight. The rubbery and liquid flow properties are markedly dependent on chain length.

Curves of the type shown in Fig. 10.1 can be characterized by two temperatures. The first is the inflection temperature defined as the value when $\log E = \frac{1}{2}(\log E_1 + \log E_2)$ where E_1 and E_2 are respectively the glassy and rubbery plateau moduli. The second temperature is defined by the modulus falling below 10^5 N m^{-2}. The glass transition is characterized by the onset of motion of chain segments which, below the transition, are essentially frozen into fixed positions on the sites of a disordered quasi-lattice. The segments vibrate around these fixed points just as molecules do about their lattice spacings, and undergo little or no diffusional motion from one lattice position to another.

10.2. Linear viscoelastic behaviour

Before proceeding with the interpretation of mechanical observations of polymeric materials, it is necessary to discuss the concept of viscoelastic relaxation.[12–17] The arguments parallel those used in the discussion of fluids, but emphasis will be given here to the slow relaxation of polymers. Many of the published mathematical treatments are more sophisticated than is necessary for the reader of this text, and so to provide the necessary formulae in §§ 10.2–10.5 we follow closely the simpler exposition presented by Ward.[17]

10.2.1. Linear viscoelastic condition

Application of Newton's Law of viscosity leads to a definition in terms of the stress, σ, and the resultant velocity gradient, dv/dy,

$$\sigma = \eta \, dv/dy \qquad (10.1)$$

where y is the direction of the velocity gradient. For a velocity gradient in the xy-plane, (10.1) is written as

$$\sigma_{xy} = \eta \frac{de_{xy}}{dt}. \qquad (10.2)$$

The shear stress, σ_{xy}, is directly proportional to the rate of change of shear strain, e_{xy}. For elastic behaviour at small strains $(\sigma_{xy})_E = Ge_{xy}$ where G is the shear modulus. We can combine the equations to give

$$\sigma_{xy} = (\sigma_{xy})_E + (\sigma_{xy})_V = Ge_{xy} + \eta \frac{de_{xy}}{dt}. \qquad (10.3)$$

In polymer solids the phenomenon of creep under load is of interest, and the following patterns of stress–strain behaviour are observed (Fig. 10.2). For an ideal elastic solid the strain follows exactly the loading cycle and is proportional to the load. However in the general case the total strain, e, is the sum of three separate parts e_1, e_2, and e_3. In the analysis e_1 and e_2 are often termed respectively the immediate and the delayed elastic deformations; e_3 is the Newtonian flow, i.e. that part of the deformation which is identical with the deformation of a viscous liquid obeying Newton's law of viscosity. Thus the simple loading experiment defines a creep compliance

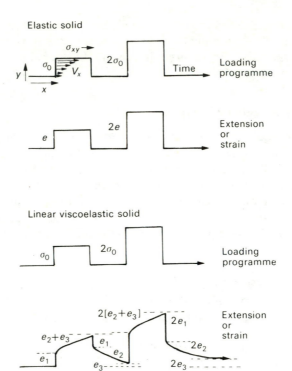

Fig. 10.2. Linear viscoelastic behaviour.

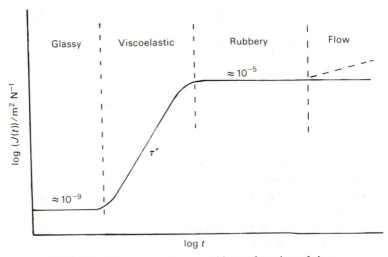

Fig. 10.3. Creep compliance $J(t)$ as a function of time.

$J(t)$, which is a function of time (Fig. 10.3)

$$e(t)/\sigma = J(t) = J_1 + J_2 + J_3 \qquad (10.4)$$

where J_1, J_2, and J_3 correspond to e_1, e_2, and e_3, respectively.

The term J_3, which defines the Newtonian flow, can be neglected for rigid polymers at ordinary temperatures, because their flow viscosities are very large and their behaviour is dominated by J_1 and J_2. However linear amorphous polymers do show a finite J_3 at temperatures above their glass transitions, which is not exhibited by cross-linked or very crystalline materials.

It is not always possible to separate J_1 and J_2. However an investigation of the over-all response of the system as a function of time does provide a useful insight into the dynamic behaviour. Thus Fig. 10.3 illustrates the variation of compliance with time at constant temperature for an idealized amorphous polymer with only one relaxation time. For short-time experiments the observed compliance has a value of about $10^{-9}\,\mathrm{m^2\,N^{-1}}$ (that of a glassy solid) and is time-independent. At very long times the compliance drops to a value of about $10^{-5}\,\mathrm{m^2\,N^{-1}}$ (that of a rubbery solid) and is again time-independent. At intermediate times the compliance lies between these values and is time-independent.

10.2.2. Stress relaxation

Creep is the manifestation of strain change under constant stress, and a related phenomenon is stress reduction at constant strain (Fig. 10.4). The time-dependent relaxation of stress is expressed in the modulus $G(t) = \sigma(t)/e$ (Fig. 10.5).

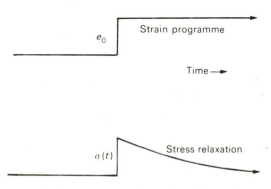

Fig. 10.4. Stress relaxation behaviour.

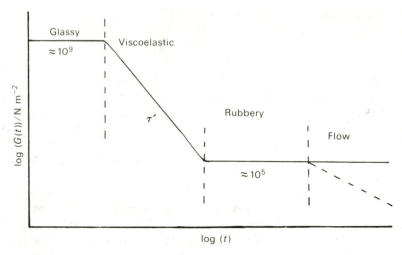

Fig. 10.5. Stress relaxation modulus $G(t)$ as a function of time.

10.2.3. Mathematical analysis of viscoelastic behaviour

The total creep, $e(t)$, can be defined in terms of the incremental stress and the creep-compliance function

$$e(t) = \int_{-\infty}^{t} J(t - \tau) \, \mathrm{d}\sigma(\tau). \tag{10.5}$$

A typical experiment can be modelled by considering the following stress cycle for a viscoelastic solid. Three specific cases are of interest (Fig. 10.6):

(i) Single-step loading of a stress σ_0 at time $\tau = 0$. For this case

$$J(t - \tau) = J(t) \quad \text{and} \quad e(t) = \sigma_0 J(t). \tag{10.6}$$

(ii) Two-step loading of a stress σ_0 at time $\tau = 0$ followed by an additional stress σ_0 at time $\tau = t_1$ (Fig. 10.6). Then the strains

$$e_1 = \sigma_0 J(t) \qquad e_2 = \sigma_0 J(t - t_1) \tag{10.7}$$

give the creep deformations produced by the two loading steps, and

$$e(t) = e_1 + e_2 = \sigma_0 J(t) + \sigma_0 J(t - t_1). \tag{10.8}$$

This shows that the 'extra creep' $e_c'(t - t_1)$ produced by the second loading step is given by

$$e_c'(t - t_1) = \sigma_0 J(t) + \sigma_0 J(t - t_1) - \sigma_0 J(t)$$

$$= \sigma_0 J(t - t_1). \tag{10.9}$$

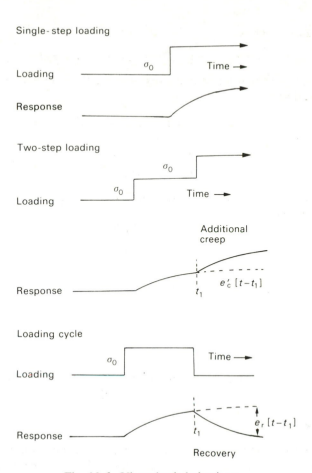

Fig. 10.6. Viscoelastic behaviour.

Equation (10.9) illustrates the linearity principle that the additional creep $e'_c(t - t_1)$ produced by adding the stress σ_0 is identical with the creep which would have occurred had this stress σ_0 been applied without any previous loading at the same instant in time t_1.

(iii) Creep and recovery. In this case the stress σ_0 is applied at time $\tau = 0$ and removed at time $\tau = t_1$. The deformation $e(t)$ at a time t greater than t_1 is given by the addition of two terms $e_1 = \sigma_0 J(t)$ and $e_2 = -\sigma_0 J(t - t_1)$ which describe the application and removal of the stress σ_0 respectively

$$e(t) = \sigma_0 J(t) - \sigma_0 J(t - t_1). \qquad (10.10)$$

The recovery $e_r(t - t_1)$ will be defined as the difference between the predicted creep under the initial stress and the àctual measured response. Thus

$$e_r(t - t_1) = \sigma_0 J(t) - \{\sigma_0 J(t) - \sigma_0 J(t - t_1)\} = \sigma_0 J(t - t_1). \quad (10.11)$$

In this way the time-dependent stress and the time-dependent strain can be expressed as

$$\sigma(t) = [G_r e] + \int_{-\infty}^{t} G(t - \tau) \frac{de(\tau)}{d\tau} \, \partial \tau. \quad (10.12)$$

$$e(t) = \int_{-\infty}^{t} J(t - \tau) \frac{d\sigma}{d\tau} (\tau) \, \partial \tau. \quad (10.13)$$

In an experiment conducted so that the strain is constant and the relaxing stress follows $G(\tau)$

$$\frac{d\sigma(\tau)}{d\tau} = \frac{dG(\tau)}{d\tau}, \quad (10.14)$$

and

$$\int_0^t \frac{dG(\tau)}{d\tau} J(t - \tau) \, d\tau = \text{constant} \quad (10.15)$$

or simply

$$\int_0^t G(\tau) J(t - \tau) \, d\tau = e(t). \quad (10.16)$$

Equation (10.16) provides the mathematical basis relating creep and stress relaxation observations.

10.3. Maxwell model

It is often useful to model macroscopic viscoelastic behaviour by picturing the solid as some combination of idealized elastic and viscous components such as a Hookean spring and a Newtonian dash pot. Of these the simplest is the Maxwell model, which considers the additive strains in a series combination of these two elements.

The stress in the spring is

$$\sigma_1 = E_m e_1 \quad (10.17)$$

and that in the dash pot is

$$\sigma_2 = \eta_m \frac{de_2}{dt}. \tag{10.18}$$

So

$$\sigma = \sigma_1 + \sigma_2 \tag{10.19}$$

$$e = e_1 + e_2, \tag{10.20}$$

and

$$\frac{d\sigma}{dt} = E_m \frac{de_1}{dt}, \tag{10.21}$$

giving

$$\frac{de}{dt} = \frac{1}{E_m} \frac{d\sigma}{dt} + \frac{\sigma}{\eta_m}. \tag{10.22}$$

The Maxwell model is useful for stress relaxation.

In this case

$$\frac{de}{dt} = 0 \quad \text{and} \quad \frac{1}{E_m} \frac{d\sigma}{dt} + \frac{\sigma}{\eta_m} = 0 \tag{10.23}$$

and

$$\frac{d\sigma}{\sigma} = \frac{E_m}{\eta_m} dt. \tag{10.24}$$

If σ_0 is the initial stress

$$\sigma = \sigma_0 \exp \frac{-E_m}{\eta_m} t. \tag{10.25}$$

This shows that the stress decays exponentially with a characteristic time constant $\tau = \eta_m / E_m$

$$\sigma = \sigma_0 \exp -t/\tau. \tag{10.26}$$

10.4. Kelvin or Voigt model

This model is similar to the Maxwell model except that the spring and dashpot are combined in parallel. Then

$$\sigma_1 = E_v e_1 \tag{10.27}$$

and

$$\sigma_2 = \eta_v \frac{de_2}{dt}$$

for the spring and dashpot, respectively.

Combining the stress and strain using (10.19) and (10.20)

$$\sigma = E_v e + \eta_v \frac{de}{dt}. \tag{10.28}$$

When the strain is constant this parallel combination gives a constant limiting stress (expansion of the spring), which often will not be shown by real systems.

When the stress is constant

$$e = \frac{\sigma_0}{E_v}\left\{1 - \exp\left(\frac{-E_v}{\eta_v}t\right)\right\} \tag{10.29}$$

and when the stress is removed the recovery is obtained by setting $\sigma = 0$ in (10.28)

$$E_v e = -\eta_v\left(\frac{de}{dt}\right) \tag{10.30}$$

or

$$e = e_0 \exp(-t/\tau) \tag{10.31}$$

where $\tau = \eta_v/E_v$ is called the 'retardation time'.

10.4.1. Relaxation and retardation time spectra

Stress relaxation with time dependent strain is described by

$$\sigma(t) = [G_r e] + \int_{-\infty}^{t} G(t-\tau)\frac{de(\tau)}{d\tau}d\tau \tag{10.32}$$

which at constant strain, 'e', gives (Maxwell model)

$$\sigma(t) = E_m'e' \exp{-t/\tau}. \tag{10.33}$$

When a series of such elements are joined in parallel,

$$\sigma(t) = 'e' \sum^{n} E_n \exp(-t/\tau_n) \tag{10.34}$$

where E_n, τ_n are the spring constants and relaxation times respectively of the nth Maxwell element. The spring constants, E_n can be replaced by the relaxation time spectrum, $f(\tau)$ giving the concentration of elements with relaxation time between τ and $\tau + d\tau$

$$\sigma(t) = [G_r'e'] + 'e' \int_0^\infty f(\tau) \exp(-t/\tau) \, d\tau \qquad (10.35)$$

and

$$G(t) = [G_r] + \int_0^\infty f(\tau) \exp(-t/\tau) \, d\tau. \qquad (10.36)$$

Generally it is found more convenient to use a logarithmic time scale, as was introduced in an analogous fashion for dielectric behaviour in Chapter 5.

$$G(t) = [G_r] + \int_{-\infty}^{+\infty} H(\tau) \exp(-t/\tau) \, d \ln \tau \qquad (10.37)$$

where $H(\tau)$ is the contribution to stress relaxation from the constituents of $f(\tau)$.

In exactly the same way a series combination of Kelvin elements yields the generalized compliance in creep,

$$J(t) = [J_u] + \int_{-\infty}^{+\infty} L(\tau)(1 - \exp(-t/\tau) \, d \ln \tau \qquad (10.38)$$

where J_u is the instantaneous elastic compliance and $L(\tau)$ is the retardation time spectrum.

The various idealized functions derived in this way are summarized in Table 4.3.

10.5. Dynamic mechanical measurements

For linear viscoelastic behaviour with strain lagging in phase behind a sinusoidal stress

$$e = e_0 \sin \omega t$$
$$\sigma = \sigma_0 \sin(\omega t + \delta) \qquad (10.39)$$

where ω is the angular frequency and δ is the phase lag.

The periodic stress can be written in terms of components in phase with, and 90° out of phase with, the periodic strain

$$\sigma = \sigma_0 \sin \omega t \cos \delta + \sigma_0 \cos \omega t \sin \delta \qquad (10.40)$$

which in terms of a complex modulus (Fig. 10.7),

$$G^* = G_1 + iG_2 \qquad (10.41)$$

yields

$$\sigma = e_0 G_1 \sin \omega t + e_0 G_2 \cos \omega t$$

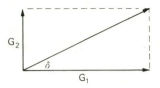

Fig. 10.7. Argand diagram of the complex modulus.

with

$$G_1 = \frac{\sigma_0}{e_0} \cos \delta \quad \text{and} \quad G_2 = \frac{\sigma_0}{e_0} \sin \delta. \tag{10.42}$$

The interrelated complex parameters are then,

$$e = e_0 \exp i\omega t \tag{10.43}$$

$$\sigma = \sigma_0 \exp i(\omega t + \delta) \tag{10.44}$$

$$\frac{\sigma}{e} = G^* = \frac{\sigma_0}{e_0} e^{i\delta} = \frac{\sigma_0}{e_0} (\cos \delta + i \sin \delta). \tag{10.45}$$

The energy dissipated per cycle, ΔE, can be obtained from the loss modulus

$$\Delta E = \int_0^{2\pi/\omega} \sigma \, de = \int_0^{2\pi/\omega} \sigma \frac{de}{dt} \, dt. \tag{10.46}$$

Substituting for σ and e

$$\Delta E = \omega e_0^2 \int_0^{2\pi/\omega} (G_1 \sin \omega t \cos \omega t + G_2 \cos^2 \omega t) \, dt$$

$$= \pi G_2 e_0^2. \tag{10.47}$$

In most cases G_2 is small compared with G_1; and the complex modulus amplitude is approximately equal to G_1.

As introduced in Chapter 1, we can consider the response of the system in the frequency plane either as modulus (Fig. 10.8), or as compliance (Fig. 10.9). For example, the Maxwell model gives

$$\frac{de}{dt} = \frac{1}{E_m} \frac{d\sigma}{dt} + \frac{\sigma}{\eta_m} \tag{10.23}$$

or

$$\sigma + \tau \frac{d\sigma}{dt} = E_m \tau \frac{de}{dt} \tag{10.48}$$

where $\tau = \eta_m / E_m$.

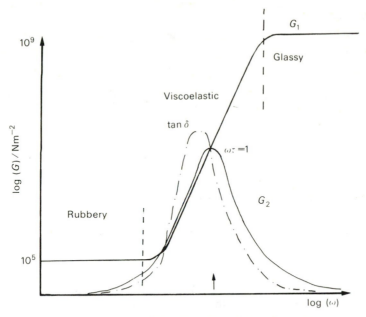

Fig. 10.8. Complex modulus $G^* = G_1 + iG_2$ as a function of frequency, ω.

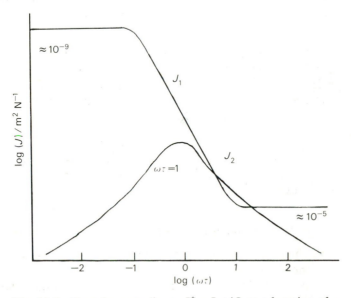

Fig. 10.9. Complex compliance $J^* = J_1 + iJ_2$ as a function of $\omega\tau$.

Putting

$$\sigma = \sigma_0\, e^{i\omega t} = (G_1 + iG_2)e \tag{10.49}$$

gives

$$\sigma_0\, e^{i\omega t} + i\omega\tau\sigma_0\, e^{i\omega t} = \frac{E\tau i\omega\sigma_0\, e^{i\omega t}}{G_1 + iG_2} \tag{10.50}$$

from which it follows that

$$G_1 + iG_2 = E_m \frac{i\omega t}{1 + i\omega\tau}. \tag{10.51}$$

Separation of real and imaginary parts gives the now familiar functions

$$G_1 = E_m \frac{\omega^2\tau^2}{1 + \omega^2\tau^2}, \quad G_2 = E_m \frac{\omega\tau}{1 + \omega^2\tau^2} \quad \text{and} \quad \tan\delta = \frac{1}{\omega\tau}. \tag{10.52}$$

These functions are general for any type of relaxation and can be associated with mechanical relaxation in the polymer material.

10.6. Methods of viscoelastic measurement

As with other experiments the mechanical relaxation properties of a polymer can be sensed either in the time or frequency domain.[18] The range covered is approximately 10^{-8}–10^8 Hz. The techniques fall into five main classes:

(1) Transient measurements: creep and stress relaxation 10^{-8}–10 Hz;
(2) Low-frequency vibrations: free oscillation methods 10^{-2}–10^2 Hz;
(3) High-frequency vibrations: resonance methods 10^2–10^4 Hz;
(4) Forced vibrations (non-resonance) methods 10^{-2}–10^2 Hz;
(5) Wave propagation methods 10^4–10^8 Hz.

The reader is referred to a number of specialist reviews for details of these techniques.[18-21] Although a large range and variety of techniques exist for the study of mechanical properties, the time–temperature superposition principle is often used.

10.7. Time–temperature superposition principle

The concept behind this approach is that temperature only shifts the relaxation on the time scale but does not alter the shape or nature of the relaxation process.[22] To illustrate this approach we shall consider the shift in relaxation of the compliance or of the peak in $\tan\delta$ (Fig. 10.10). If it is

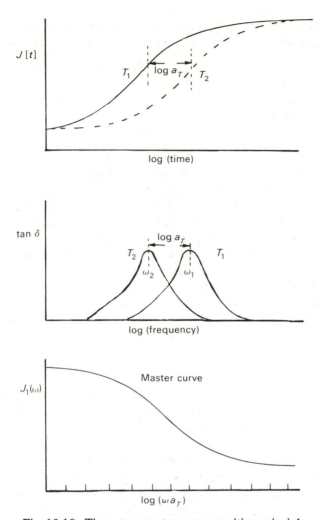

Fig. 10.10. Time–temperature superposition principle.

assumed that the relaxation is controlled by a simple rate theory then the frequency of conformational change between rotational isomeric states can be described by

$$\nu = \nu_0 \, e^{-\Delta H/RT} \tag{10.53}$$

where ν_0 is the pre-exponential factor discussed in Chapter 2.

Re-arranging (10.53) we have

$$\Delta H = -R \left[\frac{d \ln \nu}{d \, (1/T)} \right]. \tag{10.54}$$

At temperatures T_1 and T_2 the peak value of tan δ occurs at frequencies ω_1 and ω_2, respectively. The assumption is that ω_1 and ω_2 are related by the equation

$$\frac{\omega_1}{\omega_2} = \frac{e^{-\Delta H/RT_1}}{e^{-\Delta H/RT_2}}.$$

(10.55)

Substituting in eqn (10.54) yields

$$\log\left(\frac{\omega_1}{\omega_2}\right) = \log a_T = \frac{\Delta H}{R}\left\{\frac{1}{T_2} - \frac{1}{T_2}\right\}.$$

(10.56)

The factor $\log a_T$ is related to the activation energy for the relaxation process. The simplest way of applying the time-temperature superposition principle is to produce a 'master compliance curve' by choosing one particular temperature and applying only a horizontal shift on a logarithmic time scale to make the compliance curves for other temperatures join as smoothly as possible on to the curve at this chosen reference temperature. The molecular theories of viscoelasticity suggest that there should be an additional small vertical shift factor $T_0\rho_0/T\rho$ in changing from the actual temperature T(K) (at a density ρ) to the reference temperature T_0(K) (at a density ρ_0). The shift factor for an amorphous polymer may be expressed by the Williams, Landel, Ferry (WLF) theory[22-4]

$$\log a_T = \frac{C_1(T - T_s)}{C_2 + (T - T_s)}$$

(10.57)

where C and C_2 are constants and T_s is a reference temperature peculiar to a particular polymer. This holds extremely well over the temperature range $T = T_s \pm 50\,^\circ C$ for all amorphous polymers.

10.7.1. WLF theory and the glass transition process

It is impossible to discuss the form of eqn (10.57) without considering the glass transition process. In Chapter 1 the basic properties of the glass transition process were reviewed and its association with the occurrence of free volume established.[25-8] The fractional free volume $f = v_f/v$, where v_f is the difference between the total macroscopic volume and the actual volume occupied by the polymer, v_0, can therefore be written as

$$f = f_g + \alpha_f(T - T_g)$$

(10.58)

where f_g is the fractional free volume at a glass transition temperature T_g and α_f is the coefficient of expansion of the free volume.

It is possible to develop a transition state theory based on free volume by expressing the frequency of site exchange in the form

$$\nu = A \int_{f_c}^{\infty} \phi(f)\,df$$

(10.59)

which can be shown to have the form

$$\nu = \nu_s \exp\left\{-Nf_c\left(\frac{1}{f}-\frac{1}{f_c}\right)\right\} \tag{10.60}$$

where ν_s is the frequency at T_g. Using (10.58) we have

$$\ln \nu/\nu_s = \frac{(Nf_c/f_g)(T-T_g)}{T-T_g+f_g/f}. \tag{10.61}$$

Comparison of (10.61) and (10.57) indicates that $T_s \approx T_g$. However it is found that $T_s = T_g - 51\cdot6\,^\circ\text{C}$ for most practical purposes. The WLF constants C_1 and C_2 also have unique definitions on the basis of free-volume theory.

The temperature T_2 can be associated with the co-operative movements of a group of segments of the chain. If the frequency of molecular jumps is given by

$$\nu_c = A \exp-\frac{n\Delta G^*}{kT} \tag{10.62}$$

where A is a constant $= kT/h$, ΔG^* is the energy hindering the rearrangement per segment and n is the number of segments acting co-operatively as a unit.[29-31] If S is the configurational entropy of the system

$$S = \frac{N_A}{n} S_n \tag{10.63}$$

where N_A is Avogadro's constant and S_n is the entropy of a unit of n segments. Thus

$$n = \frac{N_A S_n}{S} \quad \text{and} \quad \nu_c = A \exp\frac{-N_A S_n \Delta G^*}{SkT}. \tag{10.64}$$

If it is assumed that S_n is independent of temperature, the configurational entropy of the system can be calculated directly for any temperature from the specific heat at constant pressure. Assuming the entropy is zero at T_2 we have

$$S(T) = \Delta C_p \ln\left(\frac{T}{T_2}\right) \tag{10.65}$$

where ΔC_p is the difference in specific heat of the polymer above and below the glass transition. Then eqn (10.64) becomes

$$\nu_c = A \exp\frac{N_A \Delta G^* S_n}{k\Delta C_p(T-T_2)}. \tag{10.66}$$

This gives a relaxation time equation of the form

$$\tau = \tau_0 \exp\left\{\frac{B}{(T - T_2)}\right\}. \tag{10.67}$$

In experiments this can be fitted to the WLF equation if $T_2 = T_g - 51 \cdot 6$. It is clear from the success of these apparently opposing approaches that a definition of the T_g process is not completely unambiguous.

10.8. Relaxation transitions and their relationship to molecular structure

The relaxation properties of solid polymers are conveniently divided by consideration of amorphous and crystalline polymers. The changes in mechanical relaxation associated with the highest-temperature process in amorphous polymers have been correlated with various aspects of molecular structure and lead to the conclusion that this feature is ascribable to the motion of large elements of the polymer backbone. The corresponding transition in crystalline polymers is often associated with a process more closely related to the melting transition.

10.8.1. Relaxation in amorphous polymers

A typical example of the complex spectrum of processes and mechanical responses observed in an amorphous polymer is demonstrated by

Poly (methylmethacrylate)

poly(methylmethacrylate).[32-48] This polymer exhibits four distinct relaxation features. We use the customary nomenclature of labelling the highest relaxation as the alpha-(α), and so on with decreasing temperature β, γ, δ, etc. The highest temperature relaxation is associated with the glass transition and has associated with it a large change in modulus. The β-relaxation has been shown by a combination of comparative studies on similar polymers using n.m.r. and dielectric techniques to have a major component from side-chain motion of the ester group. The γ and δ relaxations involve

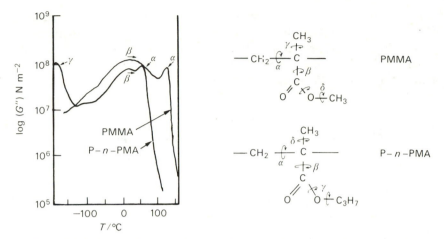

Fig. 10.11. Relaxation in poly(methylmethacrylate) and poly(n-propyl-methacrylate).

motion of the methyl groups attached to the main chain and to the side chain respectively (Fig. 10.11).

As indicated earlier the glass transition process may be considered as being controlled by the intrinsic flexibility of the backbone or as a function of the free volume available within the polymer. In the latter model, conformational change can only occur when there is sufficient free volume for the chain to move. The free volume is assumed to diffuse throughout the polymer at a rate controlled by the motion of the chains and has a critical value for the onset of large segment motion. In practice both approaches appear capable of rationalizing the data. The effects of chemical structure implicit in the flexibility approach also contain a changing intermolecular contribution and the apparent equivalence of the two approaches for the description of the glass transition lies in the fact that any change will influence both the intra- and intermolecular contributions to the over-all potential. Much of our knowledge of the effects of chemical structure is empirical due to the complexity of the interplay between these types of interaction. In spite of this some general trends do appear.

10.8.2. Main chain structure

The flexibility of the backbone is the principal feature determining the mobility of a normal straight-chain polymer. The internal rotational barrier is defined by a combination of non-bonding interactions between neighbouring atoms or groups and a basic restriction defined by the atom pair and the symmetry of the atoms involved. An additional complicating factor in the case of polymers is the effect of long-range interactions, which can lead to the formation of stable tertiary structures in the solid phase. This latter effect usually leads to the formation of crystalline materials to be discussed later in this chapter. The effects of non-bonding interactions on the internal rotational barrier can be illustrated by considering the activation energy associated with conformational changes in small molecules (Table 10.1).

Table 10.1

Typical values of the barriers to internal conformational change in small molecules

Molecule	Barrier to internal rotation $(kJ \, mol^{-1})$	Ref.
Ethane $CH_3—CH_3$	12·25	218
Ethylchloride $CH_3—CH_2Cl$	15·45	219
Dichloroethane $CH_3—CHCl_3$	22·2	220
Trichloroethane $CH_3—CCl_3$	25·1	220
Tetrachloroethane $CH_2Cl—CCl_3$	36·5	221
Pentachloroethane $CHCl_2—CCl_3$	52·4	221
Hexachloroethane $CCl_3—CCl_3$	62·8	222
1-Fluoropropene (*cis*) $CH_3—CH=CHF$	4·44	223
1-Fluoropropane (*trans*) $CH_3—CH=CHF$	9·28	224
Methylalcohol $CH_3—OH$	4·50	225
Methylmercaptan $CH_3—SH$	5·30	226
Methylamine $CH_3—NH_2$	8·25	227
Methylphosphine $CH_3—PH_2$	8·15	228
Methylsilane $CH_3—SiH_3$	7·0	229
Silane $SiH_3—SiH_3$	3·2	230

The general trends indicated from these studies tend to be paralleled in the variations observed in polymers, and the following can be identified:

(i) The flexibility will increase as the bond length is increased;

(ii) The flexibility increases for bonds formed with atoms of different symmetry, and decreases with a decrease in symmetry number for bonds formed between the same atoms;

(iii) Delocalization of electron density between unsaturated bonds increases the rigidity but this effect is reversed if steric interactions hinder the conjugation between adjacent bonds;

(iv) Although not widely studied, 1–3 non-bonding interactions have a dominant effect in defining both the activation energy for inter-conversion and the stability of the possible isomeric forms.

The above trends are reflected in the variation of the glass-transition temperature and the 'activation energy' for the alpha relaxation process.[50–2] For instance the presence of a flexible group, such as an ether link, will make the main chain more flexible and reduce the glass-transition temperature, whereas the introduction of an inflexible group such as a terephthalate residue will increase the glass-transition temperature. In reality the glass-transition temperature in both these polymers will be influenced not only by the intra- but also by the intermolecular potential and is sensitive to the thermal history, even for amorphous polymers.

10.8.3. Influence of side-groups

In general the bulkier the side-group the higher the glass-transition temperature (Table 10.2). The change in glass-transition temperature reflects both a larger barrier to internal rotation as a result of increased non-bonding interactions and larger volume required for the conformational change. The interplay of these effects is demonstrated in the change of glass-transition temperature within a series of isomeric poly(vinylbutylethers) (Table 10.2). The tertiary butyl group is more compact and creates a greater steric barrier to internal motion. The normal butyl group will be spatially more extended, occupying more volume. However it is flexible enough to adopt, rapidly, a low-energy structure. In an analogous manner the length of the flexible side-group reduces the glass transition temperature in the poly(n-alkylethers) (Table 10.2). The longer the rapidly moving chain, the greater the free volume available for relaxing the main chain.

10.8.4. Effect of main-chain polarity

Dipolar interactions, being long-range, influence both the local potential profile and the over-all energy of the lattice. The variation of T_g for a variety of polymers with differing polarity is illustrated in Fig. 10.12. The reduced mobility with increasing polarity is seen by comparing the behaviour of poly(methylacrylicesters) with poly(chloroacrylicesters), the higher values of T_g being observed in the more polar polymers. The trends within a series of polymers, and also between polymers with similar structures, parallel the expected variation in the magnitude of the intermolecular interactions.

10.8.5. Effect of molecular weight

The molecular weight will markedly influence the value of the glass-transition temperature if it is shorter than the critical entanglement value M_c. In this case the spectrum of the relaxation processes is significantly changed, and whole-molecule motions and rotational isomerism of the end-groups become dominant. They create a large extra free volume, and so lower the value of T_g. This effect of molecular weight is illustrated in Fig. 10.13 for vinylacetate polymers.

Table 10.2

Glass transitions of some common polymers[49]

Polymer	Segment structure	Glass transition temperature measured at 1 Hz (°C)
Polyethylene	$-CH_2-$	−20 to −60
Polypropylene	$-CH_2-CH-$ $\quad\quad\quad\mid$ $\quad\quad\quad CH_3$	0
Polystyrene	$-CH_2-CH-$ $\quad\quad\quad\mid$ $\quad\quad\quad C_6H_5$	116
Poly(N-vinylcarbazole)	$-CH_2-CH-$ (carbazole N structure)	211
Poly(vinyl-n-butylether)	$-CH_2-CH-$ $\quad\quad\quad\mid$ $\quad\quad\quad O-CH_2-CH_2-CH_2-CH_3$	−32
Poly(vinyl-iso-butylether)	$-CH_2-CH-$ $\quad\quad\quad\mid$ $\quad\quad\quad O-CH_2CH(CH_3)_2$	−1
Poly(vinyl-t-butylether)	$-CH_2-CH-$ $\quad\quad\quad\mid$ $\quad\quad\quad O-C(CH_3)_3$	83
Poly(vinylmethylether)	$-CH_2-CH-$ $\quad\quad\quad\mid$ $\quad\quad\quad O-CH_3$	−10
Poly(vinylethylether)	$-CH_2-CH-$ $\quad\quad\quad\mid$ $\quad\quad\quad O-C_2H_5$	−17
Poly(vinyl-n-propylether)	$-CH_2-CH-$ $\quad\quad\quad\mid$ $\quad\quad\quad O-C_3H_7$	−27
Poly(methylmethacrylate) (isotactic) (syndiotactic)	$\quad\quad\quad CH_3$ $\quad\quad\quad\mid$ $-CH_2-C$ $\quad\quad\quad\parallel$ $O=C-OCH_3$	104 40
Poly(ethylmethacrylate)	$\quad\quad\quad CH_3$ $\quad\quad\quad\mid$ $-CH_2-C-$ $\quad\quad\quad\parallel$ $O=C-OC_2H_5$	67
Poly(n-butylmethacrylate) Poly(iso-butylmethacrylate)	$\quad\quad\quad CH_3$ $\quad\quad\quad\mid$ $-CH_2-C-$ $\quad\quad\quad\parallel$ $O=C-OC_4H_9$	20 41

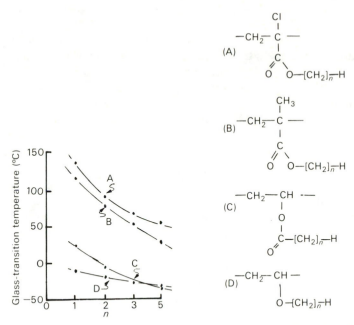

Fig. 10.12. Variation of glass transition temperature with chain length in the side-group.

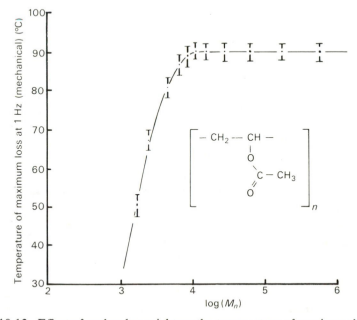

Fig. 10.13. Effect of molecular weight on the temperature of maximum loss.

10.8.6. Effect of plasticizers

In general, plasticizers are low molecular-weight materials which are added to rigid polymers to soften them. Plasticizers must be soluble in the polymer and are added at high temperature. The plasticizer lowers the glass-transition temperature and makes it easier for molecular conformational change. It has been proposed that the mode of action is through the creation of free volume. Alternatively the plasticizer alters the morphology of the solid and hence can create voids. It appears that both mechanisms involve the creation of available volume. This approach is also compatible with the observation that a small amount of a low molecular-weight fraction of a polymer can lead to a considerable loss in mechanical properties. The low molecular-weight polymer here acts as a plasticizer for the higher molecular-weight polymer, since it will have a lower T_g and hence an associated greater free volume at a particular temperature.

10.8.7. Lower temperature relaxation processes

A number of minor changes in the mechanical properties of a polymer are associated with the relaxation of specific molecular groupings. These relaxations are of considerable importance for they provide a mechanism of energy storage and dissipation which infers on a polymer a certain ductility and resilience to impact not found in other glassy materials. These lower-temperature relaxations are usually associated with the motions of small elements of the backbone and are principally controlled by intramolecular contributions to the rotational potential. The difference between the alpha and the beta process in many polymers may be one of the size of the group which undergoes conformational change. If the element involves a large number, typically six or more monomer units, the relaxation is volume-limited and associated with the glass-transition process. On the other hand, if the process involves only a small section of the chain (or a small-amplitude libration of the larger unit), relaxation can occur without any significant volume change being required. The effect of the local environment is observed, however, in the breadth of the beta relaxation. In many amorphous polymers this extends over several decades and is sensitive to changes in the morphology. It has also been shown in Chapter 1 that, on a time–temperature plot, the alpha and beta relaxation processes can merge. Intuitively, this suggests that these relaxations have a common origin.

Some of these ideas have been rationalized using a concept of correlation lengths.

10.8.8. The α- and β-relaxations as correlated molecular processes

From the above discussion it is clear that there may exist a logical connection between the beta and alpha relaxation processes.[53] If it is assumed that the beta process is associated with short-range conformational changes and that the glass transition involves large elements of the chain, then the logical connection parameter is one of correlation in the energy of motion.

Relaxation may change both the energy of the element and the entropy of the ensemble being investigated. In a condensed phase it will be assumed that the energy and relaxation time can be related through the equation

$$\tau^{-1} = 2\nu_0 \exp(-V/kT) \tag{10.68}$$

where V is the mean barrier height restricting conformational change and will include inter- as well as intramolecular contributions. The barrier will slow down the exchange between states in the region of the glass-transition temperature and 'freeze in' specific conformational structures. One way of introducing correlation is to consider that the energy difference between conformational states is determined in part by the relative occupation of the two states

$$W = W_0 + kT_c(n_1 - n_2)/(n_1 + n_2) \tag{10.69}$$

where T_c is a constant, with the dimensions of temperature, which describes the strength of the correlation between contributions. W_0 represents all other contributions to the energy difference, usually considered to be of an intermolecular origin. The occupation numbers n_1 and n_2 are defined for the two states between which relaxation occurs. Eqn (10.68) can then be rewritten as

$$\tau^{-1} = (\nu_0/2) \exp(V/kT)(\alpha(T)/(1 - (T_c/T)\alpha^2(T)) \tag{10.70}$$

where $\alpha^2(T)$ is defined by

$$\alpha^2(T) = 1 - m_0^2 \tag{10.71}$$

and m_0 is determined at any temperature by the equation

$$m_0 = \tanh(m_0 T_c/T). \tag{10.72}$$

The usual rate equations for the occupation numbers n_1 and n_2 of sites 1 and 2 have the form

$$dn_1/dt = \nu_c[n_2 \exp(-E_2/kT) - n_1 \exp(-E_1/kT)] \tag{10.73}$$

where $2V - E_1$ and $2W = E_1 - E_2$ and $N = n_1 + n_2$ with $m = (n_1 - n_2)/(n_1 + n_2)$. Then the above equation becomes

$$dm/dt = -2\nu_0 \exp(-V{-}kT)\{m \cosh(W/kT) - \sinh(W/kT)\}. \tag{10.74}$$

The energy difference then has the form of (10.69). For simplicity, if W_0 is put equal to zero, then (10.74) results. For $T > T_c$ this equation has only one solution, $m_0 = 0$. However, for $T < T_c$ it develops two stable solutions $m_0 \to \pm 1$ as $T \to 0$.

The response of the model to an applied field can be written as

$$m(t) = m_0 + M_1(t). \tag{10.75}$$

To find the linear response of this system combine (10.75) with (10.74) and expand in terms of M_1

$$\mathrm{d}m_1/\mathrm{d}t = \frac{-2\nu_0 \exp(-V/kT)}{(1 - m_0^2)^{\frac{1}{2}}} M_1\{1 - (T_c/T)(1 - m_0^2)\} + F(t)\}$$

$$\tag{10.76}$$

where $F(t)$ represents the interaction of the perturbation with the distribution between states.

The important conclusions which arise from the above analysis are that for glass-forming systems, the low-frequency relaxation will usually reveal two processes.[54-8] A singularity will however appear at $T = T_c$ and is a feature of the 'mean field' method of dealing with the interactions between the original two-site model and the rest of the system. It is clear that more sophisticated modelling will be required in order to remove the apparent singularity which arises at T_c.

A realistic comparison of the theory with data can be obtained if $V/kT_c = 25$. A single molecular mechanism is therefore invoked for the entire range of temperatures and frequencies, i.e. the activated hopping over a conformational barrier of constant height, V. In the absence of correlated motion this leads to simple Arrhenius behaviour for the general locus of the loss process in the entire temperature and frequency plane. Introducing a correlated term into the energy difference between the two conformations produces a marked distortion of the locus of the loss process. Towards lower frequencies this produces two processes—one located near T_c which is strong and identified as an α-process, and the other at lower temperatures, the β-process. This latter process is essentially the original Arrhenius process, but is modified to some extent by the correlated interactions. The existence of some degree of correlated motion, or the variation in the contributions (due to W_0) of long-range origin, lead to the considerable breadth of the β-process.

A factor of approximately $\frac{1}{25}$ for kT_c/V appears consistent with experimental observation.[54-8] The correlated motion does not radically change the barrier structure, which is largely determined by the short-range order.

The nature of the correlation in a particular polymer will depend very much on the detail of the conformational potential profile. Studies of poly(methylmethacrylate) in solution[59] have indicated that the relaxation involves both motion of the $-\text{COOCH}_3$ group and the C—C bond which links the side-group into the main chain. An inspection of molecular models shows that the steric hindrance to this rotation comes largely from the main-chain methyl substituents of the two adjacent repeat units. In this situation the relaxing group will, even in the case of the beta process, involve a substantial degree of co-operative motion. The alpha process will involve the correlated motion of a number of methylmethacrylate groups. Changes in the stereochemistry of the backbone will also influence the extent to which correlation occurs, as illustrated in Fig. 10.14.

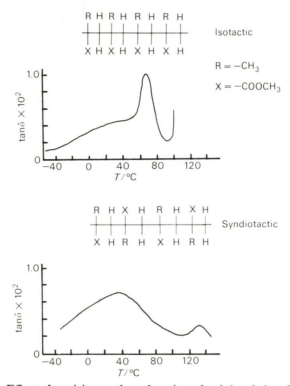

Fig. 10.14. Effect of tacticity on the relaxation of poly(methylmethacrylate).

Polyvinyl esters have a similar structure to the acrylates and exhibit a number of similarities in their relaxation properties.[60-71] The beta relaxation in this case is associated with an independent rotation of the side-groups ($-OCOCH_3$) with only local distortion of the chain. The side-group motions appear to be slightly more restricted when the side-chain oxygen atom exists adjacent to the main chain. A similar conclusion applies to the alpha relaxation process. Studies of other vinyl esters indicate that the motion is little affected by changes from methyl to benzyl, which supports the assignment of the beta relaxation to a process localized on the $-O-CO-$ group.

In the case of polyvinylchloride the precise nature of the beta relaxation has been open to some controversy, but does appear to involve quasi-librational motion of small elements of the chain backbone.[72-80]

This type of approach can be used to rationalize the beta relaxation for a large number of polymers, some of which are summarized at the end of this chapter.

In many polymers γ and even δ relaxation features have been observed. The occurrence of these features depends on the nature of the polymer structure and also on the presence of small molecule impurities. In the case of poly(methylmethacrylate) the γ-relaxation may be attributed to the rotational relaxation of the α-methyl groups.[55,56] It is difficult to see how the rotational motion of the threefold symmetrical methyl group attached to the main chain backbone can give rise to a mechanical loss. One possibility is that the onset of rotation of the α-methyl group gives rise to some local co-operative movement of the chain. Another possibility is that the energy profile contains multiples of three potential minima each of which are energetically non-equivalent.

Rotational relaxation of the methyl group in the ester side chain also gives rise to a relaxation process which occurs at considerably lower temperatures than the relaxation associated with the α-methyl group. These processes give rise to distinct n.m.r., neutron, and dielectric relaxation features, as discussed in other chapters.

The motion of groups attached to the side-chain of methacrylate polymers has been the object of attention of a number of workers for some considerable length of time. Using a combination of mechanical and dielectric observations it has been demonstrated that the motions of such pendant groups are often essentially independent of the dynamics of the main polymer matrix. Motion of the side-chain below the beta relaxation temperature will involve very little change in the over-all free volume of the matrix. As a result, the relaxation behaviour often appears to be independent of the matrix and is dominated by the intramolecular contributions to the potential profile, resembling the intramolecular relaxation behaviour of analagous small molecules. Above the beta relaxation temperature the pendant group moves with the rest of the side-chain and can significantly increase the available free volume, decreasing markedly the temperature of the glass transition.

A large number of substituted methacrylates and acrylates have been studied and a summary of the results is presented at the end of this chapter. Since the motion is similar to that observed in small molecules we shall not consider the interpretation of the data in any detail. The reader is referred to a number of specialized reviews in which the stereochemistry of the analogous small molecules are discussed in detail. The cyclic pendant groups are generally related to cyclohexane and exhibit relaxation behaviour associated with a combination of processes involving boat-twist and chair forms of the ring structure. Analogous relaxation features have been observed in other polymers and associated with very local relaxation processes (Table 10.3).

10.9. Relaxation in crystalline polymers

The precise molecular assignment of the viscoelastic behaviour in crystalline polymers is considerably more speculative than for the corresponding amorphous polymers.

X-ray diffraction studies of many crystalline polymers can be rationalized on the basis of the so-called fringe micelle model.[81] The scattering pattern is composed of discrete reflections from crystallites and a diffuse scattering from the amorphous regions. The fringe micelle model can be represented diagrammatically (Fig. 10.15). This model has undergone considerable

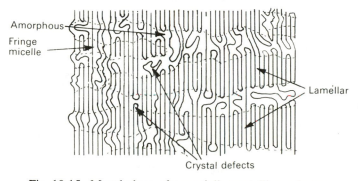

Amorphous

Fringe micelle

Lamellar

Crystal defects

Fig. 10.15. Morphology of a partially crystalline polymer.

revision since its original inception and now includes the possibility of the crystalline regions containing chain folds. Recent X-ray and electron diffraction studies of crystalline polymers have established the existence of regions of the order of 10–20 μm in area and 100 Å thick.[82-5] The thickness of the lamellae has been shown to be sensitive to the thermal history of the sample and lamellar thickening of the crystalline regions can be readily demonstrated. Electron diffraction shows that in polyethylene the polymer chains lie approximately normal to the lamellar surface and since the molecules are usually of the order of 10 000 Å in length, it can be deduced that they must be folded back and forth within the crystals. The relaxation behaviour of such a solid will be a composite of the motions of the crystalline and defective (pseudo-amorphous) phases. Support for the fringe micelle model is found in the empirical correlation of the magnitude of loss processes associated with the crystalline and amorphous regions with changes in the crystallinity as determined by X-ray or density methods. Two examples of polymers which fall within this class are poly(tetrafluoroethylene) (PTFE) and poly(ethyleneterephthalate) (PET).

10.9.1. Poly(tetrafluoroethylene) (PTFE)

The relaxation properties of a series of samples of PTFE of different crystallinity are illustrated in Fig. 10.16, displaying the temperature dependence of the logarithmic decrement for three transitions in PTFE as a

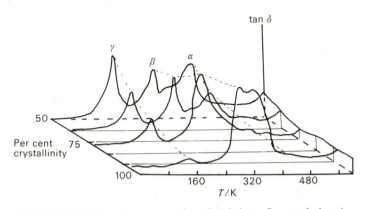

Fig. 10.16. Relaxation properties of poly(tetrafluoroethylene).

function of the degree of crystallinity.[86,87] The lowest temperature relaxation decreases in magnitude with increasing crystallinity in a very clear manner. On the basis of a two-phase model it can be identified with a transition in the amorphous regions of the polymer. The β-relaxation, on the other hand, increases in magnitude with increasing crystallinity and is therefore associated with the crystalline regions. Analysis of the α-relaxation is somewhat dependent on the method of resolving the loss peak, and of determining the strength of a relaxation process, but the consensus of opinion would seem to confirm that it decreases in magnitude with increasing crystallinity and is therefore associated with the amorphous regions.

10.9.2. Poly(ethyleneterephthalate) (PET)

In PET[88] (Fig. 10.17) the effect of crystallinity on the β-relaxation is very small and has led to a very complex interpretation in terms of this loss peak being composed of several relaxation processes. No clear distinction can be drawn between relaxation occurring in the crystalline and the amorphous regions.

In the case of the α-relaxation the loss peak decreases with increasing crystallinity, broadens, becomes very asymmetric, and moves to higher

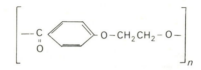

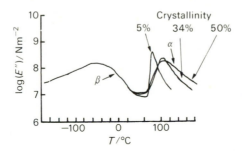

Fig. 10.17. Relaxation properties of poly(ethyleneterephthalate).

temperatures. Such behaviour may be explained on the basis of the motion occurring in the amorphous region; the presence of crystallites places constraints on the amorphous region and influences the mobility of the chains. Support for this hypothesis has been obtained from n.m.r. measurements.

In crystalline polymers, relaxations may be ascribed to the following processes:

(1) Motions which occur in the amorphous phase. These will include the glass transition, an example of which is the alpha relaxation in PET or the gamma transition in PTFE.

(2) Motions which occur in the crystalline phase. These may be of two types: (a) those involving co-operative motions of the molecular chains along the chain length of the crystallite—these relaxations will be related to the lamellar thickness; (b) those associated with defects such as end-groups in the crystal and involved chain motion.

(3) Motions of a local nature involving only one or two carbon atoms moving. This type of motion will be discussed later in this chapter.

(4) Motions of large elements of the polymer and associated with inter-lammellar shear or interfibrillar shear.

10.9.3. Relaxation in hydrocarbon polymers

The temperature dependence of the loss in low- and high-density polyethylene (Fig. 10.18) illustrates the complexity of interpretation and assignment of the molecular origins of the observed relaxation processes.[89–94] Low-density polyethylene although basically consisting of long chains of

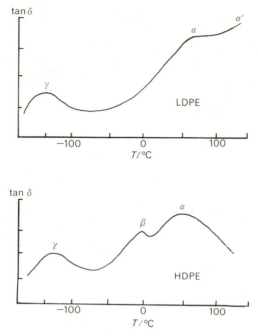

Fig. 10.18. Relaxation in polyethylene—high- and low-density polymers.

—CH_2— groups, contains a significant number of short side-branches (of the order of three per 100 carbon atoms in a typical commercial polymer) together with a few long branches (about one per molecule). High-density polyethylene, on the other hand, is very much closer to being pure $(CH_2)_n$ polymer, and the number of branches is often less than five per 1000 carbon atoms.

Low-density polyethylene exhibits three clearly distinguishable loss peaks whereas high-density polyethylene exhibits only two peaks. Comparison of the relaxation spectra indicates that the lowest temperature peak (not shown in Fig. 10.18) is apparently independent of the detailed structure of the polymer, but decreases in intensity as the crystallinity increases. It is therefore associated with relaxation in the amorphous region and has been ascribed to 'local twisting of the molecular chains'. Annealing of crystallized mats of polyethylene leads to an increase in amplitude of the peak which is attributed to the generation of defects within the lamellae, perhaps due to motion of dislocations and point defects.

From studies of polyethylenes of widely varying densities, the β-relaxation has been attributed to relaxation of the branch points. The α-peak and the α' peak in the high-density polymer are sensitive to the thermal treatment of the polymer, irradiation and polymer structure. The α-process has been assigned to translational motion of chain segments along the chain axis within the crystal lattice. Alternatively, the relaxation has been ascribed to reorientation of chain folds at the lamellae surfaces. Support for this assignment is obtained from the observation that the magnitude of the α-relaxation is inversely proportional to the lamellar thickness. The decrease in intensity on annealing is therefore directly proportional to the decrease in the number of chain folds as the lamellae thicken. The α'-peak can be interpreted as slip at the lamellar boundary. It is envisaged that the lamellae bend under the applied stress like an elastic beam in a viscous liquid. To explain the fact that the creep is totally recoverable, it is assumed that the lamellae are pinned along their length.

Further information on the molecular origins of the relaxation processes in polyethylene can be obtained by extending the studies to orientated samples of the polymer. The α-relaxation in low-density polymer is observed to develop anisotropy and this is attributed to translational motion of the chains within the lamellae. It seems probable that the chains involved form interlamellar ties so that the stresses are transmitted throughout the bulk polymer.

It is clear from this discussion that the assignment of a unique relaxation mechanism to an observed loss peak in a crystalline polymer can be a very difficult exercise.

10.10. Copolymers, blends, and grafts

If two polymers in a mixture are insoluble they will exist as two separate phases and each phase will exhibit a glass transition characteristic of the homopolymer.[95-7] This type of situation is illustrated by the behaviour of poly(styrene-butadiene-styrene) triblock polymers. Two loss peaks are observed, the lower close to that for polybutadiene and the higher close to that of polystyrene. If the two polymers are completely soluble in one another, the properties of the mixture can become almost identical to those of a random copolymer of the same composition.

The properties of a composite can be interpreted on the basis of a free-volume model. The glass transition is associated with the occurrence of a certain critical value of the free volume. In an ideal copolymer, the partial specific volumes of the two components are constant and equal to the specific volumes of the two homopolymers. If it is further assumed that the specific volume–temperature coefficients for the two copolymers are as in the homopolymers and are independent of temperature, the glass transition temperature T_g for the copolymer is given by

$$\frac{1}{T_g} = \frac{1}{(w_1 + Bw_2)}\left\{\frac{w_1}{T_{g1}} + \frac{Bw_2}{T_{g2}}\right\} \qquad (10.78)$$

where w_1 and w_2 are the weight fractions of the two monomers whose homopolymers have transitions at temperature T_{g1} and T_{g2} respectively and B is a constant which is close to unity.

The mechanical properties of a composite depend very much on the nature and ratio of the components. Mixing of two mutually insoluble polymers causes one phase to form dispersed regions, often of spherical form, surrounded by the other phase. The morphology of phase separation is determined by a number of factors, the principal being the tendency to minimize the contact area of the interface. Changing the composition of the phases alters the volume fraction of the spherical phase until hexagonal close packing is achieved at approximately 70 per cent concentration. Further increase leads to phase inversion.

The complex dynamic modulus of a composite may be represented in a number of ways none of which is totally satisfactory.[98-105] One of the simplest models is obtained by assuming that the hard phase is continuous and that the two elements are arranged in parallel.

$$E^* = (1 - V)E_1 + VE_2. \tag{10.79}$$

Where E^* is the complex Young's modulus of the system, E_1 and E_2 are the complex moduli of the two phases, and V is the volume fraction of phase 2.

The expressions for the real and imaginary parts are

$$\begin{aligned} E' &= (1 - V)E_1' + VE_2' \\ E'' &= (1 - V)E_1'' + VE_2'' \end{aligned} \tag{10.80}$$

which leads to the loss

$$\tan \delta = \frac{(1 - V)E_1' \tan \delta_1 + VE_2' \tan \delta_2}{(1 - V)E_1' + VE_1'}. \tag{10.81}$$

This model for the mechanical behaviour of a composite will describe adequately the behaviour of systems such as the tri-block styrene-butadiene-styrene, at large styrene contents.

On the other hand, the behaviour of mutually insoluble polymers where the rubbery phase is predominant appears to be best modelled using a series combination of the elastic elements. The dynamic modulus and loss tangent now have the form

$$E_1' = [V/E_1' + (1 - V)/E_2']^{-1} \tag{10.82}$$

$$\tan \delta = \left(\frac{1}{E_1'} + \frac{(1 - V)}{E_2'}\right) \left(\frac{V}{\tan {}_1E_1'} + \frac{(1 - V)}{\tan {}_2E_2'}\right). \tag{10.83}$$

In detail neither of these simple models is adequate and more complex combinations of parallel and series combinations may be introduced. The molecular origins of the more complex stress fields in these materials lies in the interpenetrating nature of the networks formed. The ideal spherical phase separation is rarely observed in practice and the composite often approximates more closely to an array of interpenetrating networks and small occlusions of the minor component.

10.11. Oriented polymer films

Partially oriented polymer films exhibit anisotropy in their mechanical properties.[106] The process of orientation is believed to be associated with the formation of fibrillar structures, oriented in the direction of draw. The process of drawing can be envisaged as the elongation of pre-existing crystallites as a result of the stresses set up when molecular ties constrain the relative movement of neighbouring lamellae. The polymers, when stressed, form a fibrillar structure and align in the direction of draw. The anisotropy in the mechanical properties arises from the difference in the nature of the forces opposing deformation. In the draw direction elongation involves distortion of the fibril structures, which can in turn be associated with deformation of the backbone structure of the polymer. Mechanical elongation perpendicular to the draw direction will involve relative movement of the fibres, in which case the structure of ties and Van der Waals forces are involved.

This type of behaviour can be demonstrated by observation of the propagation velocity of ultrasonic waves.[107] The increase of modulus in the draw direction and drop of modulus in the perpendicular direction is typical of a number of partially crystalline polymers. This topic has been extensively reviewed elsewhere and the development of anisotropy in electrical and magnetic properties has been widely used as a probe for the orientation of drawn molecules.[108]

In general the formation of fibres leads to a restriction of the motion of the polymer chains. Shifts of the relaxation peaks to higher temperature on orientation of isotropic materials can be attributed to restrictions by a combination of intra- and intermolecular forces. These constraints often lead to helix formation and so to fibre structure.

One important consequence of this fibre structure is an increase in the barrier properties to small molecular diffusion. Unfortunately drawn films often suffer from loss of mechanical rigidity and strength due to fibrillation. To avoid this occurring, polymer film is often prepared by biaxial drawing. This avoids development of a high degree of anisotropy, but does create an increase in the barrier properties and modulus of the film.

10.12. Effect of relaxation on mechanical properties[109–24]

Two of the most important properties of a polymeric material are ease of fabrication and ability to withstand impact. An impact can be idealized as a

Table 10.3

Summary of the relaxation characteristics of some common polymers

Polymer	Monomer structure	Glass-transition temperature (K)	Label	Relaxation process molecular description	Temp. of Mechanical loss peak (K)	Refs.
Polyethylene	$-(CH_2)_n-$	213	α	Translational motion of chain segments within the crystalline regions	333	125–44
			β	Relaxation of chains at branch points	273	
				Very local motion of the polymer in amorphous regions	150	
Polypropylene	$-(CH_2-CH)_n-$ with CH_3	288	α	Sensitive to tacticity of the polymer and thermal history, attributed to motion within crystalline domains	373	145–53
			β	Glass–rubber relaxation associated with large-scale motion of the chains in the amorphous region	288	
				Local motion of the polymer chain	193	
Polyisobutene	$-(CH_2C(CH_3)_2)_n-$	202	α	Glass–rubber relaxation associated with large-scale motion of the chain	202	154–6
Polystyrene	$-(CH_2CH)_n-$	373	α	Glass–rubber relaxation	398	157–71
			β	Local motions of the backbone	293	
				End-group motion and coupled local motions	163	

Polymer	Structure		Description		
Poly(vinylchloride)	$-(CH_2CHCl)_n-$	α	Glass–rubber relaxation	373	172–88
		β	Local motion of the backbone	243	
Poly(tetrafluoroethylene)	$-(CF_2CF_2)_n-$	α	Relaxation in the amorphous region	400	189–98
		α	Torsional oscillators of chain segments around the chain axis within the crystalline domain	320	
		β	Local motions in the amorphous phase	180	
Poly(methylmethacrylate)	$-(CH_2C)_n-$ with CH_3, $C=O$, $O-CH_3$ side group	α	Glass–rubber transition	383	199–207
		β	Local motions of the acrylate side-group	293	
Poly(ethylmethacrylate)	$-(CH_2C)_n-$ with CH_3, $C=O$, $O-C_2H_5$ side group	α	Glass–rubber transition	378	199–207
		β	Local motion of the acrylate side-group	283	
Poly(vinylacetate)	$-(CH_2CH)_n-$ with $O-C(=O)CH_3$ side group	α	Glass–rubber transition	325	208–17
		β	Side-group relaxation	172	

force acting for a short period at a point. Mathematically the impact is a delta function and its Fourier spectrum will normally extend over a frequency range from a few tens of hertz to approximately a kilohertz. In plastics which are capable of withstanding high impacts there must exist a loss process which matches the Fourier power spectrum generated by the impulse, which itself is capable of storing and dissipating the energy. In the absence of a viscoelastic process, the energy cannot be dissipated at the point of impact, large stresses are created, and fracture follows.

Matching the polymer power spectrum to the Fourier transform of the impulse function appears to rationalize the impact properties of a variety of polymers. In practice, since mechanical rigidity is an essential property of a plastic, it is the beta relaxation which is matched to the impulse function.

The basic dynamic spectrum influences not only the impact properties but also the fracture properties of a polymer. A fracture is considered as a propagating stress region. The polymer chains in the fracture zone experience a high stress which can be relieved partially by motion of the chains. Two extreme conditions can be identified. In a glassy material no relaxation processes exist for the removal of energy from the stress zone and brittle fracture results. If a relaxation exists with a relaxation spectrum which matches the stress rate in the fracture zone, polymer deformation and flow can occur with plastic fracture.

Simple crack propagation rates under conditions of plane strain have been shown to exhibit a reasonable correlation with the frequency spectrum and in particular again with the beta relaxation. For further discussion on this topic the reader is referred to specialized reviews on the yield and fracture properties of polymers.[106,109,124]

10.13. Review of mechanical relaxation data

To help our appreciation of the relation between chemical structure, molecular weight, and polymer mobility the following resumé of mechanical and related relaxation data for solid polymers has been prepared (Table 10.3).

References

1. Schmieder, K. and Wolf, K. *Kolloid Z.* **134**, 149 (1953).
2. Sauer, J. A. and Kline, D. E. *J. polym. Sci.* **18**, 491 (1955).
3. Buchdahl, R. and Nielson, L. E. *J. polym. Sci.* **15**, 1 (1955).
4. Becker, G. W. *Kolloid Z.* **140**, 1 (1955).
5. Merz, E. H., Nielsen, L. E. and Buchdahl, R. *Ind. Engng Chem.* **43**, 1396 (1951).
6. Jenckel, E. *Kolloid Z.* **136**, 142 (1954).
7. Illers, K. H. and Jenckel, E. *Rheol. Acta* **1**, 322 (1958).

8. Illers, K. H. and Jenckel, E. *J. polym. Sci.* **41**, 528 (1959).
9. Illers, K. H. *Z. Electrochem.* **65**, 679 (1961).
10. Sinnett, K. M. *Soc. plastics Engr Trans.* **2**, 65 (1962).
11. Tanayanagi, M. *Mem. Fac. Engng Kyushu Univ.* **23**, 1 (1963).
12. Lodge, A. A. *Elastic liquids.* Academic Press, New York (1964).
13. Coleman, B. D. and Noll, W. *Rev. mod. Phys.* **33**, 239 (1961).
14. Timoshenko and Goodier, J. N. *Theory of elasticity.* McGraw-Hill, New York (1951).
15. Love, A. E. H. *A treatise on the mathematical theory of elasticity* (4th edn). Macmillan, New York (1944).
16. Cross, B. *Mathematical structure of the theories of viscoelasticity.* Hermann, Paris (1953).
17. Ward, I. M. *Mechanical properties of solid polymers*, Wiley, New York, London (1971).
18. Ferry, J. D. *Viscoelastic properties of polymers* (2nd edn). Wiley, New York (1970).
19. McCrum, N. G., Read, B. E. and Williams, G. *Anelastic and dielectric effects in polymeric solids.* Wiley, New York (1967).
20. Hiller, K. W. *Progress in solid mechanics*, Vol. 5. North-Holland, Amsterdam (1961).
21. Leaderman, H. *Elastic and creep properties of filamentous materials and other high polymers.* Textile Foundation, Washington, DC (1943).
22. Williams, M. L., Landel, R. F. and Ferry, J. D. *J. Amer. chem. Soc.* **77**, 3701 (1955).
23. McCrum, N. G. and Morris, E. L. *Proc. Roy. Soc. A* **281**, 258 (1964).
24. McCrum, N. G. and Morris, E. L. *J. appl. Phys.* (*Jpn*) **4**, 542 (1965).
25. Doolittle, A. K. *J. appl. Phys.* **22**, 1471 (1951).
26. Doolittle, A. K. *J. appl. Phys.* **23**, 236 (1952).
27. Glasstone, S. K., Laider, K. J. and Eyring, H. *The theory of rate processes.* McGraw-Hill, New York (1941).
28. Work, R. N. *J. appl. Phys.* **27**, 69 (1956).
29. Gibbs, J. H. and DiMarzio, E. *J. chem. Phys.* **28**, 373, 807 (1958).
30. Gibbs, J. H. and DiMarzio, E. *J. polym. Sci.* **40**, 121 (1959).
31. Gibbs, J. H. In *Modern aspects of the vitreous state* (ed. J. D. MacKenzie), Chapter 7. Butterworths, Washington.
32. Deutsch, K., Hoff, E. A.and Reddish, W. *J. polym. Sci.* **13**, 365 (1954).
33. Powles, J. G., Hunt, B. I. and Sandiford, D. J. *Polymer* **5**, 505 (1964).
34. Sinnot, K. M. *J. polym. Sci.* **42**, 3 (1960).
35. Heijboer, J. *Physics of non-crystalline solids*, p. 231. North-Holland, Amsterdam (1965).
36. Thompson, A. B. and Woods, D. A. *Trans. Faraday Soc.* **52**, 1383 (1956).
37. Reddish, W. *Trans. Faraday Soc.* **46**, 459 (1950).
38. Farrow, G., McIntosh, J. and Ward, I. *Makromol. Chem.* **38**, 147 (1960).
39. Illers, K. H. and Breuer, H. *J. colloid Sci.* **18**, 1 (1963).
40. Heijboer, J. *Chem. Weekblad* **48**, 264 (1952).
41. Loskaek, S. *J. polym. Sci.* **15**, 391 (1955).
42. Rogers, S. S. and Mandelkern, L. *J. phys. Chem.* **61**, 985 (1957).
43. Alexandrov, A. P. and Lazurkin, J. *Acta phys. Chem., USSR*, **12**, 647 (1940).
44. McLaughlin, J. R. and Tobolsky, A. V. *J. colloid Sci.*, **7**, 555 (1952).
45. Tobolsky, A. V. *Properties and structure of polymers.* Wiley, New York (1960).
46. Heijboer, J. *Kolloid Z.* **134**, 149 (1956).

47. Mikhailov, G. P. and Boriseva, T. I. *Sov. Phys. Tech. Phys.* **3**, 120 (1958).
48. Ishida, Y. and Yamafuji, K. *Kolloid Z.* **177**, 97 (1961).
49. Vincent, P. I. *Physics of plastics.* Iliffe Books, London (1965).
50. Neilson, L. E. *Mechanical properties of polymers.* Reinhold, New York (1962).
51. Willbourn, A. H. *Trans. Faraday Soc.* **54**, 717 (1958).
52. Boyer, R. F., *Rubber Rev.* **34**, 1303 (1963).
53. Brereton, M. and Davis, G. R. *Polymer* **18**, 765 (1977).
54. Brereton, M., Davis, G. R., Rushworth, A. and Spence, J. *J. polym. Sci. A-2* **15**, 583 (1977).
55. Ishida, Y., Watanabe, M. and Yamafuji, K. *Kolloid Z.* **200**, 48 (1954).
56. Ishida, Y. *Kolloid Z.* **168**, 29 (1960).
57. Reddish, W. *J. polym. Sci.* Part C **14**, 123 (1966).
58. Yano, S. *J. polym. Sci.* (*Phys.*) **14**, 1877 (1976).
59. North, A. M. and Block, H. *Advn. mol. Relaxation Processes.* **4**, 1 (1970).
60. Ishida, Y., Matsuo, M. and Yamafuji, K. *Kolloid. Z.* **180**, 108 (1962).
61. Veselovskii, P. F. and Slusker, A. *Zh. Tech.* **25**, 939 (1955); **25**, 1204 (1955).
62. Mikhailov, G. P. *Makromol. Chem.* **35**, 26 (1960).
63. Saito, S. *Kolloid Z.* **189**, 116 (1963).
64. Mead, D. J. and Fuoss, R. M. *J. Amer. chem. Soc.* **63**, 2832 (1941).
65. Broens, O. and Muller, F. H. *Kolloid Z.* **140**, 121 (1955); **141**, 20 (1955).
66. Thurn, H. and Wolf, K. *Kolloid Z.* **148**, 6 (1956).
67. Hikichi, K. and Furuichi, J. Rep. Progr. polymer. Phys. (Jpn) **4**, 69 (1961).
68. Thurn, H. and Wolf, K. *Kolloid Z.* **156**, 21 (1968).
69. Schmeider, K. and Wolf, K. *Kolloid Z.* **127**, 65 (1952).
70. McKinney, J. E. and Belcher, H. V. *J. Res. nat. Bur. Stds.* **67A**, 43 (1963).
71. Williams, M. L. and Ferry, J. D. *J. kolloid Sci.* **9**, 479 (1954); **10**, 474 (1955).
72. Dyson, A. *J. polym. Sci.* **7**, 133 (1951).
73. Reddish, W. *Pure appl. Chem.* **5**, 723 (1962).
74. Ishida, Y. *Kolloid Z.* **168**, 29 (1960).
75. Koppelmann, J. and Gielesson, J. *Kolloid Z.* **176**, 97 (1961).
76. Becker, G. W. *Kolloid Z.* **140**, 1 (1955).
77. Woodward, A. E. and Sauer, J. A. *Advan. polym. Sci.* **1**, 114 (1958).
78. Tanaka, K. *Rep. Progr. polym. Phys.* (*Jpn*) **5**, 138 (1962).
79. Okuyama, M. and Yanagida, T. *Rep. Progr. polym. Phys.* (*Jpn*) **6**, 125 (1963).
80. Takayanagi, M. *High Polymers* (*Jpn*) **10**, 289 (1961).
81. Geil, P. H. *Polymer single crystals.* Interscience, New York (1963).
82. Fisher, E. W. *Naturforsch.* **12a**, 753 (1957).
83. Keller, A. *Phil. Mag.* **2**, 1171 (1957).
84. Till, P. H. *J. polym. Sci.* **24**, 301 (1957).
85. Keller, A. and Sawada, S. *Makromol Chem.* **74**, 190 (1964).
86. McCrum, N. G. *J. polym. Sci.* **34**, 355 (1959).
87. Ward, I. M. *Trans. Faraday. Soc.* **56**, 648 (1960).
88. Takayanagi, M., Neki, K., Nagai, A. and Minami, S. Zairyo **14**, 343 (1965).
89. Fischer, E. W. and Schmidt, S. C. *Z. angew. Chem.* **74**, 551 (1962).
90. McCrum, N. G. and Morris, E. L. *Proc. Roy. Soc. A.* **292**, 506 (1966).
91. Kline, D. E., Sauer, J. A. and Woodward, A. E. *J. polym. Sci.* **22**, 455 (1956).
92. Illers, K. H. *Rheol. Acta* **3**, 194 (1964).
93. Stachurski, Z. H. and Ward, I. M. *J. polym. Sci. A-2* **6**, 1083, 1817 (1968).
94. Stachurski, Z. H. and Ward, I. M. *J. macromol Sci.*, (*Phys*), **B3**(3), 445 (1969).
95. Gordon, M. and Taylor, J. S. *J. appl. Chem.* **2**, 493 (1952).

96. Mandelkern, L., Martin, G. M. and Quinn, F. A. *J. Res. nat. Bur. Stand.* **58**, 137 (1959).
97. Fox, T. G. and Leshack, A. *J. polym. Sci.* **15**, 371 (1955).
98. Smallwood, H. M. *J. appl. Phys.* **15**, 758 (1944).
99. Guth, E. and Hold, O. *Phys. Rev.* **53**, 322 (1958).
100. Dewey, I. M. *J. appl. Phys.* **18**, 578 (1947).
101. Fokin, A. G. and Shermerger, G. D. *Prikl. Mekh. Tech. Fiz.* **13**, 37 (1968).
102. Chaban, I. A. Acoustics Institute of the Academy of Sciences, USSR (1963).
103. Ninomya, K. *J. kolloid Sci.* **14**, 49 (1959).
104. Uemura, S. and Takayanagi, M. *J. appl. polym. Sci.* **10**, 113 (1966).
105. Kavai, H. and Ogava, M. *Kebunshi* **12** (10), 752 (1963).
106. Ward, I. M. *Mechanical properties of solid polymers.* Wiley (1971).
107. Datta, P. K. and Pethrick, R. A. *Polymer* **19**, 145 (1978).
108. Ward, I. M. *Structure and properties of orientated polymers.* Applied Science Publishers (1975).
109. Nadai, A. *Theory of flow and fracture of solids.* McGraw-Hill, New York (1950).
110. Bridgeman, P. W. *Studies of large plastic flow and fracture.* McGraw-Hill, New York (1952).
111. Cottrell, A. H. *Dislocation and plastic flow in crystals.* Clarendon Press, Oxford (1953).
112. Thorkildsen, R. L. *Engineering design for plastics*, p. 322. Reinhold, New York (1964).
113. Lezurkin, Y. S. and Fegelson, R. A. *Zh. Tech. Fiz.* **21**, 267 (1951).
114. Robertson, R. E. *J. appl. polym. Sci.* **7**, 443 (1963).
115. Crowet, C. and Homes, G. A. *Appl. Mat. Res.* **3**, 1 (1964).
116. Bauwens-Crowet, C., Bauwens, J. A. and Homes, G. A. *J. polym. Sci. A-2* **7**, 1745, 735 (1969).
117. Reetlin, J. A. *Polymer* **6**, 311 (1965).
118. Haward, R. N. and Thackray, G. *Proc. Roy. Soc.* **A302**, 453 (1968).
119. Andrews, E. H. *Proc. Roy. Soc.* **A277**, 562 (1964).
120. Brown, N. and Ward, I. M. *Phil. Mag.* **17**, 961 (1968).
121. Brown, N., Duckett, R. A. and Ward, I. M. *J. Phys., appl. Phys.* **1**, 1369 (1968).
122. Frank, F. C., Keller, A. and Conner, A. O. *Phil. Mag.* **3**, 64 (1958).
123. Ward, I. M. *Text. Res. J.* **31**, 650 (1961).
124. Andrews, E. H. Developments in polymer fracture, Vol. 1 (ed. E. H. Andrews), Applied Science Publishers, London (1979).
125. Schmieder, K. and Wolf, K. *Kolloid Z.* **134**, 149 (1953).
126. Flocke, H. A. *Kolloid Z.* **188**, 114 (1963).
127. Oakes, W. G. and Robinson, D. W. *J. polym. Sci.* **14**, 505 (1954).
128. Mikhailov, G. P., Lobanov, A. M. and Sazhin, B. I. *J. tech. Phys.* (*USSR*) **24**, 1553 (1954).
129. Mikhailov, G. P. and Borisov, T. I. *J. tech. Phys.* (*USSR*) **23**, 2159 (1953).
130. Kabin, S. P. *Zh. tekhn. Fiz.* **26**, 2628 (1956).
131. Sandiford, D. J. H. and Willbourn, A. H. In *Polyethylene* (ed. A. Renfrew and P. Morgan), Chapter 8. Iliffe, London (1960).
132. Nakayasu, H. H., Markovitz, H. and Plazek, D. J. *Trans. Soc. Rheol.* **5**, 261 (1961).
133. Takayanagi, M. *High Polymers* (*Jpn*) **10**, 289 (1961).
134. Iwayanagi, S. *Rep. Progr. polym. Phys.* (*Jpn*) **5**, 131 (1962).

135. Rempel, R. C. *J. appl. Phys.* **28**, 1082 (1957).
136. Thurn, H. *Kolloid Z.* **173**, 72 (1960).
137. Wada, Y. and Tsuge, K., *J. appl. Phys. (Jpn)* **1**, 64 (1962).
138. Reddish, W. and Barrie, J. T. *IUPAC Symp. Macromol. Chem. Wiesbaden.* Short communication I.A.3 (1959).
139. Tsuge, K. *J. appl. Phys. (Jpn)* **1**, 270 (1962).
140. Pechold, W. S., Blasenbrey, S. and Woerner, S. *Kolloid Z.* **189**, 14 (1963).
141. Kline, D. E., Sauer, J. A. and Woodward, A. E. *J. polym. Sci.* **22**, 455 (1956).
142. Reding, F. P., Faucher, J. A. and Whitman, R. D. *J. polym. Sci.* **57**, 483 (1962).
143. Boyer, R. F. *Rubber Rev.* **34**, 1303 (1963).
144. Baccaredda, M. and Butta, E. *J. polym. Sci.* **31**, 189 (1958).
145. Sauer, J. A., Wall, R. A., Fuschillo, N. and Woodward, A. E. *J. appl. Phys.* **29**, 1385 (1958).
146. Willbourn, A. H. *Trans. Faraday Soc.* **54**, 717 (1958).
147. McCrum, N. G. *J. polym. Sci.* **34**, 355 (1959).
148. Newman, S. and Cox., W. P. *J. polym. Sci.* **46**, 29 (1960).
149. Von Schweider, E. *Ann. Phys.* **24**, 711 (1907).
150. Passaglia, E. and Martin, G. M. *J. Res. Nat. Bur. Stds.* **68**, 519 (1964).
151. Ferry, J. D. *Viscoelastic properties of polymers.* Wiley, New York (1961).
152. Tobolsky, A. V. *Properties and structure of polymers.* Wiley, New York (1960).
153. Turner, A. and Bailey, F. E. *J. polym. Sci.* **1B**, 601 (1963).
154. Ferry, J. D. and Fitzgerald, E. R. *Proc. 2nd Int. Congr. Rheol.*, p. 140. (1953). *J. Colloid Sci.* **8**, 224 (1953).
155. Kabin, S. P. and Mikhailov, G. P. *Zh. Tekhn. Fiz.* **26**, 511 (1956).
156. Catsiff, E. and Tobolsky, A. V. *J. Kolloid Sci.* **10**, 375 (1955).
157. Schmieder, K. and Wolf, K. *Kolloid Z.* **134**, 149 (1953).
158. Sauer, J. A. and Kline, D. E. *J. polym. Sci.* **18**, 491 (1955).
159. Buchdahl, R. and Nielsen, L. E. *J. polym. Sci.* **15**, 1 (1955).
160. Becker, G. W. *Kolloid Z.* **140**, 1 (1955).
161. Merz, E. H., Nielsen, L. E. and Buchdahl, R. *Ind. engng Chem.* **43**, 1396 (1951).
162. Jenckel, E. *Kolloid Z.* **136**, 142 (1954).
163. Illers, K. H. and Jenckel, E. *Rheol. Acta* **1**, 322 (1958).
164. Illers, K. H., *Z. Elektrochem.* **65**, 679 (1961).
165. Sinnott, K. M. *Soc. plastics Engr Trans.* **2**, 65 (1962).
166. Broens, O. and Muller, F. H., *Kolloid Z.* **140**, 121 (1955); **141**, 20 (1955).
167. Baker, E. B., *Rev. sci. Instrum.* **20**, 716 (1949).
168. Kastner, S. *Kolloid Z.* **187**, 27 (1963).
169. Barb, W. G., *J. polym. Sci.* **37**, 515 (1959).
170. Overberger, C., Frazier, C. J., Mandelman, J. and Smith, H. *J. Amer. chem. Soc.* **75**, 3326 (1953).
171. Curtis, A. J. *Soc. plastics Engr. Trans.* **18**, 82 (1962).
172. Fordham, J. W., McCairn, G. H. and Alexander, L. H. *J. polym. Sci.* **39**, 335 (1959).
173. George, M. R. H., Grisenthwaite, R. J. and Hunter, R. F. *Chem. Ind. (Lond.)* 719, 1114 (1958).
174. Reding, F. P., Faucher, J. A. and Whitman, R. D. *J. polym. Sci.* **57**, 483 (1962).
175. Davis, J. M., Miller, R. F. and Busse, W. F. *J. Amer. chem. Soc.* **63**, 361 (1941).
176. Fuoss, R. M. and Kirkwood, J. G. *J. Amer. chem. Soc.* **63**, 385 (1941).
177. Wurstlin, F. *Kolloid Z.* **110**, 71 (1948).
178. Nielsen, L. E., Buchdahl, R. and Lavreault, R. *J. appl. Phys.* **21**, 607 (1950).

179. Dyson, A. *J. polym. Sci.* **7**, 133 (1951).
180. Wolf, K. *Kunstoffe, Plastics* **41**, 89 (1951).
181. Thurn, H. *Z. angew. Phys.* **7**, 44 (1955).
182. Thurn, H. and Wurstlin, F. *Kolloid Z.* **156**, 21 (1958).
183. Reddish, W. *Soc. Chem. Ind. Symp.* (*London*) *Monograph 5*, 138 (1959).
184. *Ishida, Y. M. Kolloid Z.* **190**, 16 (1963).
185. Kiessling, D. *Kolloid Z.* **176** 119 (1961).
186. Koppelman, J. and Gielessen, J., *Kolloid Z.* **175**, 97 (1961).
187. Saito, S. *Kolloid Z.* **189**, 116 (1963).
188. Sommer, W. *Kolloid Z.* **167**, 97 (1959).
189. Sperati, C. A. and Starkweather, H. W. *Advan. polym. Sci.* **2**, 465 (1961).
190. Bunn, C. W. and Howells, E. R. *Nature* **174**, 549 (1954).
191. Pierce, R. H., Clark, E. S., Whitney, J. F. and Bryant, W. M. D. Meeting of Amer. Chem. Soc. Atlantic City, New Jersey (September 1956).
192. Clark, E. S. and Muus, L. T. Meeting Chem. Soc. Assoc. Milwaukee, Wisconsin (June 1958).
193. Rigby, H. A. and Bunn, C. W. *Nature* **164**, 583 (1949).
194. Schulz, A. K. *J. chem. Phys.* **53**, 933 (1956).
195. McCrum, N. G. *J. polym. Sci.* **27**, 555 (1958).
196. Takayanagi, Y. *J. appl. polym. Sci.* **5**, 468 (1961).
197. Ohzawa, Y. and Wada, Y. *Rep. Progr. polym. Phys.* (*Jpn*) **3**, 436 (1964).
198. Krum, F. and Muller, F. H. *Kolloid Z.* **164**, 8 (1959).
199. Mikhailov, G. P. and Borisova, T. I. *Vysokomolekul Soedin* **2**, 1772 (1961).
200. Nagata, N. K., Hikichi, M., Kaneri, M. and Furuichi, J. *Rep. Progr. polym. Phys.* (*Jpn*) **6**, 235 (1963).
201. Gall, W. G. and McCrum, N. G. *J. polym. Sci.* **50**, 489 (1961).
202. Heijboer, J. Physics of non-crystalline solids, p. 231. North-Holland, Amsterdam (1965).
203. Powles, J. G., Strange, J. H. and Sandiford, D. J. *Polymer* **4**, 401 (1963).
204. Odajima, A., Woodward, A. E. and Sauer, J. A. *J. polym. Sci.* **55**, 181 (1961).
205. Ishida, Y. and Yamafuji, K. *Kolloid Z.* **177**, 97 (1961).
206. Rogers, S. and Mandelkern, L. *J. phys. Chem.* **61**, 985 (1957).
207. Shetter, J. A. *J. polym. Sci.* **1B**, 209 (1963).
208. Ferry, J. D. and Strella, S. *J. colloid Sci.* **13**, 459 (1958).
209. Bueche, F. *J. appl. Phys.* **26**, 738 (1955).
210. Fujita, H. and Kishimoto, A. *J. colloid Sci.* **13**, 418 (1958).
211. Child, W. C. and Ferry, J. D. *J. colloid Sci.* **12**, 327, 389 (1957).
212. Dannhauser, W., Child, W. and Ferry, J. D. *J. colloid Sci.* **13**, 103 (1958).
213. Wolf, K. and Schmeider, K. *Symp. Inter. Chim. Macromol. Suppl. Ric.* 3 (1955).
214. Thurn, H. and Wolf, K. *Kolloid Z.* **173**, 72 (1960).
215. Jenckel, E. *Kolloid Z.* **100**, 163 (1942).
216. Kovacs, A., Stratton, R. A. and Ferry, J. D. *J. phys. Chem.* **67**, 152 (1963).
217. Matsuo, S. and Yamafuji, K. *J. phys. Soc.* (*Jpn*) **15**, 2295 (1962).
218. Weiss, S. and Leroi, G. *J. chem. Phys.* **48**, 962 (1968).
219. Fateley, W. G. and Miller, F. A. *Spectrochim. Acta.* **19**, 611 (1963).
220. Brier, P. N., Higgins, S. S. and Bradley, R. D. *Mol. Phys.* **21**, 72 (1971).
221. Allen, G., Brier, P. N. and Lane, G. *Trans. Faraday Soc.* **63**, 824 (1967).
222. Morino, Y. and Hirota, E. *J. chem. Phys.* **28**, 185 (1958).
223. Beudet, R. A. and Wilson, E. B. *J. chem. Phys.* **37**, 1133 (1962).
224. Dale, J. *Spectrochim. Acta* **22**, 3373 (1966).
225. Seigel, S. *J. chem. Phys.* **27**, 989 (1957).

226. Lees, R. M. and Baker, J. G. *J. chem. Phys.* **48**, 5299 (1968).
227. Itoh, T. *J. Phys. Soc. Jpn* **11**, 264 (1956).
228. Kojima, T., Brieg, E. L. and Lin, C. C. *J. chem. Phys.* **35**, 2139 (1961).
229. Herschbach, D. R. *J. chem. Phys.* **25**, 358 (1956).
230. Lowe, J. P. In *Progress in physical organic chemistry*, Vol. 6 (ed. A. Streitweiser and R. W. Taft). Interscience, London and New York (1968).

SPECTROSCOPIC AND
SCATTERING PHENOMENA

11.1. Introduction

Infrared absorption spectroscopy has been used for many years both for the characterization of macromolecules and for obtaining structural information regarding the arrangement and conformation of polymer chains. In this respect infrared dichroism has proved particularly valuable for the study of ordered polymer systems.[1] The elastic scattering of visible light has also been widely used for the study of time-averaged conformational and dimensional properties of polymer molecules in dilute solution.[2] However, light-scattering observation of dynamic properties has become widespread only with the introduction of the laser, and is a much more recent phenomenon in the polymer field.

All forms of light scattering arise from fluctuations in the dielectric constant of the medium and theoretically the scattering process is treated similarly whether these fluctuations arise from molecular vibrations, sound waves, or diffusional motion. A basic theoretical treatment of light scattering is presented in Chapter 2. In the case of inelastic scattering, it is now possible to measure frequency shifts over the whole of the range $10-10^{14}$ Hz at very low signal levels, so that a wide variety of dynamic phenomena in polymers can be studied.

In this chapter we shall consider the applications of Rayleigh, Brillouin, and Raman scattering of visible light, infrared absorption, and dichroism, and finally the effect of intermolecular interactions on the vibrational spectra of polymer solids.

11.2. Rayleigh linewidth (quasi-elastic) spectroscopy

In the polymer field, Rayleigh linewidth spectroscopy, (or photon-correlation spectroscopy), has mainly been applied to the study of molecular diffusion in solution. Direct measurements of diffusion coefficients are

possible and information concerning the conformational state and hydrodynamic properties of polymers is readily obtained using this technique.

11.2.1. Experimental techniques

Since the spectral widths of the scattered beam are very small (10^7 Hz)
compared with the frequency of the exciting visible laser light ($\sim 10^{15}$ Hz)
light-beating techniques are necessary. Classically, light-beating involves
the shifting of the central frequency of the scattered spectrum from 10^{15} Hz
to zero, permitting the measurement of very small shifts.

Photon-correlation spectroscopy involves the scattering of a beam of
coherent photons with a random time distribution and the detection of
departures from Poisson statistics manifested by correlations in photon
arrival times after scattering. The experiment therefore measures the
intensity correlations of the scattered photons at the cathode of the phototube and so gives the spectrum. Since in practice detection is limited to a
coherence area or less, only time correlations are measured. The technique
involves the processing of the fluctuations of the phototube current, $i(t)$, and
no frequency filtering is involved, in contrast to Brillouin and Raman
detection. The current auto-correlation function is defined as

$$G_i(\tau) = \langle i(t) . i(t+\tau) \rangle$$

and is related to the power spectrum by its Fourier transform. In the case of
Gaussian light, $G_i(\tau)$ and the auto-correlation function of the scattered field
$g^{(1)}(\tau)$, when normalized, are related by

$$G_i(\tau) = e\langle i \rangle \delta \tau + \langle i \rangle^2 (1 + |g^{(1)}(\tau)|^2).$$

The first term represents the shot noise (i.e. white noise), in the spectrum, e
is the electronic charge, the second term is a d.c. photocurrent, and the final
term contains the desired information about the correlations in the scattered
field.

Experimentally the intensity of scattered light is measured at an angle θ to
the incident laser beam. The angle θ can be varied, and the detecting
phototube has gain and noise properties which permit the detection of single
photons. The electronic processing of the fluctuations in the detector current
is usually accomplished with an auto-correlation computer which measures
$G_i(\tau)$ and which may operate directly on the single photon pulses. In this
case, the upper frequency limit is set by the speed of the electronics (10^7 Hz)
or if current integration is performed for weak signals this limit is set by the
integration time constant (10^5–10^6 Hz). The experimental time involved in
measuring $G_i(\tau)$ varies from a few seconds to about an hour depending on
the scattered signal strength. Other methods for obtaining the auto-

correlation function include the use of spectrum analysers, photon counting, or on-line computers.

11.2.2. Applications

It is well known that molecular weights can be obtained from the absolute intensity of light scattered from a polymer solution and that information concerning the size of the macromolecule, e.g. radius of gyration, can be obtained from the angular dependence of the mean intensity. Here we are concerned with the time-dependent, rather than the time-averaged, value of the scattered intensity and thus probe the random Brownian motion of the macromolecule.

Polymers small compared with $\lambda \simeq K^{-1}$: *monodisperse non-interacting molecules.* For sufficiently small scatterers, the factor $\exp(i\mathbf{K} \cdot \mathbf{r})$ in the scattering amplitude term (eqn 2.70) is the same for all $\mathbf{r}$ within one scatterer. Thus the electric fields scattered by the different elementary dipole radiators within *one* scatterer are virtually in phase at the detector. The scatterers are then essentially point scatterers and only motions of the centre of mass will contribute to the time dependence of the scattered light. If we further assume the scatterers to be non-interacting and identical, then

$$|g^{(1)}(\tau)| = \exp(-\Gamma\tau) \tag{11.1}$$

where $\Gamma = D_T K^2$ and D_T is the translational diffusion coefficient. If Gaussian statistics apply, then the time-dependent part of the measured correlation function decays according to $\exp(-2\Gamma\tau)$. In a photon correlation measurement, D_T can frequently be obtained to an accuracy of 1 per cent in a measurement lasting one minute.

From the Einstein relationship

$$D_T = kT/\xi, \tag{11.2}$$

a direct measure of the frictional coefficient can be obtained. For spherical molecules, ξ is given by the Stokes equation,

$$\xi = 6\pi\eta R_H \tag{11.3}$$

where R_H is the hydrodynamic radius and η the viscosity. Thus a rapid measure of the size of spherical macromolecules can be obtained in solutions. Regardless of shape, the molecular weight can be determined from the Svedberg equation,

$$M = \frac{N_A kTS}{D_T(1 - \bar{V}\rho_s)}. \tag{11.4}$$

Here k is Boltzmann's constant and N_A is Avogadro's constant. The sedimentation coefficient, S, and partial specific volume, $\bar{V}$, must be determined independently. For spherical particles the 'hydrodynamic' volume $(4/3)\pi R_H^3$ can be compared with the dry volume $(M/N_A)\bar{V}$ to obtain the degree of solvation.

In the case of random-coil synthetic polymers, experimental problems arise. In addition to poly-dispersity it is difficult to work at concentrations low enough to neglect the interparticle interactions due to their extended conformation. Nevertheless, it has been possible to confirm the relationship

$$D_T \propto M^{-\alpha}$$

for fairly monodisperse polymers of a certain type (e.g. polystyrene in cyclohexane) and to obtain α with a fair degree of accuracy.[3]

Polydisperse, non-interacting macromolecules. Where there is a spread of particle sizes and/or shapes in a sample, the correlation function of the scattered electric field becomes a sum of exponentials

$$|g^{(1)}(\tau)| = \int G(\Gamma) \exp(-\Gamma\tau)\, d\Gamma \qquad (11.5)$$

where $G(\Gamma)$ is the normalized distribution of decay rates. In principle, the desired $G(\Gamma)$ could be obtained by Laplace inversion of the experimental $|g^{(1)}(\tau)|$ obtained from the measured intensity correlation function. In practice, this approach is very sensitive to statistical error in the results and is not generally used. One technique used to overcome this problem is to assume some parametric form for $G(\Gamma)$ and then to calculate $|g^{(1)}(\tau)|$ for various values of the parameters until a best fit with the experimental $|g^{(1)}(\tau)|$ is obtained. This approach is useful if one has a precise knowledge of the form of $G(\Gamma)$ but is of little value for arbitrary samples.[4] Various polydisperse samples have been studied in this way.[5]

Monodisperse interacting polymers. At finite concentrations the assumption of particle independence is no longer valid and the polymer molecules will interact to some degree. This interaction occurs both indirectly through the hydrodynamic velocity fields set up in the solvent by the motion of each particle and, for example, by direct hard sphere shielded or coulombic interactions between the macromolecules. When the range of interaction is short compared with $1/|K|$ the effects of interactions can be included in a generalized Einstein equation,[6]

$$D_T = \frac{c(\partial\mu/\partial c)}{N_A\xi} \qquad (11.6)$$

where c is the polymer concentration, $(\partial\mu/\partial c)$ the 'osmotic compressibility', and ξ is the friction coefficient.

The effects of interaction on ξ and $(\partial\mu/\partial c)$ can then be calculated separately. For hard spheres a simple result for the first-order concentration dependence can be obtained

$$\frac{c}{N_A}\left(\frac{\partial\mu}{\partial c}\right) = kT(1+8\Phi+\cdots)$$

$$\xi = \xi_0(1+K_S\Phi+\cdots) \tag{11.7}$$

where Φ, $(\propto c)$ is the volume fraction of the solvated macromolecules in solution and K_S is a constant of value 6–8. For small Φ

$$D_T \simeq \frac{kT}{\xi_0}\{1+(8-K_S)\Phi+\cdots\}. \tag{11.8}$$

Thus D_T is expected to show a considerably smaller first-order concentration dependence than osmotic pressure or conventional light scattering whose magnitude depends on $\partial\mu/\partial c$ or the sedimentation coefficient, determined by ξ.

For fairly concentrated macromolecular solutions with long-range forces, any one molecule will, most of the time be interacting strongly with others. A complicated many-body problem must then be solved to obtain theoretical expressions for the scattered field-correlation function in terms of the interparticle potential.

Monodisperse non-interacting polymers comparable or large compared with $1/K$. For large polymers, studied at finite scattering angle θ, the fields scattered by elementary dipoles within one scatterer will no longer be in phase at the detector. There will be some degree of destructive interference giving rise to particle scattering factors $P(\theta)$ less than unity. If the particles are not spherical, the resultant scattered field will be a function of particle orientation and will thus fluctuate in time as the particle undergoes rotational Brownian motion. The electric field-correlation function for this situation may be written[7]

$$|g^{(1)}(\tau)| = A_0(K)\exp(-D_TK^2\tau)+A_1(K)\exp[-(D_TK^2+6D_R)\tau]$$

$$+\text{other terms} \tag{11.9}$$

where A_0, A_1, etc. are K-dependent amplitudes and D_R is the rotational diffusion coefficient. Translational effects dominate $(A_0 \gg A_1)$ at small K, but rotational contributions become increasingly important as θ is increased. Thus if measurements are made for a range of θ, the different angular dependences of these two contributions allows simultaneous measurement of both translational and rotational diffusion coefficients.

As an example, consider tobacco mosaic virus, TMV, a fairly rigid rod-shaped macromolecule about 300 nm long and 18 nm wide. When the molecule is oriented parallel to the scattering vector, $\mathbf{K}$, there will be a significant phase difference $(\mathbf{K}.\mathbf{r})$ of about π between light scattered from the two ends of the molecule so considerable destructive interference will occur at the detector. When the molecule is oriented perpendicular to **K** the phase shifts will be small, (**K . r**) of about $\pi/20$, and the particle will scatter essentially as a point scatterer. Several measurements of D_R for TMV have been made in this manner.

For large flexible macromolecules the scattered electric field may fluctuate similarly due to internal motions, and this in turn will contribute to the time dependence of the scattered light. In this case a formula similar to (11.9) holds, with D_R replaced by the appropriate reciprocal relaxation time. Several such studies of large flexible macromolecules have been carried out.[8,9]

In this section we have considered only optically isotropic macromolecules for which the scattered light has the same polarization as the incident light. If polymers of any size have an anisotropic polarizability term, a portion of the scattered light will be depolarized and this will reflect both rotational and translation motions. Biological macromolecules are generally fairly isotropic however, so that the intensity of depolarized scattering is less than 0·5 per cent of that of the polarized scattering. Nevertheless rotational motions of some molecules have been obtained from the time dependence of the depolarized scattered light. In the following section, the theory of scattering from anisotropic macromolecules will be considered.

11.2.3. Classical scattering from anisotropic fluctuations

In the general anisotropic case, the off-diagonal components in the polarizability tensor are no longer zero, and the induced dipole is no longer parallel to the incident beam polarization. The scattered beam now has components parallel and perpendicular to the scattering plane. If I_{VV} defines the scattered light polarized in the same plane as the exciting radiation and I_{HV} radiation polarized at 90° to the plane of the incident beam, the total intensity of scattered light is given by,

$$I_S(\mathbf{k}, \omega) = I_{VV}(\mathbf{k}, \omega) + I_{HV}(\mathbf{k}, \omega) \qquad (11.10)$$

where

$$I_{VV} = (A/2\pi) \int dt \int d^3r \, e^{i(\mathbf{k}.\mathbf{r}-\omega t)} \int d^3r' \langle \alpha_{zz}(\mathbf{r}+\mathbf{r}'t)\alpha_{zz}(\mathbf{r}', 0)\rangle, \qquad (11.11a)$$

$$I_{HV} = (A/2\pi) \int dt \int d^3r \, e^{i(\mathbf{k}.\mathbf{r}-\omega t)} \int d^3r' \langle \alpha_{yz}(\mathbf{r}+\mathbf{r}'t)\alpha_{yz}(\mathbf{r}', 0)\rangle, \qquad (11.11b)$$

$\alpha_{zz}(\mathbf{r}, t)$ and $\alpha_{yz}(\mathbf{r}, t)$ are the zz and yz elements of the polarizability tensor at point **r** and time t, and A is a constant depending on the scattering geometry. These results are limited to molecules smaller than the wavelength of light, λ, and to small scattering angles θ. Equation (11.11b) represent the anisotropic or depolarized scattering and the measurement of this should in principle provide information on the asymmetry of the scattering segment via the off-diagonal elements of the polarizability tensor.

Consider a polymer molecule to be divided into N_S segments each of which are regarded as scattering units. Segment i has a polarizability tensor α^i in the laboratory coordinate system. We assume that the polymer polarizability tensor can be expanded as the sum of the individual segmental contributions

$$\alpha_{yz}(\mathbf{r}, t) = \sum_{i=1}^{N_S} \alpha^i_{yz}(t)\delta(\mathbf{r}_i(t) - \mathbf{r}) \tag{11.12}$$

where $\mathbf{r}_i(t)$ is the position of segment i at time t. From this we obtain the contribution to the depolarized spectrum from one molecule. Using the appropriate transformation matrices, the laboratory and segmental coordinate systems can be related, and explicit forms of I_{HV} worked out for particular cases. In the case of rigid rods[10]

$$I_{HV}(\mathbf{k}, \omega) \simeq (\alpha_{zz} - \alpha_{yy})^2 \frac{k^2 D_T + 6D_R}{\omega^2 + (k^2 D_T + 6D_R)^2} \tag{11.13}$$

where D_T is the translational diffusion coefficient and D_R the rotational diffusion coefficient. Thus, the rotational motion can be studied using depolarized scattering measurements. This technique is particularly useful at small scattering angles when it is important to discriminate against the incident radiation. In this limit K becomes small and (11.13) reduces to

$$I_{HV}(\mathbf{k}, \omega) \simeq (\alpha_{zz} - \alpha_{yy}) \frac{6D_R}{\omega^2 + (6D_R)^2} \tag{11.14}$$

where the linewidth is due solely to rotational diffusion.

11.3. Brillouin spectroscopy

As we have seen, Brillouin spectroscopy is the scattering of light from acoustic modes which are periodic and coherent fluctuations in the density of the medium. The scattered light is shifted from the incident frequency by an amount corresponding to the acoustic velocity, with a linewidth related to the lifetime of the excitation. Since accessible sound frequencies using visible sources are in the region 5×10^8–2×10^{10} Hz, complex moduli in the hypersonic range can be measured. In addition, studies of the total scattered intensity, particularly when compared with the contribution from non-propagating excitations, provide information about the thermodynamic parameters and structural features of the system.

11.3.1. Experimental techniques

The spectral range that we should like to cover lies between the upper-frequency limit of the photon correlation technique, 10^7 Hz, and the lower limit of diffraction grating instruments, about 10^{11} Hz. Also, since visible lasers are normally used, the resolving power of the system must be of the order of 10^7.

To achieve these objectives a Fabry–Perot etalon must be used which achieves a high resolving power by the interference of beams multiply reflected between two highly reflective plates. Even then the lower limit is difficult to achieve in practice. A typical experimental arrangement would involve an ionized gas laser as source with the scattered light collected at an angle θ to the incident beam and collimated into the etalon. The emerging signal is detected by a phototube and recorded by usual techniques or, at low intensities, by multichannel scaling. The accessible region of the spectrum free from overlap between orders is determined by the free spectral range, $\Delta\nu$, of the etalon. This is determined by the plate separation and in frequency units is given by $\Delta\nu = c/2d$. It can be adjusted between the rough limits of 10^8 and 10^{12} Hz to satisfy experimental requirements. At large plate separations however, mechanical stability problems are encountered and lower values of $\Delta\nu$ are very difficult to obtain. Greater stability can be achieved using a confocal arrangement if an adjustable free spectral range is not required.

The reflectivity finesse, F_R, is defined in terms of the reflection coefficients, r, of the plates, $F_R = \pi r^{\frac{1}{2}}(1 - r)^{-1}$, but in practice other factors such as plate flatness tend to limit the finesse obtainable. Normally finesse values of 40 and 50 are attainable corresponding to reflectivities of about 95 per cent with plates flat to about $\lambda/100$. Contrast ratios, $C \simeq 4F^2\pi^{-2}$ of about 10^3 are then obtained, and these are adequate for samples of good optical quality. Many polymers, however, are of poor optical quality and the Brillouin lines are lost in the random scatter. This problem can be alleviated by operating the etalon in a multipass mode in which the beam is reflected back through the sample m times where optimum values of m range from 3 to 5. In this way contrast ratios of up to 10^8 may be obtained.

The etalon may be scanned by varying the gas pressure leaving the plates stationary or by moving the plates by applying a voltage to piezo-electric stacks fixed to the plates. In pressure scanning, the optical distance between the plates is changed by altering the refractive index of the medium and is useful for slow scans where a high degree of linearity is required. Alignment problems are minimized using this technique. The piezo-electric scanning technique has the advantage that when signal levels are small, scans can be repeated at frequencies up to several hundred hertz and the signal stored in a multichannel analyser.

11.3.2. Applications

Brillouin spectroscopy is the only technique for measuring viscoelastic and thermodynamic properties of polymers in the frequency range 10^8–

10^{11} Hz. This frequency range lies between the region of resonant molecular vibrations studied in infrared and Raman spectra and lower frequencies arising from diffusional motion of the polymer chains as a whole. Motions in this frequency region therefore, are likely to involve significant lengths of polymer chain which are difficult to analyse in terms of the motion of single or small groups of atoms. Nevertheless progress has been made in this area.

Glass transitions in polymers have received considerable attention using the Brillouin technique. The frequency dependence of the glass transition has been studied over a wide range. These measurements follow the change in sound velocity with temperature, and the transition is observed as a discontinuity in slope.

In the case of poly(methylmethacrylate) (PMMA),[11,12] the sound velocity has been measured at a series of temperatures over a frequency range extending from 8·9 to 11·3 GHz. The main glass transition is indicated by a clear discontinuity in the gradient of the temperature dependence at around 110 °C. This transition is thought to be associated with the main-chain motions in some way, but why the temperature corresponds to the low-frequency transition and the reason for the change in gradient are not clearly understood. The glass transitions in poly(vinylchloride) (PVC),[11] polystyrene,[12] poly(ethylmethacrylate) (PEMA),[12] and polybutadiene[13] have also been studied, but again the detailed nature of the transitions at these temperatures and frequencies remains obscure.

The Brillouin linewidths, and hence phonon attenuation coefficients, have also been measured for PMMA and PVC over a range of temperatures. The attenuation coefficients were found to be almost constant up to the low-frequency glass transition (110 °C for PMMA and 65 °C for PVC) and then to increase rapidly. The increase was attributed to the main α mechanical loss process whose characteristic frequency rises rapidly with temperature.

The measurement of scattered light intensity also provides a method of probing the ordered regions which may exist in amorphous polymers. In the case of PMMA, a typical amorphous synthetic polymer, the Landau–Placzek ratio is expected to be about 0·1 below the glass transition, rising to about 0·3 above. Experimental values are found to be roughly two orders of magnitude larger than those predicted by theory. The lowest measured value on a carefully annealed sample of PMMA is 7. The discrepancy between theory and experiment suggests the presence of static inhomogeneities in the polymer which enhance the intensity of the central peak. The nature of these inhomogeneities is not clear at the present time. 'Frozen in' density fluctuations below the glass transition have been suggested as a possible scattering mechanism. The presence of structural features in the amorphous state, such as ordered regions of up to 40 Å in size has also been postulated. These suppositions remain speculative at the present time.

Finally, the Brillouin spectra of certain copolymers of PEMA and PMMA[12] contain two distinct transitions corresponding to the glass transition temperatures of the constituents. This implies a degree of separation in the sample sufficient to give rise to blocks of each constituent at least 200 nm in size. Thus Brillouin scattering may be used to study the microscopic composition of copolymers.

11.4. Raman scattering

The detailed quantum mechanical treatment of Raman scattering can be found in a number of texts dealing with vibrational spectroscopy.[14,15] In the study of polymer systems observations of Raman scattering have developed towards two rather different applications. On the one hand, the polarization characteristics of the scattered light have been used to yield information on the orientation of the Raman-active scattering centres. On the other hand recent advances in instrumentation have permitted the observation of Raman bands very close to the exciting line, which bands yield information on low-frequency skeletal, torsional, and related modes of motion.

Studies of static orientation in polymers fall outwith the scope of this volume, and the interested reader is referred to a recent review in this field.[16]

11.4.1. Low-frequency vibrations

The recent advances in Raman instrumentation have permitted the observation of Raman bands very close to the exciting line, in some cases as close as a few cm^{-1}. Studies in this low-frequency region are of special importance for the observation of low-energy excitations, such as skeletal torsional modes and longitudinal modes which involve the whole polymer chain.

An example of the second type of motion occurs in crystals of polyethylene. In lamellar crystals of polyethylene, the polymer chain folds back upon itself many times, with the length of the straight sections varying between 10 and 20 nm, depending on the crystallization parameters. The straight section is similar to that of an extended n-paraffin chain for which Raman modes in the region 26–150 cm^{-1} have been reported, frequencies which were inversely proportional to chain length. If the chain is treated as a continuum, the vibrations may be attributed to longitudinal acoustic modes of the whole chain whose frequency is given by,

$$\nu = (1/2l)(E/\rho)^{\frac{1}{2}} \qquad (11.15)$$

where l is the chain length, ρ the density, and E is Young's modulus. The analogous low-frequency modes in crystalline polyethylene have been correlated with the lamellar thickness as measured by low-angle X-ray scattering.[17,18] The good agreement between the predictions of the Raman and X-ray results suggests that the crystalline and amorphous (folded region) have the same modulus. However, normal mode calculations for the polyethylene unit cell indicate mixing between a torsional component and the longitudinal mode, which mixing results from intermolecular forces.[19] These forces also give rise to transverse modes which propagate at right angles to the chain axis.[20] Keller and his colleagues[21] have studied samples in which the straight chain is inclined at various angles to the crystal surface, and conclude that the spectra are characteristic of the inclined chain length rather than of the crystal thickness. Thus these measurements are characteristic of the chain rather than the crystal platelet. Longitudinal acoustic modes have also recently been observed in bulk crystallized polyethylene.[21,22]

11.5. Infrared absorption

11.5.1. Infrared dichroism

The interaction energy of electromagnetic radiation of electric field strength $\mathbf{E}$ and an electric dipole $\boldsymbol{\mu}$ is proportional to $\boldsymbol{\mu} \cdot \mathbf{E}$. The effect of such a perturbation term in the hamiltonian, $\mathscr{H}$, is to permit, in an electromagnetic field, induced transitions between vibrational states v'' and v' in a molecule.

The energy absorbed by a molecule is proportional to $[(\partial\boldsymbol{\mu}/\partial Q) \cdot \mathbf{E}]^2$ where $\partial\boldsymbol{\mu}/\partial Q$ is the change in dipole moment with respect to the normal coordinate, Q, and is thus proportional to the transition moment for this vibration. This is a scalar product, so that radiation with an electric vector parallel to the transition moment can be adsorbed while radiation with an electric vector perpendicular to the transition moment will not be adsorbed. In a crystal, or oriented polymer, such moments are uniquely oriented in space, so specific interaction with polarized radiation is possible.

The anisotropic absorption of such an oriented transition moment is characterized by the dichroic ratio of the absorption band, defined as

$$R_0 = \int_0^\infty \varepsilon_\pi(\nu)\,\mathrm{d}\nu \Big/ \int_0^\infty \varepsilon_\sigma(\nu)\,\mathrm{d}\nu \qquad (11.16)$$

where $\varepsilon_\pi(\nu)$ and $\varepsilon_\sigma(\nu)$ are extinction coefficients for plane-polarized radiation incident normal to the transition moment and with the electric vector oriented respectively parallel and perpendicular to this direction. The extinction coefficient $\varepsilon(\nu)$ is defined by Beer's law,

$$I(\nu) = I_0(\nu) \exp(-\varepsilon(\nu)l) \qquad (11.17)$$

where I_0 and I are the incident and transmitted intensities respectively and l

is the sample thickness. If the absorption band can be represented by a Lorentz equation, viz.

$$\varepsilon(\nu) = \frac{a^2}{(\nu - \nu_0)^2 + b^2} \tag{11.18}$$

where ν_0 is the frequency of maximum absorption and a and b are constants,

$$R_0 = \frac{\varepsilon_\pi(\nu_0)}{\varepsilon_\sigma(\nu_0)}. \tag{11.19}$$

The experimentally determined dichroic ratio is

$$R = \frac{D_\pi(\nu_0)}{D_\sigma(\nu_0)} \tag{11.20}$$

where D is the measured optical density, i.e. $\log(I_0/I)$. Various experimental factors can lead to R being different from R_0. These include overlapping bands, imperfect polarization of the incident radiation, the use of convergent rather than parallel radiation, etc. However, the qualitative characteristics of R will be unchanged.

So far we have considered a single oriented molecule. The simplest model for a molecular system is the oriented-gas approximation, in which intermolecular interactions are neglected and the observed dichroic ratio is interpreted as the sum of contributions from independent fixed molecular transition moments. This is a useful approximation since interactions between molecules (and often between identical units of a single long-chain molecule) are often weak and can be neglected. This approximation, however, is inadequate for detailed study since factor-group symmetry and correlation field effects should then be included. The oriented-gas model should only be used when it has been established that effects due to the crystalline field can be neglected.

11.5.2. Models for oriented polymers

Since oriented polymers or fibres usually contain regions of greater and less regularity, with the chain direction varying from place to place but tending on the whole to follow one direction, it is not possible to infer uniquely the orientation distribution of transition moments from the dichroic ratio of a single band. Recourse must then be made to reasonable postulated models for the distribution function, calculating the dichroic ratio in terms of the parameters of this distribution and characterizing the polymer in terms of these parameters.

Consider the case of a uniaxially-oriented set of transition moments, all making an angle θ with the fibre axis. The extinction coefficient for parallel radiation is given by

$$\varepsilon_\pi(\nu_0) \propto \int_0^{2\pi} \left(\frac{\partial \mathbf{\mu}}{\partial Q}\right)^2 \cos^2 \theta \frac{m}{2\pi} \, d\phi \tag{11.21}$$

where m is the total number of transition moments. The corresponding coefficient for perpendicular radiation is

$$\varepsilon_\sigma(\nu_0) \propto \int_0^{2\pi} \left(\frac{\partial \mathbf{\mu}}{\partial Q}\right)^2 \sin^2 \theta \cos^2 \theta \frac{m}{2\pi} \, d\phi. \tag{11.22}$$

The dichroic ratio is then,

$$R_0 = \frac{\varepsilon_\pi(\nu_0)}{\varepsilon_\sigma(\nu_0)} = 2 \cot^2 \theta. \tag{11.23}$$

If, on the other hand, the molecular chain axis, makes an angle θ with the fibre axis, the transition moments being distributed uniformly at an angle α to the chain axis, then

$$R_0 = \frac{2 \cot^2 \alpha \cos^2 \theta + \sin^2 \theta}{\cot^2 \alpha \sin^2 \theta + (1 + \cos^2 \theta)/2}. \tag{11.24}$$

Similarly, more complex distribution functions can be assumed and their dichroic properties evaluated. However, unless the assumption of a specific orientation distribution can be verified by other techniques, there is little to be gained from this approach.

11.5.3. Analysis of vibrational spectra of high polymers

Two theoretical methods of analysis are used for the prediction of the vibrational spectrum of a polymer: symmetry analysis and normal coordinate analysis. The first seeks to provide general statements concerning the nature of the spectrum based on the symmetry of the polymer molecular structure. This includes derivation of the selection rules and possible influence of the symmetry on modes of vibration. The second, which has been outlined in Chapter 2, attempts to calculate the normal frequencies and modes of vibration of the polymer.

To illustrate the symmetry analysis, consider a single infinitely long, planar, zig-zag polyethylene chain illustrated in Fig. 11.1. Its structure is defined by a set of symmetry elements. These are C_2, a two-fold rotation axis; C_2^s, a two-fold screw axis; i, a centre of inversion; σ, a mirror plane; and σ_g, a glide plane. In addition, there are the identity operation E, and the infinite number of translations by multiples of the unit-cell repeat distance. All of these leave the configuration of the molecule unchanged. This set of symmetry elements, in which the translations are considered as equivalent to

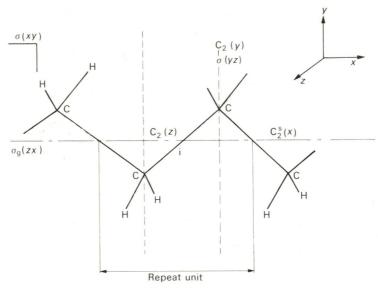

Fig. 11.1. Symmetry elements of a single, infinite, planar, zig-zag polyethylene chain.

the identity operation, form the factor or unit-cell group. They constitute a group which is isomorphic with the point group D_{2h}.

The character table for D_{2h} is included in Table 11.1. This designates the symmetry species into which the vibrations divide. Each species consists of characters $+1$ or -1 in the present case, which describe the behaviour of the normal mode with respect to the various symmetry operations. From this character table, the number of normal modes of various types which are found in each species and their infrared and Raman activity can be determined. Thus the total number of normal modes under a symmetry species i is given by,

$$n_i = \frac{1}{N} \sum_R U_R(\pm 1 + 2 \cos \phi_R)\chi_i(R) \qquad (11.25)$$

where N is the order of the group, U_R is the number of atoms which remain invariant under the operation R, the plus or minus sign depends on whether R is a pure rotation through ϕ or a rotation through ϕ followed by a reflection in a plane perpendicular to the axis, $\chi_i(R)$ is the character of the symmetry element in the ith species, and the summation extends over all the symmetry elements.

The number of pure translations under a given symmetry species is given by

$$n_i(T) = \frac{1}{N} \sum_R (\pm 1 + 2 \cos \phi_R)\chi_i(R). \qquad (11.26)$$

Table 11.1

Character table and selection rules for polyethylene chain

D_{2h}	E	$C_2(z)$	$C_2(y)$	$C_2^s(x)$	i	$\sigma(xy)$	$\sigma_g(xz)$	$\sigma(yz)$	n_i	T	T'	R'	n_i'	IR	Raman
A_g	1	1	1	1	1	1	1	1	3	0	1	0	2	ia	Pol
B_{1g}	1	1	-1	-1	1	1	-1	-1	2	0	1	1	0	ia	Dep
B_{2g}	1	-1	1	-1	1	-1	1	-1	1	0	0	1	0	ia	Dep
B_{3g}	1	-1	-1	1	1	-1	-1	1	3	0	R_x	1	1	ia	Dep
A_u	1	1	1	1	-1	-1	-1	-1	1	0	0	1	0	ia	ia
B_{1u}	1	1	-1	-1	-1	-1	1	1	3	T_z	0	1	1	a	ia
B_{2u}	1	-1	1	-1	-1	1	-1	1	3	T_y	0	0	2	a	ia
B_{3u}	1	-1	-1	1	-1	1	1	-1	2	T_x	0	1	0	a	ia
U_R	6	0	2	0	0	2	0	6							
$\phi(°)$	0	180	180	180	180	0	0	0							
$\pm 1 + 2\cos\phi$	3	-1	-1	-1	-3	1	1	1							
$U_R(\pm 1 + 2\cos\phi)$	18	0	-2	0	0	2	0	6							
$U_R(s)$	2	0	0	0	0	2	2	2							
$(U_R(s) - 1)(\pm 1 + 2\cos\phi)$	3	1	-1	1	3	1	-1	1							
$U_R(s - v)$	2	0	2	0	0	2	0	2							
$(U_R(s - v))(1 \pm 2\cos\phi)$	6	0	-2	0	0	-2	0	2							
$2\cos\phi(\pm 1 + 2\cos\phi)$	6	2	2	2	6	2	2	2							

Activity columns: IR, Raman

The difference between n_i and $n_i(T)$ is the number of internal vibrational modes, except for the one rotation about the chain axis. These can in turn be subdivided into modes within chemical groups (e.g. CH_2) and translatory or rotatory types of oscillations between groups. These can be determined from the equations,

$$n_i(T') = \frac{1}{N} \sum_R [U_R(s) - 1][\pm 1 + 2 \cos \phi_R] \chi_i(R) \tag{11.27}$$

and

$$n_i(R') = \frac{1}{N} \sum_R [U_R(s - v)][1 \pm 2 \cos \phi_R] \chi_i(R) \tag{11.28}$$

where $U_R(s)$ is the number of chemical groups which remain invariant under the symmetry operation R, v represents the number of groups which consist of a single atom only, and $U_R(s - v)$ is the number of chemical groups other than single atoms which are invariant to the symmetry operation R. The number of modes internal to a chemical group $n_i' = n_i - n_i(T) - n_i(T') - n_i(R')$. These quantities are included in Table 11.1.

The selection rules for infrared and Raman activity can also be determined. Modes under a given species can only be infrared-active if the quantity,

$$n_i(\mu) = \frac{1}{N} \sum_R (\pm 1 + 2 \cos \phi_R) \chi_i(R) \tag{11.29}$$

is different from zero. This equation is identical with that for $n_i(T)$ indicating that the polarization properties of the modes will be the same as the translational direction associated with that particular species. Modes of a given species can be Raman-active if the corresponding quantity $n_i(\mathbf{a})$, associated with the polarizability is not zero.

$$n_i(\mathbf{a}) = \frac{1}{N} \sum_R 2 \cos \phi_R (\pm 1 + 2 \cos \phi_R) \chi_i(R) \tag{11.30}$$

The activities of the various symmetry species are shown in Table 11.1. Recently several detailed vibrational studies of polyethylene have been reported.[23,24]

11.5.4. Crystallinity

Although in general the vibrational spectrum of a polymer is not greatly influenced by changes in the degree of crystallinity, usually there are, nevertheless, a few bands whose intensity depends on the state of the polymer. A well-known example is the 728 cm^{-1} band in molten polyethylene which becomes a doublet when the polymer crystallizes. The majority of bands in a polymer spectrum come from both crystalline and amorphous regions, but some are characteristic of the amorphous regions only. These are probably bands which are normally inactive for symmetry reasons in the ordered regions of the polymer, but which become active when the symmetry is lost in the disordered amorphous regions. The measurement of crystallinity in polymers is an important technological application of vibrational spectroscopy. The infrared method, in particular, is often more rapid and convenient than other ways, especially if a series of measurements is to be carried out under different conditions.

Use of crystalline bands. Crystalline bands can be used to measure crystallinity but suffer from the disadvantage that they cannot be used independently of other methods. Usually, the extinction coefficient for a crystal band must be measured for a specimen of known crystallinity. Density measurements can be used to establish the degree of crystallinity. If d_s is the density of the completely crystalline form (from X-ray determination of the size and shape of the unit cell), d_1 the density of the amorphous form, and d the density of the samples whose crystalline fraction is X, then X is given by the relation

$$X = (d - d_1)/(d_s - d_1).$$

Use of amorphous bands. Measurements on amorphous bands are less effected by orientation effects, which may be significant with crystalline bands, and can therefore yield more accurate results. Having identified a suitable amorphous band, the absorption must be measured for a completely amorphous specimen of known thickness. The amorphous content of other specimens can then be determined. Quenching the molten polymer in liquid nitrogen or partial cross-linking with β-rays are often effective in producing an amorphous specimen.

Infrared dichroism has been used to study the orientation of chain molecules in linear and ethyl-branched polyethylene for both the amorphous and crystalline regions of the polymer during the drawing and annealing process.[25] The bands chosen as typical of the crystalline and amorphous regions of the spectrum respectively had frequencies of 1894 and 1368 cm^{-1}. The degree of orientation was determined from the dichroic ratio and plotted as a function of draw ratio and annealing temperature. The results favoured a two-phase model for partially oriented crystalline polyethylene. It was also shown that the chains in the crystalline phase became oriented earlier than those in the amorphous phase, in agreement with models of plastic deformation in polymers. The vinyl and methyl end-groups were also shown to have become incorporated in the non-crystalline phase in agreement with earlier theoretical predictions.[26] From the mean orientation of the chain molecules, in the non-crystalline regions, for one sample of polyethylene it was estimated that about 25 per cent form tie molecules, 52 percent form folds, and 23 per cent contain end-groups. One chain-fold was also estimated to contain 24 methylene groups.

The temperature variation of the infrared bands at 1303, 1352, and 1368 cm^{-1}, characteristic of the amorphous region of polyethylene have also been followed.[27] The frequencies of these bands, arising from methylene vibrations, as a function of temperature, favoured the conclusion that the 1368 and 1303 cm^{-1} bands were closely related to the concentration of *trans*-linkages and 1352 cm^{-1} to the concentration of *gauche*-linkages. An apparent discontinuity in the intensity curves at $-130\,^{\circ}$C was associated with the glass transition.

11.5.5. Strength of polymers

The measurement of dichroic ratio as a function of temperature of an oriented polymer gives information on the relaxation of the ordered structure and thus on the backbone chain motions. In the case of polycarbonate polymers these measurements have shown that the chains were in motion below the glass-transition temperature and supported the theories on the impact strength of glassy polycarbonates.[28] The relationship between stress and infrared dichroism has also been studied by measuring dichroism as a function of elongation.[29] In the case of polychloroprene by studying bands sensitive to crystallinity, it was possible to follow changes in crystallinity during mechanical treatment. The results suggested that in contrast to vulcanized natural rubber, the sample orientation in polychloroprene in the crystalline phase was almost the same as in the amorphous phase. In the case of rubber,[30] orientation of the crystalline phase was evidently completed almost immediately upon elongation, whereas in the amorphous phase the molecular chains oriented gradually during the course of stress relaxation.

The relation of tensile strength of a polymer to molecular orientation of the chains within the amorphous part of oriented fibres has also been studied.[31] In the case of polyvinyl alcohol and its copolymers with *N*-vinylpyrrolidone, the dichroic ratio of the 916 cm^{-1} band in both polymers, was found to be a linear function of the extent of elongation at several different temperatures. This band was characteristic of the amorphous region, while in contrast a crystalline band at 1141 cm^{-1} gave an infinitely large ratio at threefold elongation. The breaking strengths of the polymers were also a linear function of elongation. It follows therefore that the dichroic ratio is also a linear function of the breaking strength and could therefore be used to estimate the fracturing strength of polymers.

11.5.6. Linewidths in vibrational spectra of polymers

Very few papers have been devoted to the study of band half-widths of polymers, in contrast to the extensive study of the half-widths in liquids and molecular crystals. In solid polymers particularly at low temperatures, the band shape or half-width is governed principally by intermolecular interactions rather than molecular reorientation. Although there has been no satisfactory theory to explain half-widths, it has been found that they can be related in an empirical way to polymer intermolecular interactions. For instance, half-widths of infrared bands in polypropylene, poly(ethyleneterephthalate), and poly(oxymethylene) films were all found to increase with increasing temperature and decreasing crystallinity.

Anharmonic interactions have been used to interpret the half-widths of crystalline polymers.[32] Anharmonicities derived for one-dimensional crystal models of both zig-zag and linear chains predicted a particular relationship between half-width and temperature. Experimental half-widths for the $975\ cm^{-1}$ poly(ethyleneterephthalate) band, the 809, 899, and $975\ cm^{-1}$ polypropylene bands, and the $1146\ cm^{-1}$ poly(vinylalcohol) band were used to test the predictions of the theory. A plot of $\log(\Gamma - \Gamma_0)/\Gamma$ against $1/T$, where Γ is the half-width of the band at T K, and Γ_0 is the extrapolated half-width at 0 K, was almost linear. This linear dependence supported the validity of the calculation and the underlying assumption that the half-widths were related to third-order anharmonicity. Deviation from linearity at higher temperatures indicated the participation of even higher-order anharmonicities.

A comparative test of the effectiveness of Raman intensities and half-widths as measures of crystallinity in polymers has also been made.[33] The half-bandwidths of the $1730\text{-}cm^{-1}$ C—O stretching mode in a variety of samples of poly(ethyleneterephthalate) was plotted against density. The resultant linear relationship, however, must be interpreted with caution since the assumption of constant amorphous density in the oriented samples may not be valid. However, considering the diversity of samples, which included powders, unoriented heat-crystallized filaments, drawn yarns, and high-pressure crystallized material, a relationship between bandwidth and crystallinity seems reasonable. The variation in half-width is explained in terms of rotation of the carbonyl groups out of the plane of the benzene ring in the amorphous phase. The increase in the number of 'local environments' of the carbonyl group tends to increase the bandwidth.

11.6. Intermolecular interactions in solid polymers

Intermolecular interactions in solids manifest themselves in the vibrational spectra of solids in two ways. They determine the lattice vibrational spectra of molecular solids and also are responsible for the splitting and shifting of intramolecular vibrational modes.

11.6.1. Lattice vibrational spectra

The basic theory is best illustrated by application first to a simple system, a linear triatomic molecule such as CO_2 and then extension to more complex polymeric systems. The displacement of the qth molecule in the pth cell is denoted by $r_\alpha(p/q)$ where α can assume the x, y, or z designations. The spatial direction of the linear molecule is denoted by the unit vector,

$$\Lambda_\alpha(p/q) = \Lambda_\alpha^0(p/q) + \lambda_\alpha(p/q) \qquad (11.31)$$

where Λ_α^0 is the equilibrium value and $\lambda_\alpha(p/q)$ is the displacement coordinate. Each molecule has only two degrees of rotational freedom and therefore a redundancy is included if α is allowed to assume x, y, and z. The redundancy condition in this case is

$$\sum_\alpha [\Lambda_\alpha(p/q)]^2 = \sum_\alpha [\Lambda_\alpha^0(p/q)]^2 = 1 \qquad (11.32)$$

which simply signifies that the unit vector has retained its length and only changed its orientation. Symmetry coordinates of the crystal space groups are next formed by a transformation of the molecular displacement coordinates to linear combinations which belong to irreducible representations of the translation group. These symmetry coordinates are of the form,

$$r_\alpha^{\mathbf{k}}(q) = N^{-\frac{1}{2}} \sum_p \exp[2\pi i \mathbf{k} . \mathbf{R}(p)] r_\alpha(p/q) \qquad (11.33)$$

$$\lambda_\alpha^{\mathbf{k}}(q) = N^{-\frac{1}{2}} \sum_p \exp[2\pi i \mathbf{k} . \mathbf{R}(p)] \lambda_\alpha(p/q) \qquad (11.34)$$

where N is the number of unit cells in the crystal, $\mathbf{k}$ is a wave vector, and $\mathbf{R}(p)$ is the position vector of the pth cell. The use of such coordinates will enable factorization of the energy problem or the normal vibrational frequencies into problems concerning a particular point in the Brillouin zone designated by the wave vector $\mathbf{k}$. The problem can then be formulated generally in terms of $\mathbf{k}$ and solved for any point in the zone. However, the transition probability from the ground state to a single phonon state (i.e. excitation of one quantum of lattice vibration) will vanish both for the Raman effect and for infrared absorption except for $\mathbf{k} = 0$. In principle, the vibrational degree of freedom excited will have $\mathbf{k}$ equal to the wave vector of the radiation interacting with the solid. However, this wave vector will be very close to zero since the wavelengths of the light involved are very long compared with the lattice spacing.

For $\mathbf{k} = 0$, further simplification of the problem is possible by making use of the symmetry at this point, which is the full factor group symmetry. For solid CO_2 this factor group is T_h, and therefore symmetry coordinates can be formed which belong to irreducible representations of the factor group by forming the proper linear combinations of $\mathbf{k} = 0$ functions.

The resulting symmetry coordinates are of the form

$$S_i = \sum_{\alpha,q} \Gamma'_{i,\alpha q} r^0_\alpha(q) \qquad (i = 1, \ldots, 12) \tag{11.35}$$

$$S_i = \sum_{\alpha,p} \Gamma''_{i,\alpha q} \lambda^0_\alpha(q) \qquad (i = 13, \ldots, 24) \tag{11.36}$$

where

$$r^0_\alpha(q) = N^{-\frac{1}{2}} \sum_p r_\alpha(p/q) \tag{11.37}$$

$$\lambda^0_\alpha(q) = N^{-\frac{1}{2}} \sum_p \lambda_\alpha(p/q). \tag{11.38}$$

For space group T^6_h, the molecular centres are located at face-centred cubic (fcc) sites and the molecular axes are oriented parallel to the body diagonals of the cube. There are four such directions and accordingly, four equivalent molecules in the primitive unit cell. We therefore expect 4×3 translational degrees of freedom for each point $\mathbf{k}$ in the Brillouin zone. At the centre of the zone, $\mathbf{k} = 0$, the three acoustic frequencies are zero and represent the translational motion of the molecule as a whole. Thus, we expect nine translational optical modes. Group theoretical analysis predicts that this 9×9 reducible representation contains the following irreducible representations of the factor group T_h; $A_u + E_u + 2T_u$. Since the translations x, y, and z belong to the representation T_u we expect two absorption bands in the far-infrared.

To solve the frequency problem, the potential energy in the harmonic approximation is expanded up to second-order terms in the displacements. For $\mathbf{k} = 0$ this is best done in terms of those linear combinations of the molecular displacements which form the symmetry coordinates S_i. The potential energy then has the following form,

$$2U = \sum_{ij=1}^{12} F_{ij} S_i S_j \tag{11.39}$$

where the force constants F_{ij} are the second derivatives of the crystal potential with respect to the symmetry displacements. The frequencies are then given by the roots of the secular equation of dimension 12

$$|F_{ij} - M\omega^2 \delta_{ij}| = 0 \tag{11.40}$$

where M is the mass of the molecule. Since the potential energy has been written in terms of symmetry coordinates, this determinant will factor into smaller blocks, each for an irreducible representation of the factor group.

The librational lattice modes are treated in a similar manner. They span a reducible representation of order 8, which reduces into irreducible representations of the factor group T_h according to $E_g + 2T_g$. Thus we expect three Raman lines in the low-frequency region.

To obtain solutions for $\mathbf{k} \neq 0$, it is of no advantage to use symmetry coordinates. The potential energy is expanded in terms of displacement coordinates. Since for $\mathbf{k} \neq 0$, the translational motions are no longer separable from the librations the problem must be solved in full, which implies a secular determinant of order 20 for each value of $\mathbf{k}$. Three of these solutions will be acoustic branches while the remainder are optical branches.

Polyethylene. A detailed calculation of the lattice modes in crystalline polyethylene was carried out by Tasumi and Shimanouchi.[34] Two translational lattice modes are expected and these were calculated to be 76 and 105 cm^{-1}. One of these modes was observed at 76 cm^{-1} in the far-infrared spectrum.[35,36] The second expected infrared active mode was found later by Dean and Martin[37] who reported a weak absorption at 109 cm^{-1} in a polyethylene sample cooled to 20 K. In view of the difference in lattice dimensions expected for the different temperatures, the agreement between calculated and observed frequencies is very satisfactory. These authors[37] also reported a weak absorption at 110 cm^{-1} for paraffin samples cooled to 2 K. The effects of cell-dimension changes with temperature on the frequencies of the lattice motions were also investigated by Tasumi and Shimanouchi. They were able to explain the variation in frequency of the 76-cm^{-1} mode with temperature on the basis of changes in the unit cell dimensions.

As we have seen in Chapter 2, Zerbi[38,39] has considered in some detail the vibrational spectra of chain molecules with conformational disorder. His technique enables the exact calculation of the dynamical matrix of a polymer chain which contains a random distribution of conformational defects. The main difficulty lies in the solution of the eigenvalue equation of the very large dynamical matrices which result. The negative eigenvalue theorem was used to calculate the density of vibrational states $g(\nu)$. Zerbi applied this theoretical treatment to polyethylene and found that although most of the observed features of the infrared and Raman spectra could be explained in terms of $\mathbf{k} = 0$ one phonon modes, some could not. These included bands at 198, 252, 386, 538, 1075, 1128, 1300, 1350, 1367, and 1440 cm^{-1} in the infrared of the solid polymer and bands in the Raman spectrum at 720 and 1465 cm^{-1}. The model adopted consisted of 200 methylene units joined in an all trans conformation. Kinks and fold defects were introduced into the chain in a random and in a regular fashion. The density of states was calculated and compared with the experimental anomalies. The whole of the density of states of the perfect crystal became spectroscopically active. The infrared bands at 1440, 1300, 1128, and 1075 cm^{-1} coincided with singularities in the density of states. The broad features at 538 and 252 cm^{-1} in the infrared were accounted for as activation of the ν_5 and ν_9 acoustical branches. The Raman bands at 1465 and 720 cm^{-1} resulted from singularities which were normally only infrared active for the ideal chain but became Raman-active for the imperfect chain. The bands at 1350 and 1367 cm^{-1} were thought to be localized defect modes, but not all the observed bands could be accounted for in this way.

Several theoretical calculations of the phonon dispersion curves of polyethylene in the low-frequency region have been carried out because of its relevance to specific heats and neutron inelastic scattering data. For example, the frequency distribution, specific heat, and Young's modulus have been calculated for orthorhombic polyethylene. Since specific heats are dependent upon the frequency distribution in crystals, the calculated values are dependent on the force field selected. In one such calculation,[40] of the frequency distribution of chain molecules in polymer crystals, methylene groups were approximated to point masses since their internal frequencies all lie above 700 cm^{-1}. The force field was divided into interchain and intrachain interactions. The intrachain force field was of the Urey–Bradley type and the interchain force field was expressed in terms of the three types of nearest-neighbour interactions. In the orthorhombic case, in the limit, there are three acoustic vibrations equivalent to translations along the a-, b-, and c-axes, and five $\mathbf{k} = 0$ vibrations of the complete lattice.

11.6.2. Intramolecular vibrational spectra

The ground-state wavefunction for a crystal is given by the product of ground-state vibrational molecular wavefunctions

$$\boldsymbol{\Psi}^0 = \prod_{q,p} \boldsymbol{\psi}^0_{q,p} \tag{11.41}$$

where q designates the site in the unit cell and p numbers the unit cells. An excited state function for the crystal might be written as a product representing all molecules but one in the ground state and the r, s molecule in an excited state i,

$$\mathscr{L}^i_{r,s} = \boldsymbol{\psi}^i_{r,s} \prod_{q,p} \boldsymbol{\psi}^0_{q,p}. \tag{11.42}$$

This function describes a localized excitation on the molecule r, s and does not possess the translational symmetry of the crystal. The correct crystal function belonging to representation $\mathbf{k}$ of the infinite translation group is a Bloch-type function,

$$\boldsymbol{\Phi}^i_r = \frac{1}{\sqrt{N}} \sum_s \exp(i\mathbf{k} \cdot \mathbf{R}_s)\mathscr{L}^i_{r,s}. \tag{11.43}$$

If the crystal structure contains a primitive unit cell with more than one distinguishable molecule, i.e. if $r > 1$, further linear combinations between functions of this type have to be formed in order to diagonalize the problem.

A set of solutions are found by diagonalizing the matrix of the crystal problem for each value of $\mathbf{k}$ in the first Brillouin zone. It is easily shown that only the states of $\mathbf{k} = 0$, can interact with light since the wavelength is very much longer than the unit-cell dimension. The potential V is written as a sum whose first term is the sum of the intramolecular potentials V_0, and the additional term represents a sum of two-molecule interactions

$$V = V_0 + \sum_{u,v} V_{r,s;u,v}. \tag{11.44}$$

It is usually justified to treat the sum over two-molecule interaction terms as a perturbation for molecular crystals. The perturbed energies obtained by first order perturbation theory are then given by the roots of the secular determinant

$$|\mathbf{H}' - \mathbf{E}(\lambda - \lambda_0)| = 0 \tag{11.45}$$

where λ is the matrix of the perturbed roots

$$\lambda_i = 4\pi^2 c^2 \nu_i^2 \quad \text{and} \quad H'_{i,\alpha} = \int \psi^{i,\alpha} \sum_{u,v} V_{r,s;u,v} \psi^{i,\alpha} \, d\tau. \tag{11.46}$$

Here $\psi^{i,\alpha}$ is one of the linear combinations of terms (11.43) which now belongs to one of the irreducible representations of the factor group of the crystal. In this case, the determinant, eqn (11.45) factors into blocks according to the irreducible representations. The first-order solution for the frequency of the αth component of the multiplet associated with the ith non-degenerate vibration of the free molecule is given by

$$hc\nu_{i,\alpha} = hc\nu_i^0 + \mathbf{D}^i + \mathbf{M}^{i,\alpha} \tag{11.47}$$

with

$$\mathbf{D}^i = \sum_{u,v} \int \left[|\psi_{r,s}^i \psi_{u,v}^0|^2 - |\psi_{r,s}^0 \psi_{u,v}^0|^2 \right] V_{r,s;u,v} \, d\tau$$

and

$$\mathbf{M}^{i,\alpha} = \sum B_{\alpha r}^* B_{\alpha u} \int (\psi_{r,s}^i \psi_{u,v}^0)^* V_{r,s;u,v} (\psi_{r,s}^0 \psi_{u,v}^i) \, d\tau. \tag{11.48}$$

Here the $B_{\alpha r}$ are the coefficients in the symmetry combinations α of functions Φ_r^i where the index r refers to a particular site in the unit cell or, in other words, a sublattice consisting of translationary equivalent molecules. The term $\mathbf{D}^i$ is an energy-shift term and gives the frequency shift of the vibrational level in the crystal as compared to that in the free molecule. This term is analogous to a Coulomb term. The term $\mathbf{M}^{i,\alpha}$ represents a splitting term and has a different value for each $\mathbf{k} = 0$ component in the crystal exciton band. If there are Z molecules in the primitive unit cell, then there will be Z

different $\mathbf{k} = 0$ states. Each such state is characterized by a different symmetry combination of functions and for each state $\mathbf{M}^{i,\alpha}$ will have a different value depending on the coefficients $B_{\alpha r}$ which appear in the symmetry function. The selection rules are then found by considering the final crystal states of a given symmetry. These states are of course classified according to the factor group of the crystal, which consists of operations as in a point group except that it is coupled with translations and as a result may contain screw axes instead of pure rotation axes and glide planes instead of reflection planes.

11.6.3. Interaction potentials

The first attempts at quantitative calculation were made for an intermolecular potential which was expanded in a multipole series retaining the first term, the dipole–dipole term. Thus the transition dipole moment which could be obtained from absorption intensities was used to carry out these calculations. Varying success was obtained with this approximation and it seems that the dipole–dipole interaction is only important in the case of very strong infrared absorption bands, and even here the splitting energies observed are too large to be accounted for by the dipole approximation. This interaction potential is thus generally unsatisfactory.

Another type of intermolecular interaction potential is expressed in terms of a sum over non-bonded pairs of atoms of neighbouring molecules in the crystal. This formulation implies that an intermolecular pair potential can be expressed as a sum of atom–atom interaction terms. A number of atom–atom interaction potentials have been proposed, all of which are of the form

$$-A/R^6 + B \exp(-CR) \tag{11.49}$$

where the constants A, B, and C for C–C, C–H, and H–H interaction terms have been derived from the crystal structures and physical properties of hydrocarbons.[41,42]

Polyethylene. The splitting observed in the infrared-active internal vibrations of crystalline polyethylene (term $\mathbf{M}^{i,\alpha}$) has been treated by Tasumi and Shimanouchi.[34] Crystalline polyethylene is orthorhombic and contains infinite polymethylene chains. The factor group is D_{2h}. The force field used was a Urey–Bradley type, containing non-bonded H–H interaction terms, based on a previous normal coordinate analysis of polyethylene. The intermolecular perturbation potential contained only short-range H–H interactions. In the infrared spectra, the CH_2 scissors and rocking vibrations at 1468 and 725 cm^{-1} respectively are split by about 10 cm^{-1} due to intermolecular interaction. The intermolecular potential included four H—H non-bonded distances of two CH_2 groups and the force constants were adjusted to reproduce the observed splittings in the infrared spectra. The final set of four constants were compared with the force constants calculated from exponential-6 potentials of the type $-A/R^6 + B \exp(-CR)$ and also to the potential proposed by de Boer.[43] Reasonable agreement was obtained with these potentials with the latter being the more successful. The relative importance of transition dipole coupling and inter-molecular H–H interactions has also been studied.[44] The former was found to be of minor importance in determining the crystal field splittings in crystalline polyethylene.

The crystal field-splitting effect in the Raman spectra of polyethylene and deuterated polyethylene have also been measured at low temperature.[45] At $-160\,^\circ$C splitting of the bands at $1296\ \mathrm{cm}^{-1}$ (1295 and $1297\ \mathrm{cm}^{-1}$) and $1066\ \mathrm{cm}^{-1}$ (1065 and $1068\ \mathrm{cm}^{-1}$) in polyethylene and $826\ \mathrm{cm}^{-1}$ (827 and $830\ \mathrm{cm}^{-1}$) in deuterated polyethylene were observed. These splittings were attributed to repulsive forces between hydrogen atoms on neighbouring chains. A similar explanation was used to explain band-splitting in poly-tetrafluoroethylene.[46] Studies of mixed crystals of polyethylene and deu-terated polyethylene provided evidence of chain segregation associated with crystallization rates and molecular weight. This was suggested by band-splitting in the infrared in the methylene rocking region arising from the association of similar chains rather than mixing of chains of the opposite species.

Crystal field splitting at low temperatures has also been observed in polytetrafluoroethylene.[47] Doublets observed at 575–$595\ \mathrm{cm}^{-1}$ and at $1215\ \mathrm{cm}^{-1}$ at $-180\,^\circ$C were attributed to crystal field splitting.

Recently, further detailed work on the vibrational spectra of polyethylene in the C—H stretching region has been carried out.[48] In the case of Raman spectra, the lineshapes were found to be dependent on the environment of the polymethylene chain. This was interpreted in terms of Fermi resonance interactions between the methylene C—H stretching mode and appropriate binary combinations involving the methylene bending mode. The shapes of these bands depend on the dispersion of the bending mode fundamental throughout the Brillouin zone. For the isolated chain, only parallel dis-persion is involved, but for the crystal, perpendicular dispersion is equally important and leads to the observed dependence on crystal structure.

References

1. Elliot, A. *Infrared spectra and structure of organic long-chain polymers.* Arnold, London (1969).
2. Fabelinskii, I. L. *Molecular scattering of light.* Plenum, New York (1968). Fluery, P. A. and Boon, J. P. *Advan. chem. Phys.* **24**, 1 (1973).
3. King, T. A., Knox, A., Lee, W. I. and McAdam, J. D. G. *Polymer*, **14**, 151 (1973).
4. Tagami, Y. and Pecora, R. *J. chem. Phys.* **51**, 3298 (1969).

5. Reid, T. F. *Macromolecules* **5**, 771 (1972).
6. Phillies, G. D. J. *J. chem. Phys.* **60**, 976 (1974).
7. Pusey, P. N., Vaughan, J. M. and Williams, G. *J. chem. Soc., Faraday Trans. II* **70**, 1696 (1974).
8. Fugime, S. *Advan. Biophys.* **1**, 3 (1972).
9. King, T. A., Knox, A. and McAdam, J. D. G. *Chem. Phys. Lett.* **19**, 351 (1973).
10. Pecora, R. *J. chem. Phys.* **49**, 1036 (1968).
11. Jackson, D. A., Pentecost, H. T. A. and Powles, J. G. *Mol. Phys.* **23**, 425 (1972).
12. Mitchell, R. S. and Guillet, J. E. *J. polym. Sci.* A2 **12**, 713 (1974).
13. Huang, Y. Y. and Wang, C. H. *J. chem. Phys.* **61**, 1868 (1974).
14. Koningstein, J. A. *Introduction to the theory of the Raman effect.* Reidel, Dordrecht (1972).
15. Long, D. A. *Raman spectroscopy.* McGraw-Hill, New York (1977).
16. Shepherd, I. W. In *Advances in infrared and Raman spectroscopy* (ed. R. J. H. Clark and R. E. Hester), Vol. 3. Heyden, London (1977).
17. Peterlin, A., Olf, H. G., Peticolas, W. L., Hibler, G. W. and Lippert, J. L. *J. polymer Sci., polymer Letters*, **9**, 583 (1971).
18. Peticolas, W. L., Hibler, G. W., Lippert, J. L., Peterlin, A. and Olf, H., *Appl. Phys. Lett.* **18**, 87 (1971).
19. Wu, C. K. and Nicol, M. *J. chem. Phys.* **58**, 5150 (1973).
20. Marsh, D. I. and Martin, D. H. *J. Phys. C* **5**, 2309 (1972).
21. Folkes, M. K., Keller, A., Stejny, J., Coggin, P. L., Fraser, G. V. and Hendra, P. J. *Colloid polym. Sci.* **253**, 354 (1975).
22. Cutler, D. J., Hendra, P. J., Walker, J. H., Cudby, M. E. A. and Willis, H. A. *Spectrochim. Acta* **34A**, 391 (1978).
23. Piseri, L. and Zerbi, G. *J. chem. Phys.* **48**, 3561 (1968); Tasumi, M. and Krimm, S. *J. chem. Phys.* **46**, 755 (1967).
24. Bailey, R. T., Hyde, A. J., Kim, J. J. and McLeish, J. *Spectrochim. Acta* **33A**, 1053 (1977); Hendra, P. J., Jobic, H. P., Marsden, E. P. and Bloor, D. *Spectrochim. Acta* **33A**, 445 (1977).
25. Glenz, W., Peterlin, A. and Wilkie, W. *J. polym. Sci.* A2 **9**, 1191 (1971); Glenz, W. and Peterlin, A. *Kolloid Z.* **247**, 786 (1971); *J. macromol. Sci.* **4B**, 473 (1970).
26. Kilian, H. G. *Kolloid Z.* **231**, 534 (1969).
27. Jackson, J. F. and Hsu, T. S. *Polym. Preprints* **12**, 726 (1971).
28. Yannas, V. I. and Lunn, A. C. *J. polym. Sci. B polym. Lett.* **9**, 611 (1971).
29. Takenaka, T., Shimura, Y. and Gotoh, R. *Kolloid Z.* **237**, 193 (1970).
30. Gotoh, R., Takenaka, T. and Hayama, N. *Kolloid Z.* **205**, 18 (1965).
31. Savitskaya, A. N., Klimenko, I. B., Vol'f, L. A. and Androsov, V. F. *Polym. Sci.* (*USSR*) **12**, 894 (1970).
32. Vettigren, V. I. and Kosobukin, V. A. *Optics and Spectrosc.* **31**, 311 (1971).
33. Melveger, A. J. *J. polym. Sci.* A2 **10**, 317 (1972).
34. Tasumi, M. and Shimanouchi, T. *J. chem. Phys.* **43**, 1245 (1965).
35. McKnight, R. V. and Moeller, K. D. *J. opt. Soc. Amer.* **54**, 132 (1964).
36. Frenzel, A. O. and Bulter, J. P. *J. opt. Soc. Amer.* **54**, 1059 (1964).
37. Dean, G. D. and Martin, D. H. *Chem. Phys. Lett.* **1**, 415 (1967).
38. Zerbi, G., Piseri, L. and Cabassi, F. *Mol. Phys.* **22**, 241 (1971).
39. Zerbi, G. *Pure appl. Chem.* **26**, 499 (1971).
40. Kitagawa, T. and Miyazawa, T. *Bull. Chem. Soc. Jpn* **43**, 372 (1970).
41. Kitaigorodskii, A. *J. chim. Phys.* **63**, 9 (1966).
42. Williams, D. E. *J. chem. Phys.* **47**, 4680 (1967).

43. de Boer, J. *Physica* **9**, 363 (1942).
44. Tasumi, M. and Krimm, S. *J. chem. Phys.* **46**, 755 (1967).
45. Boerio, F. J. and Koenig, J. L. *J. chem. Phys.* **52**, 3425 (1970).
46. Bank, M. I. and Krimm, S. *J. polym. Sci. B, polym. Lett.* **8**, 143 (1970).
47. Boerio, F. J. and Koenig, J. L. *J. chem. Phys.* **54**, 3667 (1971).
48. Snyder, R. G., Hsu, S. L. and Krimm, S. *Spectrochim. Acta* **34A**, 395 (1978).

NEUTRON SCATTERING STUDIES

12.1. Introduction

Neutron scattering as a technique has been available for just over ten years. Neutron sources are very expensive and so funding for such work usually involves a central facility, often with multinational co-operation (for example, the European facility at Grenoble). Nevertheless, although neutron scattering spectrometers are not to be found in university or industrial laboratories, it is possible for a wide variety of research workers to obtain access to these central facilities.

A number of excellent books and review articles have been published[1-3,9] in which the theory and practice of thermal neutron scattering have been discussed in detail. The present discussion will be limited to the concepts and parameters necessary for an understanding of the data which can be obtained for polymer solids and solutions.

Neutrons are capable of two distinct interactions with an atom. The dominant scattering process is between the neutron and the nucleus, involving particularly an interaction with the spin states of the nucleus. Interaction of the neutron with unpaired electrons can also occur, in which case the scattering process involves the magnetic moment of the unpaired electrons. This gives rise to magnetic neutron scattering. For the large majority of organic systems we are concerned only with normal neutron–nucleus scattering processes.

Neutron scattering can be separated into elastic and inelastic components. In the former, the energy of the scattered neutron remains unchanged. Such a process closely resembles the elastic scattering of photons (visible light and X-ray) and so provides information on the spatial arrangement of the scattering centres. It can, therefore, be used to study the equilibrium conformations of polymer chains. However, a very important difference from photon scattering arises from the wavelengths (or more correctly wavevectors) of the scattered particles. Neutrons have effective wavelengths of the order of nanometres, whereas visible photons have wavelengths approximately a thousand times greater. Thus neutrons are better able to 'probe' low amplitude or localized phenomena, and this is particularly important when interest is centred on the small-scale motions of groups in polymer chains.

Inelastic scattering, when the neutron energy changes during scattering, can occur when the scattering centre is itself in motion and capable of either losing or gaining energy from the neutron. Since the energy exchange with thermal motion is usually small, this leads simply to a broadening of the scattering peak and gives rise to what is called quasi-elastic scattering. On the other hand, if the atom producing the scattering is undergoing quantized rotational or vibrational motion, then the energy exchange leads to a discrete energy spectrum similar to that obtained from infrared or Raman studies as discussed previously.

In order to appreciate the potential of the technique, it is necessary to consider briefly the details of the scattering process. The basic concepts parallel those of photon scattering discussed previously, and can be found in detail in the reference texts[1-3]

12.2. Momentum and energy transfer

12.2.1. Scattering lengths

As indicated in the discussion of photon scattering, the resultant intensity of a scattered beam is a function of both the energy and the angle of scattering. The wave vector of the incident beam, $\mathbf{k}_i$, is related to that of the scattered beam, $\mathbf{k}_s$, via $\mathbf{K} = \mathbf{k}_s - \mathbf{k}_i$. For neutrons the vector difference, $\mathbf{K}$, is called simply the Q-vector, $\mathbf{Q}$ of magnitude Q (Fig. 12.1). If the incident

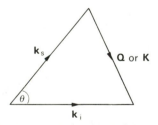

Fig. 12.1. Neutron scattering vectorial representation.

energy is E_0 and the energy of the scattered neutron is E, then the energy transfer is

$$
\begin{aligned}
E - E_0 &= \Delta E \\
&= \hbar\omega \\
&= \tfrac{1}{2}m(v^2 - v_0^2) \\
&= \frac{\hbar^2}{2m}(\mathbf{k}_s^2 - \mathbf{k}_i^2)
\end{aligned}
\tag{12.1}
$$

where v_0 and v are the velocities of the incident and scattered neutron. In the case of elastic scattering, no energy change occurs ($\Delta E = 0$) and $\mathbf{k}_s = \mathbf{k}_i$. Alternatively, for inelastic scattering $\Delta E \neq 0$. In practice the energy change can be positive or negative depending upon whether the neutron has gained or lost energy during the scattering event. This in turn depends on the incident energy of the neutron.

The momentum transfer is $\hbar \mathbf{Q}$ which can be written as

$$\hbar \mathbf{Q} = \hbar(\mathbf{k}_s^2 + \mathbf{k}_i^2 - 2k_s k_i \cos \theta)^{\frac{1}{2}} \tag{12.2}$$

and if $\mathbf{k}_s = \mathbf{k}_i$ (i.e. the event is elastic) then (12.2) reduces to

$$Q = |\mathbf{Q}| = \frac{4\pi}{\lambda} \sin \frac{\theta}{2}. \tag{12.3}$$

Quasi-elastic scattering occurs as the limit $\Delta E \to 0$ is approached.

The amplitude of the isotropic scattered wave can be described in terms of a scattering length, b. This quantity is complex, the real part can be positive or negative, and the imaginary part is a measure of the neutron absorption, which depends on the atom involved. The neutron can interact with a nucleus of spin I leading to spin states $I + \frac{1}{2}$ and $I - \frac{1}{2}$. Associated with these interactions are the scattering lengths b^+ and b^-. In general there may be a total of T spin states, in which $2(I + 1)$ states are associated with b^+ and $2(I - \frac{1}{2}) + 1 = 2I$ states are associated with b^-. The total number of states is thus $2(2I + 1)$, and since the probability of each state occurring is identical, then the mean scattering length is

$$\bar{b} = \left(\frac{I+1}{2I+1}\right)b^+ + \left(\frac{I}{2I+1)}\right)b^- \tag{12.4}$$

and

$$\bar{b}^2 = \left(\frac{I+1}{2I+1}\right)|b^+|^2 + \left(\frac{I}{2I+1}\right)|b^-|^2. \tag{12.5}$$

Eqns (12.4) and (12.5) refer to the rather artificial situation of scattering from an array of identical nuclei. In the usual case where there are a number of isotopes present, each with different spin, the b for a particular atom is written as:

$$\bar{b} = \sum_n \left\{ C_n \frac{I_n+1}{2I_n+1} b_n^+ + C_n \frac{I_n}{2I_n+1} b_n^- \right\} \tag{12.6}$$

where C_n is the concentration of the nth isotope with spin state I_n and which has scattering lengths b_n^+ and b_n^- associated with it.

12.2.2. Scattering cross-sections

The quantity with which we are concerned is the double differential scattering cross-section, $d^2\sigma/d\Omega\, dE$, which is the probability that neutrons will be scattered by an array of N atoms with cross-section σ into a solid angle Ω with energy change dE. The quantity $d^2\sigma/d\Omega\, dE$ can be calculated by including probabilities descriptive of the following processes:

  (i) A neutron with wave vector $\mathbf{k_i}$ is scattered to give a vector $\mathbf{k_s}$ in the presence of an interaction potential V;
 (ii) The nucleus changes from an initial state $\mathbf{l_0}$ to final state $\mathbf{l}$. For this case energy is conserved, i.e. the energy gained by the nucleus $= E_{\lambda'} - E_\lambda = \hbar\omega$ and is described in the calculation of the cross-section by a Dirac delta function;
(iii) A neutron in initial spin state, s, is finally found to have a spin state of s'.

The combined expression has the form[4]

$$\frac{d^2\sigma}{d\Omega\, dE} = \frac{k_s}{k_i}\left(\frac{m}{2\pi\hbar^2}\right)^2 \sum_{\lambda s} P_\lambda P_s \sum_{s'\lambda'} \langle \mathbf{l}s'\lambda'|\hat{V}|\mathbf{l_0}s\lambda\rangle^2 \delta(\hbar\omega + E_\lambda - E_{\lambda'}) \tag{12.7}$$

which the reader will recognize as the Born approximation and is of similar form to that for photon scattering. $\hat{V}$ is given as

$$\hat{V}(r) = \frac{2\hbar^2}{\pi m}\sum_J b_J \delta(\mathbf{r} - \mathbf{R}_J) \tag{12.8}$$

where R_J is the position vector of the Jth nucleus which has a scattering length b_J. Eqn (12.7) becomes

$$\frac{d^2\sigma}{d\Omega\, dE} = \frac{k_s}{k_i}\sum_{\lambda s} P_s P_\lambda \sum_{s'\lambda'} \langle s'\lambda'|\sum_J b_J \exp(i\mathbf{Q}\cdot\mathbf{R}_J)s\lambda|^2\rangle \delta(\hbar\omega + E_\lambda - E_{\lambda'}). \tag{12.9}$$

This equation is the starting point from which scattering theory is usually developed, and by which observations of scattering are related to molecular properties.

It is possible, under certain conditions, to simplify (12.9) into parts descriptive of coherent and incoherent scattering, when

$$\frac{d\sigma}{d\Omega} = \left(\frac{d\sigma}{d\Omega}\right)_{coh} + \left(\frac{d\sigma}{d\Omega}\right)_{incoh} \tag{12.10}$$

where the coherent cross-section

$$\left(\frac{d\sigma}{d\Omega}\right)_{coh} = \overline{b}^2 \left|\sum_J \exp(i\mathbf{Q}\cdot\mathbf{R}_J)\right|^2 \tag{12.11}$$

and the incoherent cross-section

$$\left(\frac{d\sigma}{d\Omega}\right)_{incoh} = N[\overline{b^2} - \overline{b}^2] = \overline{N(b-\overline{b})^2}. \tag{12.12}$$

The incoherent cross-section is isotropic and does not depend on a phase term, therefore no information can be obtained about the relative positions of the nuclei in an array. Information about the positions of atoms can be obtained only from coherent measurements, the observed cross-section being determined by a combination of the mean value of the scattering length and interference effects between waves arising from different nuclei.

Correspondingly the total cross-section σ_T is made up of two parts

$$\sigma_{coh} = 4\pi\overline{b}^2 \tag{12.13}$$

and

$$\sigma_{incoh} = 4\pi(\overline{b^2} - \overline{b}^2). \tag{12.14}$$

Typical values in barns for these cross-sections, plus the absorption cross-section, σ_A, are shown in Table 12.1 for nuclei encountered in common

Table 12.1

Neutron scattering cross-sections for some common elements

Element	Spin I	$b \times 10^{14}/m$	σ_T/b	σ_{coh}/b	σ_A/b
^1H	$\frac{1}{2}$	-0.374	81.5	1.76	0.19
^2H	1	0.667	7.6	5.59	0.0005
^{12}C	0	0.665	5.51	5.56	0.003
^{14}N	1	0.94	11.4	11.1	1.1
^{16}O	—	0.580	4.24	4.23	0.0001
Cl†	—	0.96	15	11.58	19.5
B†	—	0.54	4.4	3.66	430
^{28}Si	0	0.42	2.22	2.20	0.06

† Naturally occurring element with more than one isotope.

polymers. The values of $\overline{b}$ are different for ^1H and ^2H and the incoherent cross-section for ^1H is almost an order of magnitude greater than that for any other nucleus. As a result, the scattering from normal organic compounds (hydrogen-containing) will be dominated by the hydrogen incoherent scattering.

The type of molecular information available from neutron scattering can be summarized diagramatically (Fig. 12.2). The type of experiment required to collect the data depends on the part of the total scattering which is to be probed. Although it is inappropriate to consider the whole experiment in detail, some general features are worthy of note.

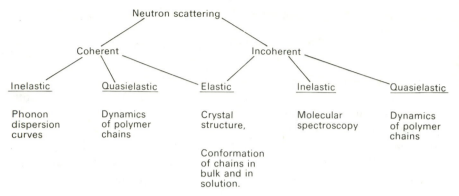

Fig. 12.2. Summary of molecular data obtainable from neutron-scattering studies.

12.3. Techniques for studying neutron scattering

The simplest spectrometer would require a source of radiation, some methods of selecting the required energy or wavelength, and a method of detecting the scattered radiation. Detection involves measuring the intensity as a function of momentum and/or energy transfer. One can therefore divide scattering instruments into two categories, those used to measure (a) elastic and (b) inelastic scattered radiation. Instruments in category (a) are used to investigate the structural or conformational properties of a system whereas these in (b) are used to study dynamic properties.

A typical source (nuclear reactor) produces a flux of neutrons near its core in the region of 10^{18}–10^{19} neutrons $m^{-1} s^{-1}$. The neutrons pass out of the reactor through beam tubes, the ends of which are placed near the reactor core. The energies of the neutrons are reduced by surrounding the core with D_2O which provides a Maxwellian distribution of neutron wavelengths peaking at about 0.1 nm with a long wavelength tail. By cooling or heating the beam tube it is possible to shift the wavelength distribution to produce a greater proportion of long- or short-wavelength neutrons respectively. Depending on the collimation and monochromation of the beam, the flux will be reduced to a value at the sample of the order of 10^9–10^{10} neutrons $m^{-1} s^{-1}$. The longer the wavelength, the smaller the flux available, and hence the more difficult it is to perform the scattering experiment.

Monochromators can be constructed in a number of ways. Single crystals are popular when neutrons of a particular wavelength are desired; selection is achieved by placing the crystal at its Bragg angle. Crystals are also used in beam tubes as energy cut-off filters, only neutrons with wavelengths greater than the Bragg cut-off pass through. For example, beryllium has a Bragg cut-off of 0.396 nm and scatters wavelengths shorter than 0.394 nm.

Velocity selection is an alternative method which has been widely used. Choppers are inserted in the beam, and only neutrons with the correct energy are able to pass through two suitably placed rotating filters.[5,6] Unfortunately, it is usual that an improvement in resolution is accompanied by a decrease in the flux intensity at the sample.

Detectors usually are designed to monitor the recoil products produced as a result of reaction with certain nuclei, e.g. ^{10}B, ^{3}He, or ^{6}Li. The reaction products can be detected either by their ionization of a gas or by light emission (scintillation counting). Detectors come in the form of tubes, cylinders, and more recently arrays of cells. The latter have been developed to facilitate spatial resolution.

12.4. Conformational studies of polymers

Although, in this volume, we are primarily concerned with the way in which the technique can provide dynamic data, it is of interest to consider how the neutron-scattering experiment can provide information on static polymer conformation.

Fundamentally the equations[7] are analogous to the Zimm equation[8] used in light-scattering studies. Plots of the intensity, determined for various values of the polymer concentration, c, and the transfer vector, $\mathbf{Q}$, are extrapolated to zero c and $\mathbf{Q}$ to allow determination of the mean square radius of gyration and the weight average molecular weight.

In light-scattering measurements the application of the above method relies on there being sufficient difference between the refractive indices of the polymer and the solvent. This is often not easily achieved. In the case of neutron scattering a significant contrast factor can be achieved by deuteration of the polymer and this is then studied in a hydrogenated solvent or vice versa. The particular advantage that this yields is the ability to study the dimensions of a deuterated polymer dispersed in a matrix of the hydrogenated polymer. In this way dimensions can be determined in bulk solid polymers. The effects of deuteration, although observable, do not markedly influence the thermodynamic properties of these mixtures. It is not appropriate to discuss here the literature on the equilibrium dimensions of polymers determined by this method, and the reader is referred to a recent review[9] for a detailed discussion of this topic. In general, results for polystyrene,[10-13], poly(ethyleneoxide),[9] poly(dimethylsiloxane),[10,14,15] poly(vinylchloride),[9] and polyethylene[16] agree with Flory's proposal that in the amorphous bulk state a polymer adopts its unperturbed dimensions with a Gaussian coil structure.

12.5. Dynamics of polymer chains—quasi-elastic scattering

It was pointed out in the previous section that molecular motion affects the observed neutron scattering spectrum in two ways: through neutron quasi-elastic scattering (NQES) and through neutron inelastic incoherent scattering (NIIS). The origins of these phenomena are slightly different and will be treated separately. It must be remembered, however, that in principle these methods sense the same molecular dynamic spectrum.

12.5.1. Neutron quasi-elastic scattering

NQES generally senses the condition $QR_g > 1$, where R_g is the root mean square radius of gyration, $\langle s^2 \rangle^{\frac{1}{2}}$, and so reflects the segmental and internal modes of the polymer chain. However, it is now possible also to measure a range of $QR_g < 1$. Further, photon correlation spectroscopy (PCS) senses the limit $QR_g \ll 1$, the data here being dominated by over-all translational motion of the polymer chain. In principle, a combination of these two techniques allows the whole relevant range of $\mathbf{Q}$-values to be studied.

12.5.2. Scattering from polymer solutions[9]

Theoretical analysis of the NQES experiments usually leads to presentation of the data either in terms of a scattering law, $S(\mathbf{Q}, \omega)$ or $I(\mathbf{Q}\, t)$, or the behaviour of the spectral width at half-maximum, $\Delta\omega$, as a function of momentum transfer. The simplest approach to modelling scattering from polymers is an analysis based on a bead and spring model.[17-20] However, the form of $I(\mathbf{Q}, t)$ so calculated does not lead to an accurate fit of the neutron scattering data. This is not completely surprising since this model is known, from hydrodynamic studies of dilute solutions of polymers in theta solvents, to hold only for the long-time behaviour. This long-time condition is equivalent to small energy and momentum transfer in neutron scattering, i.e. $QR_g \gg l > Qd$, where d is the interatomic distance and l the step length of the polymer.

Scattering in the above condition has the form, for Rouse behaviour (no hydrodynamic interaction)

$$I(\mathbf{Q}, t) = \exp\left\{ -Q^2 \frac{l^2}{3} \left(\frac{W}{\pi} |t| \right)^{\frac{1}{2}} \right\}$$ (12.15)

and in the Zimm limit (hydrodynamic behaviour)

$$I(\mathbf{Q}, t) = \exp\left\{ -Q^2 \frac{l^2}{6\pi} \Gamma(\tfrac{1}{3}) |\bar{W}t|^{\frac{2}{3}} \right\}$$ (12.16)

where $W = 3kT/\pi\eta_0 dl^3$ and $\bar{W} = 3kT/\pi\eta_0 l^3$ are reciprocal correlation times of the order of 10^{-16} to 10^{-14} s and η_0 is the solvent viscosity. The specific time dependences of (12.15) and (12.16) lead to different dependences of $\Delta\omega$ on $\mathbf{Q}$ for $S(\mathbf{Q}, \omega)$. In the Rouse limit

$$\Delta\omega_{\text{incoh}} = 0 \cdot 01\omega l^4 \mathbf{Q}^4$$

$$(12.17)$$

$$\Delta\omega_{\text{coh}} = 0 \cdot 066\omega l^4 \mathbf{Q}^4$$

whereas the Zimm limit

$$\Delta\omega_{\text{incoh}} = 0 \cdot 075\omega l^3 \mathbf{Q}^3$$

$$(12.18)$$

$$\Delta\omega_{\text{coh}} = 0 \cdot 055\omega l^3 \mathbf{Q}^3.$$

These equations are considered to be valid provided that $Ql \ll h^2$ where $h^2 = 6d^2/\pi l^2$. It should be noted that (12.17) and (12.18) imply that $I(\mathbf{Q}, t)$ depends only on η_0, the solvent viscosity and is independent of the chain chemistry, i.e. is the same for all polymers. This approximation may be valid for the long-time limit, but is unlikely to be appropriate to the description of shorter-time behaviour.

The description of the motion of a polymer chain in which rotational motion is hindered has been attempted.[21] In the long-time limit the equation has the form

$$I(\mathbf{Q}, t) \propto \exp(-At^{\frac{1}{2}})$$

$$(12.19)$$

and predicts that the change over from $\mathbf{Q}^4$ to $\mathbf{Q}^3$ behaviour with variation of $\mathbf{Q}$ should be a function of chain stiffness. An appropriate solution of the equations of motion for a chain with persistence length, a yields

$$\Delta\omega \propto (\mathbf{Q}a)^{\frac{8}{3}}.$$

$$(12.20)$$

An extension of this type of equation, considered to be valid for a semi-dilute solution of polymer chains where the radius of gyration of each coil is larger than the distance between neighbouring coils, has the form[22,23]

$$\Delta\omega = \Delta\omega_R\{1 + (\xi^2/\mathbf{Q}^2)\}$$

$$(12.21)$$

where $\Delta\omega_R$ is the simple Rouse half-width proportional to $\mathbf{Q}^4$, and ξ is the static screening length. If $\mathbf{Q} \ll \xi^{-1}$ then (12.21) becomes $\Delta\omega \propto \mathbf{Q}^2$, which is in agreement with the experimental evidence. Recently scaling arguments have been used[24,25] to extend this treatment to concentrations where entanglements are likely to be present. Two regimes have been identified: (i) when $\mathbf{Q} < \xi^{-1}$ where the diffusion is probably co-operative and (ii) when $\mathbf{Q} > \xi^{-1}$ where single-chain behaviour should be observed with $\Delta\omega$ proportional to $\mathbf{Q}^3$.

Poly(tetrahydrofuran) in CS_2,[26] polystyrene in CS_2,[27] and poly(ethyleneoxide) in water[28] have been studied. However, the general conclusion drawn from these studies is that existing techniques for dilute solutions limit the range investigated to $Ql \simeq 1$, so that interpretation of the data does not yet yield all the information in principle available.

12.5.3. Studies of bulk polymers

Although the theories described above refer to dilute solutions, for historical reasons most of the data have been obtained on bulk polymers. The first measurements[29] were made on linear dimethylsiloxanes, oligomers, and polymers, the degree of polymerization being in the range 3 to 2000. Five cyclic dimethylsiloxanes with ring sizes between 3 and 18 were also measured. The 'effective' diffusion coefficients D_{eff} obtained from eqn (12.14) were of the order of 10^{-9} m^2 s^{-1}, which is similar in magnitude to small-molecule diffusion coefficients.

The value of D_{eff} decreased as n increased, reaching an asymptotic value at about $n = 200$. It will be recalled that similar molecular-weight dependences have been noted from dielectric and ultrasonic relaxation studies discussed in the earlier chapters. The values of D_{eff} for the larger cyclic siloxanes were comparable with those for the linear polymers and cross-linking was observed to have little effect on the results. Similar asymptotic behaviour has been reported for poly(ethyleneoxide) when n is greater than 70, the steeper climb to an asymptotic result being attributed to differences in the chain stiffness.[30] The activation energy for D_{eff} is of the order of 8 kJ mol^{-1} for the siloxane polymers. It is clear that D_{eff} reflects over-all rotation and translation at low molecular weights, and is dominated by segmental motion in the higher molecular-weight polymers.

A variety of other polymers have been studied[26-8,31-4] and include poly(methylphenylsiloxane), poly(propyleneoxide), poly(ethyleneoxide), and poly(isobutylene). In the case of poly(propyleneoxide), by deuteration of the methyl group, it was possible to establish that the predominant motion does involve the polymer backbone. The activation energies for diffusive motion, observed by different techniques and calculated assuming a simple Arrhenius type of relationship indicate that

$$E_{viscosity} > E_{relaxation} > E_{neutron}.$$

The reason for this trend is not clear, and it is evident that some refinement of the analysis and data is still desirable.

12.6. Dynamics of polymer molecules—inelastic scattering

In this experiment, the energy transferred to the neutron provides information on the rotational and vibrational motions executed by the polymer backbone. The observed spectrum resembles a combination of the infrared and Raman spectra of the polymer. Whereas the photon interaction is controlled by specific selection rules, the neutron scattering is roughly proportional to the amplitude of vibration of the atom with which it interacts. As a result, the spectrum is very rich in structure and for organic materials is dominated by the hydrogen vibrations. Deuteration of the polymer is useful in assignment of vibrational modes. The particular advantage which neutron inelastic scattering has over conventional spectroscopy is its sensitivity to torsional vibrations. These are often large-amplitude vibrations and give rise to distinct neutron scattering features. However, the small dipole moment or polarization change associated with such motions makes their study by infrared or Raman spectroscopy difficult if not impossible.

The intensity of vibrational bands will be dependent on the following factors:[1,35]

(i) The momentum transfer $\mathbf{Q}$; since $\mathbf{Q}$ is dependent on angle the spectrum varies with angle of scattering;
(ii) The displacement vector and hence the amplitude of the normal vibrational mode;
(iii) The value of the incoherent scattering cross-section of the atoms involved in the motion.

Use of torsional vibrations in the estimation of activation energies for segmental motion is subject to two important constraints:[36]

(a) Analysis of an asymmetric rotor (one in which the equilibrium states differ in energy) is usually impossible, since, due to the lower population of the higher energy isomer, the associated torsional mode can rarely be observed. The analysis also requires a precise knowledge of the energy difference between states and this is rarely available for polymers;
(b) Even for symmetric rotors, where analysis is in principle possible, the observed torsional vibration may not reflect the true barrier. The first torsional transition occurs at the bottom of the potential well; rotational isomerism often involves ten or more such transitions and may involve changes in the molecular geometry on achieving the eclipsed conformation. For the effects of molecular distortion to be measured, overtones of the torsional vibration must be measured. These are usually a factor of ten or more weaker in intensity and often are masked by the higher frequency normal modes. As a result, the values obtained from the torsional analysis can either under- or overestimate the true barrier to conformational change.

As a result of the above constraints, studies to date have been restricted to observation of either methyl- or phenyl-group rotation. Neutron scattering is particularly useful for these studies as both motions are difficult to study by either infrared, Raman, or nuclear magnetic resonance spectroscopy. In the latter technique coupling between the relaxation of neighbouring nuclei complicates the analysis of the data.

12.6.1. Methyl group rotation

The methyl group has C_3 symmetry and for the zeroth approximation only V_3 is required for the description of the rotational profile. For hindered internal rotation about a single bond, the potential energy opposing rotation has the form

$$V(\phi) = (V_3/2)(1 - \cos 3\phi) \tag{12.22}$$

where ϕ is the azimuthal angle.

A variety of polymers have been investigated and values of V_3 obtained for the motion of the methyl group (Table 12.2). The apparent discrepancies

Table 12.2

Methyl group motion

Polymer	ν_{tor}/cm^{-1}	$V_3/kJ\ mol^{-1}$	$E_A/kJ\ mol^{-1}$	Ref.
Poly(methylmethacrylate)				36
—OCH_3 (all isomers)	100	4–6	very low	
α-CH_3 (isotactic)	300	23	16	
α-CH_3 (syndiotactic)	360	33	23–35	
Poly(propyleneoxide)	230	13	15·9	36
Poly(α-methylstyrene)				36
(heterotactic)	380	37		
(head-to-head)	300	13		
Poly(methylvinylether)	100	2·5		36
Poly(dimethylsiloxane)	165	6·9	9–10	37
Poly(4-methylpentene-1)	240	15	1·7	38

between values obtained from neutron scattering and from analysis of mechanical data are sometimes attributed to quantum mechanical tunnelling. This is assumed to occur when a particle passes through a region of potential energy greater than its own kinetic energy and is represented by the occurrence of a significant amplitude of the wave function inside the barrier. For symmetric rotors such as methyl groups, conditions for quantum mechanical tunnelling are favourable since the energy levels are identical in each potential well. A correction to the activation energy due to quantum mechanical effects leads to better agreement between values obtained by various methods.[39,40] An alternative (albeit less favoured) view is that the potential is not adequately represented by a simple threefold cosine potential as the methyl rotor achieves the eclipsed state.

12.6.2. Phenyl group motion

Although a variety of relaxations are possible in polystyrene[41,42] it appears that it is possible to identify in both Raman and neutron scattering a band at 60 cm^{-1} which has been attributed to phenyl-group torsional motion.[43-5] It is clear that this motion is coupled with other motions of the chain and it is not a simple task to obtain from the data an activation energy for phenyl motion. However, it appears clear that this motion is intimately connected with the low-temperature anomaly in the specific heat observed in this polymer.

12.7. Future trends

Neutron scattering may be expected to make a valuable contribution to the study of the equilibrium size of a polymer in dilute, concentrated, and bulk phases. Although early expectations that neutron scattering technique would solve most problems in polymer dynamics have not been fulfilled, a more realistic attitude now prevails. In this context studies are now advancing from dynamic phenomena which can be studied more conveniently by other techniques to those[46,47] where large transfer vectors, localized phenomena, or avoidance of selection rules make the characteristics of neutron scattering invaluable, and where it can play a complementary (rather than duplicating) role with other methods.

The recent appearance of the neutron spin-echo technique[48,49] has considerably modified the frequency-time range available for observation of dynamic processes. The technique uses the spin polarization properties of the neutron to probe the small energy changes occurring during scattering. A neutron beam is polarized by the application of a high magnetic field and then scattered by the sample. The resultant beam possesses a slightly different spin polarization from the original source, the difference being a measure of the energy transfer during the scattering process. The experiment is carried out as a null method, the beam polarization being restored to its original value by a second magnetic field.

To date, two studies have been reported[48,49] using this technique and both indicate the potential of the method for the exploration of segmental motion of polymers. Since the longest time available for observation is of the order of 10^{-6} s, the fascinating idea of investigating the reptation motion of a single polymer using neutron scattering does not at present appear feasible. There is, however, a number of interesting problems which can now be tackled with the extended frequency capability offered by this method.

References

1. Marshall, W. and Lovesey, S. W. *Theory of thermal neutron scattering*. Clarendon Press, Oxford (1971).
2. Turchin, V. E. *Slow neutrons*. Israel Program for Scientific Translation, Jerusalem (1965).
3. Allen, G. and Higgins, J. S. *Rep. Progr. Phys.* **36**, 1073 (1973).
4. Van Hove, L. *Phys. Rev.* **95**, 249 (1954).
5. Harryman, M. B. M. and Hayter, J. *AERE Harwell Report*, RRL 73/25 (1973).
6. Bunce, L. J., Harris, D. H. C. and Stirling, G. C. *UKAEA Report.* Harwell, R6246 (1971).
7. Kratky, O. *Pure appl. Chem.* **12**, 483 (1966).
8. Zimm, B. H. *J. chem. Phys.* **16**, 1093 (1948).
9. Maconnachie, A. and Richards, R. W. *Polymer* **19**, 739 (1978).
10. Ballard, D. G. H., Wignall, G. D. and Schelten, J. *Eur. polym. J.* **9**, 965 (1973).
11. Wignall, G. D., Ballard, D. G. H. and Schelten, J. *Eur. polym. J.* **10**, 801 (1974).
12. Cotton, J. P., Decker, D., Benoit, H., Farnoux, B., Higgins, J. S., Jannick, G., Ober, R., Picot, C. and de Cloiseaux, J. *Macromolecules* **7**, 863 (1974).
13. Kirste, R. G., Kruse, W. A. and Schelten, J. *J. makromol. Chem.* **162**, 299 (1973).
14. Schelten, J., Kruse, W. A. and Kriste, R. G. *Kolloid Z.* **251**, 919 (1973).
15. Gerannt, R., Pechhold, W. and Grossman, H. P. *Colloid polym. Sci.* **225**, 285 (1977).
16. Schelten, J., Ballard, D. G. H., Wignall, G. D., Longman, G. and Schmatz, W. *Polymer* **17**, 751 (1976).
17. Rouse, P. E. *J. chem. Phys.* **21**, 1272 (1953).
18. Zimm, B. H. *J. chem. Phys.* **24**, 269 (1956).
19. De Gennes, P. G. *Physics* **3**, 37 (1967).
20. Dubois Violette, E. and De Gennes, P. G. *Physics* **3**, 181 (1967).
21. Jannick, G. and Saint James, D. *J. chem. Phys.* **49**, 486 (1968).
22. Jannick, G. and De Gennes, P. G. *J. chem. Phys.* **48**, 2360 (1968).
23. Jannick, G. and Summerfield, G. C. *Harwell Report* IAEA,-SM-155/C-2(1970).
24. De Gennes, P. G. *Macromolecules* **9**, 587, 594 (1976).
25. Pincus, P. *Macromolecules* **10**, 210 (1977).
26. Allen, G., Gosh, R. E., Higgins, J. S., Cotton, J. P., Farnoux, B., Jannick, G. and Weill, G. *Chem. Phys. Lett.* **38**, 577 (1976).
27. Akcasu, Z. and Higgins, J. S. *J. polym. Sci., polym. Phys.* **15**, 1745 (1977).
28. Maconnachie, A., Vasudevan, P. and Allen, G. *Polymer* **19**, 33 (1978).
29. Allen, G., Brier, P. N., Goodyear, G. and Higgins, J. S. *Faraday Symp. Chem. Soc.* **6**, 169 (1972).
30. Allen, G., Higgins, J. S. and Wright, C. J. *J. chem. Soc. Faraday Trans. II* **70**, 348 (1974).
31. Allen, G., Gosh, R. E., Heidemann, A., Higgins, J. S. and Howells, W. S. *Chem. Phys. Lett.* **27**, 308 (1974).
32. Higgins, J. S., Gosh, R. E. and Howells, W. S. *J. chem. Soc. Faraday Trans. II* **73**, 40 (1977).
33. Larsson, K. E. *Faraday Symp. chem. Soc.* **6**, 167 (1972).
34. Singleton, R. PhD. thesis. Manchester (1977).
35. Boutin, H. and Yip, S. *Molecular spectroscopy with neutrons.* MIT Press, Cambridge, Massachusetts (1968).

36. Cunliffe A. V. In *Internal rotation in molecules* (ed. W. J. Orville-Thomas), Chapter 7. Wiley, London (1974).
37. Sauer, J. A. *J. polymer. Sci.* **C32**, 69 (1971).
38. Allen, G., Wright, C. J. and Higgins, J. S. *Polymer* **15**, 319 (1974).
39. Stejakol, E. O. and Gutowsky, H. S. *J. chem. Phys.* **28**, 388 (1958).
40. Higgins, J. S., Allen, G. and Brier, P. N. *Polymer* **13**, 157 (1972).
41. Yono, O. and Wada, Y. *J. polym. Sci. A-2* **9**, 669 (1971).
42. Froix, M. F., Williams, D. J. and Goedde, A. O. *Macromolecules* **9**, 354 (1976).
43. Spells, S. J., Shepherd, I. W. and Wright, C. J. *Polymer* **18**, 905 (1977).
44. Kim, J. J., McLeish, J., Hyde, A. J. and Bailey, R. T. *Chem. Phys. Lett.* **22**, 503 (1973).
45. Zoller, P., Fehl, D. L. and Dillinger, J. R. *J. polym. Sci., polym. Phys.* **11**, 1441 (1973).
46. Allen, G. and Fewster, S. In *Internal rotation in molecules* (ed. W. J. Orville-Thomas). Wiley, London (1974).
47. Allen, G. and Wright, C. J. *Int. Rev. Sci. Phys. Chem. Series II, Macromol. Sci.* **8**, 223 (1975).
48. Richter, D., Hayter, J. B., Mezei, F. and Ewen, B., *Phys. Rev. Lett.* **41**, 1484 (1978).
49. Higgins, J. S., Nicholson, L. K. and Hayter, J. B., *Polymer* (in press).

DIFFUSION-CONTROLLED REACTIONS
IN POLYMER SYSTEMS

13.1. The nature of diffusion-controlled reactions

The majority of chemical reactions of molecularity greater than unity can take place only when the reactants approach each other to less than a certain critical distance. In liquid and solid phases this condition usually implies that one reactant molecule must be placed in the nearest-neighbour shell of its reaction partner. Very often, too, there is the additional requirement that a reagent undergo some form of rotational diffusion into a specific orientation or conformation. In condensed phases, the time over which any two molecules exist as nearest neighbours is larger than the time characterizing the quasi-vibrational thermal collisions between molecules. Consequently, from a kinetic standpoint, the two reactant molecules remain paired for a finite time at a separation permitting chemical reaction, a situation which has been called an encounter.[1]

In the case of reactant molecules diluted in a solvent of similar molecular size, the concept of an encounter is quite precise. The radial distribution function for a liquid is such that it is meaningful to describe a 'nearest-neighbour' shell around any one molecule, and to make use of the inter-molecular separation corresponding to a maximum in the function. It is then a relatively simple matter to relate the encounter to observable macroscopic properties of the system.

When polymeric species are under consideration, the reactive positions on the molecules may occupy only a very small element in the total molecular volume, and the conformation of these positions with respect to the rest of the molecule may not be fixed. Thus it is possible for translational diffusion to have brought the centres of gravity of the macro-reactants to a separation such that reaction is possible, and yet the reactive sites could still be separated by several solvent molecules or polymer segments. It is then convenient to visualize the over-all reaction as occurring by way of three processes.[2,3] These are:

(a) Translational diffusion of two reactant species from a random separation until some segments of the two are no longer separated by solvent molecules;

(b) Rearrangement or rotation of the reactants so that the groups 'in contact' become the reactive portions of the macromolecules;

(c) chemical reaction.

These three stages can be written as a kinetic scheme

$$A + B \underset{k_{-D}}{\overset{k_D}{\rightleftharpoons}} (A.B.) \qquad \text{translation}$$

$$(A.B.) \underset{k_{-R}}{\overset{k_R}{\rightleftharpoons}} (A:B) \qquad \text{reorientation}$$

$$(A:B) \overset{k_C}{\to} \text{Product} \qquad \text{reaction.}$$

Where (A.B.) and (A:B) represent the macromolecular pair with reactive groups in positions which are, respectively, unfavourable and favourable for chemical reaction.

This treatment is justifiable so long as the unfavourably oriented pair has a lifetime sufficiently long to permit some rearrangement. Then, although the two motion processes are occurring simultaneously, we can use the scheme to obtain limiting values for an observed reaction rate constant.

Chemical control: $\quad k_{obs} = k_D k_R k_C / k_{-D} k_{-R}, \quad k_C \ll k_{-R}, \quad k_{-D} k_R / k_{-R}$

Translational diffusion control: $\quad k_{obs} = k_D, \quad k_C \gg k_{-D}, \quad k_R \gg k_{-D}$

$$(13.1)$$

Reorientation diffusion control: $\quad k_{obs} = k_D k_R / k_{-D}, \quad k_R \ll k_{-D}.$

It will be noticed that when the later steps (reorientation or chemical reaction) are rate-determining, the observed rate constants contain terms descriptive of an equilibrium concentration of encounter pairs.[4]

The types of fast reaction which exhibit diffusion-controlled behaviour in normal liquids include radical–radical reactions, ion-recombination reactions, the quenching of electronically excited species, and certain reactions involving macromolecular reactants of low diffusivity. Of course in solids, glasses, or viscous liquids, mass transport processes are very inefficient, and chemical reactions which are normally considered to be rather slow can become diffusion-controlled.[5]

So far, we have discussed only the quantitative effect of diffusion on the rate of a single chemical step. However, diffusive processes may govern which products are obtained from complex chemical reactions involving competitive reactions. Some particularly striking examples of this are to be found in photochemistry, where the diffusion-controlled quenching of excited states may compete with other relaxation processes. A competition between diffusive separation and reaction is also found when reactive entities are formed in pairs.

13.2. Calculation of diffusion-controlled reaction rates

Diffusion-controlled processes represent the only broad class of chemical reaction for which it is now possible to calculate reaction rates from fairly well authenticated first principles. The theoretical aspects of these have been reviewed by Noyes.[6] The theoretical treatments have involved two rather different lines of thought. The first applies Fick's laws of diffusion to the movement of reactant molecules across possible concentration gradients in the solution, whereas the second focuses attention on the potential reactivity of an isolated pair of molecules. Somewhere between these approaches lies a treatment based on a consideration of the reactant pair-probability distribution.

13.2.1. Simple application of Fick's laws

The problems associated with the diffusion together of reactant species are essentially the same as those dealt with in the coagulation of colloids. If one reactant, A, can remove several molecules of another, B, then the concentration of B in the immediate neighbourhood of A is reduced below the average value for the whole system. There is thus a gradient in the concentration of B about each A molecule, and there results a net flux of B molecules towards A. This net flux is equal to the rate of the reaction. Indeed, this concentration gradient can be postulated even for the case when each A molecule is capable of reacting with only one B molecule. Use of Fick's laws followed by solution of the relevant diffusion equations allows evaluation of the reaction flux, both when the concentration gradient is changing during the initial stages of the reaction, and when steady-state diffusion has been achieved.

By use of this treatment the observed rate constant when a steady state has been realized is,

$$k_{\mathrm{obs}} = \frac{4\pi r_{\mathrm{AB}} D_{\mathrm{AB}}}{1 + (4\pi r_{\mathrm{AB}} D_{\mathrm{AB}})/k} \tag{13.2}$$

in units determined by those of the quantities r_{AB}, the radius of the 'capture' sphere formed by B molecules around an A molecule, D_{AB}, the relative diffusion coefficient of the approaching reactants which is the sum of the individual coefficients, $D_{\mathrm{A}} + D_{\mathrm{B}}$ and k, the rate constant which would be observed if the concentration of B molecules in the nearest-neighbour shell of an A molecule was the same as the average over the whole system. It is normal to give rate constants in units of molecule $cm^{-3} s^{-1}$ which requires choosing r_{AB} in cm^{-1} for D_{AB} in $cm^2 s^{-1}$.

The rate constant for the extreme case when reaction takes place at every encounter is often calculated by using the simple boundary condition that the concentration of B in the nearest-neighbour shell of each A molecule is zero

$$k_{obs} = \pi r_{AB} D_{AB} .\qquad(13.3)$$

This treatment, first carried out by Smoluchowski,[7] is in fact only applicable when $k \gg 4\pi r_{AB} D_{AB}$, which inequality may not always be attained. The choice of correct boundary conditions, and the assumptions inherent in the Fick's law treatment, have been extensively covered[6,8-10] in a number of papers. In particular, the facts that a liquid is not an isotropic continuum and that a degree of correlation may exist between the movements of the various reactants, cannot be encompassed by the use of Fick's equations.

13.2.2. Diffusion and pair-probability distribution

The alternative theoretical approach to the problem of rates of diffusion-controlled reaction has involved a study of the behaviour of molecular pairs, and an evaluation of the pair-probability distributions.

It is possible to discuss perturbations in the pair-probability distribution in terms of the same parameters used in the preceding section. In this fashion the effect of diffusive flux and chemical reaction on the reactant pair-distribution function has been examined by Collins and Kimball[11] and by Waite.[10,12]

The treatment is quite general, and the simplified expression for a diffusion-controlled reaction rate constant is

$$k_{obs} = 4\pi r_{AB} D_{AB} \{1 + r_{AB}/\pi D_{AB} t)^{\frac{1}{2}}\} .\qquad(13.4)$$

This can be recognized as the time-dependent form of eqn (13.3) for which $k \gg 4\pi r_{AB} D_{AB}$, and can easily be converted to the more exact form by replacing r_{AB} by $r_{AB}/\{1 + (4\pi r_{AB} D_{AB}/k)\}$.

13.2.3. The behaviour of molecular pairs

The discussion of the potential reactivity of an isolated molecular pair is attractive, because by this approach it might be possible to avoid the approximations involved in applying macroscopic continuum equations and concepts to molecular phenomena. This line of attack has been extensively investigated by Noyes.[13]

The problem is tackled by considering an imaginary system in which no real chemical reaction takes place during an encounter of an A molecule and a B molecule, but in which it is possible to distinguish molecules that have been in a situation where they would have reacted had the system been a real one. Using this idea we can express any reaction rate constant as a primary constant (characteristic of a truly random distribution of reactants or of zero time) modified by a function which accounts for the fact that long-lived reactants must have been formed a greater-than-average distance from other reactants. While the expression is quite general in character, it does involve certain parameters which cannot be related to experimentally observable quantities without the use of assumptions similar to those inherent in Fick's laws.

13.2.4. Collisions on pseudo-crystalline lattices

One of the first to appreciate the significance of the fact that collisions between reactants in liquids occur in isolated sequences was Rabinowitch.[14] Using a pseudo-crystalline model of the liquid state, he derived the first equation which included both chemical and diffusive processes. The derivation, based on calculations of jump diffusion between lattice sites, took no account of the non-random distribution of reactants in space, and has now been largely superseded by the treatments mentioned previously.

13.2.5. Corrections for long-range interactions

When two reactants are capable of exerting long-range interactions upon each other, the rates at which they diffuse together (or apart) will be affected.

The most widely used method[6,12,15] of allowing for these interactions is to replace the encounter radius r_{AB} in the equations presented above by an 'effective' value r_{eff}. This parameter will be larger than r_{AB} for attractive forces, and less than r_{AB} for forces of repulsion.

The actual magnitude of r_{eff} can be found by integration of the long-range interaction from infinite separation up to r_{AB}, and is

$$r_{eff} = \left[\int_{r_{AB}} \exp(U/kT) \, (dr/r^2) \right]^{-1} \qquad (13.5)$$

where U is the interaction energy and is a function only of the A–B separation. Electrostatic or any other relevant interactions can be inserted for U and the integration carried out without difficulty.

It is important to realize that r_{eff} has no physical significance. Indeed for intermolecular repulsions r_{eff} may be less than any realistic molecular dimension.

13.3. Extension of theory to polymeric reagents

Although a number of papers have been published dealing with the theoretical prediction of diffusion-controlled reaction rates of polymeric species, none is totally satisfactory. The problem centres on the relative emphasis given to translational and reorientation diffusion and to the fact that these are concurrent rather than exactly consecutive.

The simplest model is simply to treat the macro-reactants as spherical coils and to assume that translational diffusion is the rate-determining step.[16] Under these circumstances eqn (13.2) can be applied. If the further substitutions are made that:

(a) In this case the encounter separation is related to the coil radii by a random-walk relationship,

$$r_{AB}^2 = R_A^2 + R_B^2;$$ (13.6)

(b) The diffusion coefficients obey a Stokes–Einstein law

$$D_A = kT/6\pi\eta R_A$$ (13.7)

where η is the solution viscosity;

(c) The coil radii are proportional to the square root of the degrees of polymerization,

$$R_A \propto N_A^{\frac{1}{2}}$$ (13.8)

then

$$k_{obs} = \frac{4kT}{6\eta}\left\{\left(1 + \frac{N_A}{N_B}\right)^{\frac{1}{2}} + \left(1 + \frac{N_B}{N_A}\right)^{\frac{1}{2}}\right\}.$$ (13.9)

The rate constant predicted by this model has an unusual dependence on the size of the reacting entities. It is a minimum, and independent of chain length, in monodisperse systems, and in heterodisperse systems depends on the size distribution rather than the average size. This means that in many processes, where the former quantity does not vary, the rate constant is independent of molecular weight. Absolute values calculated from eqn (13.8) for the radical–radical reaction observed in free-radical polymerization appear to be about a hundred times larger than those observed in practice, indicating that neglect of the reorientation process is serious.

The first thoughts[2,16,17] on accommodating the small volume of the reactive sites considered the reaction as the simple diffusive approach of two spheres, but the collision or capture radius was defined by the reactive-site geometry whereas the mobility was determined by the whole macromolecular radius. This approach, referred to rather light-heartedly as the 'ball and chain model'[2] or the 'volume swept out model'[17], predicts reaction rate constants of similar magnitude to those observed in practice.

Combination of the Smoluchowski and Stokes–Einstein relations as before now yields

$$k_{obs} = \frac{2kT}{3\eta} R_E\left(\frac{1}{N_A^{\frac{1}{2}}} + \frac{1}{N_B^{\frac{1}{2}}}\right)$$ (13.10)

where R_E is the reactive site capture radius. This expression now has an inverse square root dependence on the chain length of the reactants.

It is probable that (13.8) and (13.9) represent two extreme situations, since the former equation overestimates the segmental mobility at the reaction site (effectively assuming unit probability of reaction per polymer–polymer contact) whereas the latter underestimates it (effectively retaining the macrochain mobility right up to the moment of chemical reaction).

Because or these inadequacies in the concept of simple collision between two spheres, an alternative approach is to modify the rate constant for diffusive collision of two spheres by a probability that the reactive sites come together during the 'collision'. This probability can be considered to be similar to the 'steric factor' familiar in the collision theory of chemical reactions in gases.

The simplest model is to treat the reactants as spheres on the surface of which the reactive sites occupy a fraction of the total area. When the spheres are rigid and undergo collisions that are essentially elastic, the reaction probability requires that the contact point should lie in both reactive zones, and so is the product of the reactive area fractions. While this oversimplified picture might be applicable to reactions of certain biopolymers such as globular proteins, it is obviously inapplicable to flexible linear molecules.

The first obvious extension of this hard-sphere model, and one still placing major emphasis on translational processes, is to consider the active site as lying on the surface of a deformable soft sphere.[18] This sphere undergoes both rotational and translational diffusion, and collisions with other spheres are markedly inelastic, producing a two-dimensional contact surface. Points on either sphere move along this surface as the sphere rotates. At any instant in time the path traced by a particular point, originally on sphere A, may intersect with that traced by a point on sphere B. In this way the three-dimensional problem of contact between active sites can be reduced to a two-dimensional collision calculation. In order to find the time over which the two-dimensional surface exists, it is necessary to make some assumptions about the deformation of the spheres. With the simplest one, that the spheres approach each other essentially unretarded up to the final deformation, and then separate in an elastic fashion, it is possible to calculate a reaction probability per collision of macrospheres. This can then be combined with the collision rate for the spheres calculated from the Smoluchowski equation with the 'reflecting' boundary condition, k not greater than $4\pi r_{AB} D_{AB}$.

These models with the reactive entity on the surface of a sphere have obvious deficiencies for reactions of linear chains, where any segment (including the chain ends) has a higher probability of being in the 'interior' of the flexible coil than on its 'surface'. This means that treatment based on modification of the translational diffusion-controlled approach of two polymeric reactants must take account of interpenetration of the polymer coils. Although an expression involving reactive-site diffusion in the inter-penetration volume has been suggested in early papers[2,16] on diffusion-controlled polymer reactions, difficulties in accommodating segmental motion mean that much ensuing work on reaction in interpenetrating coils has adopted a statistical rather than a kinetic approach.

One of the problems to be overcome lies in determining just how easy interpenetration of the coils may be, and what is the accessibility of the reactive sites. One of the attempts to calculate the accessibility of chain ends in a coiled linear polymer assumed that one reactant could penetrate the second only to a certain critical distance, and then calculated the statistical probability that the reactive sites could meet in the interpenetration volume.[18] For the simplest case when interpenetration is limited by the number of chain segments of one coil lying behind the reactive site of the second molecule, the reaction probability reduces to

$$P_c = \sqrt{2}\pi \frac{N_c S_c^2}{Z} \tag{13.11}$$

where S_c is the critical interpenetration distance, N_c is the number of segments behind the reactive centre which exert the force opposing further interpenetration, and Z is the number of equivalent freely rotating links in each chain.

In fact the rigorous calculation of interpenetration geometries and lifetimes is very difficult, and a number of papers have been published in which the extreme of complete interpenetrability has been assumed.[19-21] Under these circumstances the chain end is assumed to have the same mobility as the polymer coil, and the resulting rate constant is independent of the coil size (and so also of the polydispersity). However rate constants so calculated are still higher than those observed in practice, and it is again necessary to introduce some arbitrary constraints on the localized motion of the active segments.

One of the better treatments of combined translational and segmental diffusion of chain ends is due to Burkhart.[22] The derivation starts from the kinetic scheme presented in (13.1), and initially follows the analysis of Benson and North.[16] The generalized diffusion-control condition is

$$k_{obs} = \frac{k_D k_R}{k_{-D} + k_R}, \qquad k_c \gg k_{-D}. \tag{13.12}$$

Diffusion and dimensional parameters are then introduced using the Smoluchowski equation to evaluate k_D and k_R

$$k_D = 4\pi D_{AB} r_{AB}$$

$$k_R = 4\pi D_E r_E C_E \tag{13.3}$$

where subscripts A,B refer to whole macromolecules and subscript E to the chain ends. C_E is the mean concentration of chain ends in the proximate pair. The constant k_{-D} is evaluated by assuming that the equilibrium concentration of proximate pairs is determined solely by their volume, and is relatively unperturbed by the reaction. This is equivalent to assuming that the reorientation step is significant in determining the rate of the whole process. The concentration of such pairs is the number of B chains, per unit volume, whose centres lie within a sphere of radius R_{AB} centred on an A-chain centre. So

$$\frac{k_D}{k_{-D}} = \tfrac{4}{3}\pi R_{AB}^3 \quad \text{or} \quad k_{-D} = 3D_{AB}/R_{AB}^2. \tag{13.14}$$

Then

$$k_{obs} = \frac{4\pi D_E D_{AB} R_E R_{AB}}{D_{AB}R_{AB} + D_E R_E}. \tag{13.15}$$

The analysis now considers the frictional forces acting on the chain segments so as to evaluate D_E and D_{AB}. The concept of internal viscosity[23] is introduced in the same way as in the calculations of polymer intrinsic viscosity, and a final equation (for macroreactants monodisperse in chain length) is

$$k_{obs} = \frac{12\pi R_E kT}{\lambda \eta L} \ln \frac{[9\lambda \eta L + 18B]}{[\lambda \eta L + 18B]} \tag{13.16}$$

where chains, of contour length L, can be subdivided into Z statistical units so that the internal viscosity force constant, B is given by the ratio β/Z with β a stiffness parameter. λ is a shape-resistance parameter which is $19 \cdot 2/Z^{\frac{1}{2}}$ for Gaussian chains. Values of k_{obs} calculated[22] for $R_E = 0 \cdot 4$ nm, $\eta = 5 \times 10^{-4}$ kg m^{-1} s^{-1}, and $L/Z = 0 \cdot 125$ nm at 298 K are given in Table 13.1.

The predicted rate constants are of the correct order of magnitude. Furthermore they are inversely dependent on solvent viscosity at low values of the stiffness parameter, β, becoming independent of solvent viscosity as the internal resistance to reorientation increases. On the other hand, in the most flexible and longest chains, the rate constant is very insensitive to chain length, but becomes proportional to $Z^{\frac{1}{2}}$ as the stiffness increases or chain length decreases.

One of the advantages of this formulation is that the shape-change and stiffness parameters, λ and β, can be evaluated from flow measurements such as viscosity and flow birefringence. In this respect it is more useful than treatments where D_E is evaluated using empirical or less easily obtained parameters.

The latest development in this area has been a calculation[24] of the probability that intramolecular collision can occur between the two

Table 13.1

Calculated values of k_{obs}

β	Z	$k_{obs}/\text{dm}^3\,\text{mol}^{-1}\,\text{s}^{-1}$
10^{-3}	200	$3 \cdot 3 \times 10^6$
	2000	$3 \cdot 2 \times 10^7$
	20 000	$1 \cdot 8 \times 10^8$
	200 000	$1 \cdot 4 \times 10^8$
10^{-4}	200	$3 \cdot 3 \times 10^7$
	2000	$2 \cdot 6 \times 10^8$
	20 000	$4 \cdot 0 \times 10^8$
	200 000	$1 \cdot 9 \times 10^8$
10^{-5}	200	$3 \cdot 0 \times 10^8$
	2000	$9 \cdot 6 \times 10^8$
	20 000	$5 \cdot 7 \times 10^8$
	200 000	$1 \cdot 5 \times 10^8$

(presumably reactive) ends of a polymer chain. This is, in fact, a rather old problem in polymer kinetics, having been considered many years ago by Haward[25] in the context of self-termination of biradicals. In this case the diffusion equation for the configurational distribution of the chain is examined in much the same way as described in Chapters 3 and 4, and those configurations leading to ring closure selected. It transpires that the reaction rate is strongly affected by the short-time behaviour of the segmental motion (as might be expected on an intuitive basis). Consequently the calculated rate is dependent on the model selected to represent the polymer chain. When it is assumed that the two groups are connected by a simple harmonic spring the calculated rate constant is given by

$$k_{\text{calc}} = \frac{R_E}{L\tau_m}, \tag{13.17}$$

where L is the root mean square separation of the reactive chain ends, and τ_m is the maximum relaxation time (normal mode) of the end-to-end vector.

An alternative model is to consider the chain as an array of beads and springs as discussed in Chapter 3. Under these circumstances the reaction rate becomes independent of the ratio R_E/L. It must be stressed, however, that analysis of this model strictly applies to the limit $R_E/L \to 0$, and so has little applicability to situations where the chain length is finite compared with the size of the Gaussian subunits. However the predicted rate constant is

$$k_{\text{calc}} \sim 1/\tau_m \tag{13.18}$$

where τ_m is again the maximum relaxation time of the chain.

Both (13.17) and (13.18) have the rather disturbing feature that the phenomenological dependence on the *high*-frequency segmental motions results in an equation containing only the time constant of the *lowest-frequency* motion. A wholly satisfactory treatment of this reaction rate must await the mathematical treatment of high-frequency motions in terms of parameters which can relate to 'whole-chain' modes.

13.4. Experimental observations of diffusion-controlled reactions

13.4.1. The rate-determining step in the free-radical termination reaction

By far the largest number of studies of diffusion-controlled reactions in polymer systems have involved the radical–radical reaction, and particularly the termination step in free-radical polymerization. Various chemical and physical aspects of this reaction have been reviewed[26,27] in two texts, with particular emphasis on the role of diffusive processes.

The first point to be determined in any study of such a reaction is whether the chemical or diffusive step is rate-determining. The test most commonly applied is to conduct the polymerization in solvents of differing viscosity but of similar solvent power towards polymer. Under conditions of diffusion-controlled termination the termination rate constant should vary inversely as the solvent viscosity (frequently as the first power) and consequently (though less stringently) the over-all polymerization rate should increase as the square root of the solution viscosity. In all the systems which have been tested to date (with the possible exception of butylacrylate[2,28] about which there is some doubt) a dependence of reaction rate on solution viscosity has been observed down to the lowest obtainable solution viscosities. Unfortunately, these studies have not covered a wide range of chemical types, but include acrylamide,[29] methylmethacrylate,[2,28–32] various alkyl-methacrylates,[33] various copolymers of methyl methacrylate,[34–6] and vinylacetate.[34]

The general conclusion which can be drawn from these studies is that if a free-radical polymerization is characterized by a termination rate constant with a value 10^6–10^8 dm^3 mol^{-1} s^{-1} then it is extremely probable that the reaction is diffusion-controlled. This criterion is sometimes extended to a comment on the Arrhenius activation energy observed for the process. Thus, if the activation energy is comparable with the activation energy for viscous flow of solvent, or is sufficiently small that the reaction rate must be very rapid, then there is a definite probability that the reaction will be diffusion controlled. This last criterion is not always totally reliable, especially when the internal segmental rearrangement of a polymer chain is a slow or inefficient process.

As was pointed out by Burkhart,[22] in the case of very slow segmental rearrangement of stiff chains, the reaction rate becomes less dependent on the solution viscosity. There is some indication that this is the case with higher alkylmethacrylates.[33] The dependence on solvent viscosity decreases as the ester side-group is changed from methyl, through isobutyl to 3,5,5-trimethylhexyl.

Although the observation of a solvent viscosity effect suggests that some diffusive process is the rate-determining step in termination, it does not tell us whether translational diffusion or some segmental rearrangement (which also moves the chain-end past several solvent molecules) is the important slow step. An unambiguous assessment of the importance of segmental rotation is not easy to make, and the evidence presented to date takes the form of inference rather than proof. However, every indication is that in the polymerization systems studied (and mentioned earlier) segmental rearrangement is the important rate-determining phenomenon.

The evidence takes two forms. The first is that the dependence of termination rate on polymer concentration (conversion or monomer to polymer) is not the same as the concentration dependence of self-diffusion.[31] The second is that the rate of the diffusion-controlled termination reaction varies with chemical structure over a wider range than does the rate of diffusion, more closely resembling the variation in rotational processes such as dielectric relaxation.

While it might have been thought that a study of the molecular-weight dependence of the reaction rate constant would be informative on this question, in fact this has turned out to be not the case. Part of the reason has been the difficulty in obtaining precise values for the single-step rate constant in a chain reaction, and part has been the uncertainty as to the dependence predicted by theory (§ 13.3). In addition, in most of the systems studied, propagation was occurring at the same time as termination, so that the radicals were heterodisperse in size with a number population monotonously decreasing with chain length. Consequently any kinetic effects, which might in any case be slight, can be disguised by this distribution.

A detailed and precise examination of the dependence of the termination rate constant upon final polymer (and hence free-radical) molecular weight has been made for poly(methylmethacrylate) by Fischer, Mücke, and Schulz.[30] These authors covered a twentyfold range of degree of polymerization (above 2×10^3), yet found that the termination-rate constant (at low conversion) was independent of the degree of polymerization. However O'Driscoll and Mahabadi[37] have re-examined the same system using the technique of spatially intermittent polymerization to obtain more precise values for the rate constant. At 298 K it varies as an inverse function of chain length for short chains, but becomes independent of chain length when the degree of polymerization is greater than 10^4. This is very much in line with the behaviour predicted[16,22] for chains where segmental reorientation is rate-determining.

A rather elegant examination of the chain-length effect in a radical termination reaction has been made by Borgwardt, Schnabel, and Henglein.[38] These authors studied the disappearance of poly(ethyleneoxide) radicals formed by the pulse radiolysis of dilute aqueous solutions

$$\cdot OH + \sim CH_2CH_2O\sim \rightarrow \sim \dot{C}H_2CHO\sim + H_2O.$$

They found that the bimolecular termination rate constant was dependent on the polymer degree of polymerization up to values of 10^3, but did appear to be independent of chain length above this value. These observations were made on polymers, some samples of which were almost monodisperse in molecular weight, and of course the radical chain length did not increase during the radical lifetime. The length of 10^3 together with the actual rate constants, indicate that this chain is roughly ten times as flexible as that of poly(methylmethacrylate), a result very much in line with that obtained from dielectric studies (Chapter 5).

For chains of degree of polymerization less than 10^3 the reaction rate constant varies almost as the inverse square root of the chain length. This is in accord with a model in which translational diffusion is rate-determining, but the collision volume involved is the size of the reactive chain segment, not the over-all polymer chain. In other words the important diffusive motion is migration of the radical segment retarded to the low diffusivity of the whole chain.

Whether a change from translation-controlled to segmental-controlled reaction occurs for all chains cannot yet be stated. However, the poly(ethyleneoxide) chain is one of the most flexible (most rapid backbone rotation) known, so perhaps it is not surprising that segmental rearrangement is not so clearly the slow process as in sterically hindered chains like poly(methylmethacrylate).

13.4.2. Structural effects on the free-radical termination reaction

Perhaps the strongest evidence for the importance of segmental reorientation comes from an examination of the effect of steric constraints caused by groups substituted on the polymer backbone. Thus almost all vinyl polymer chains of molecular weight around 10^5 have diffusion coefficients 2–3×10^{-11} m^2 s^{-1}, yet the kinetic rate constants vary over a thousandfold range. When comparison is made with some other measurement of segmental mobility, the parallel is quite striking. Thus structural effects on the termination rate constant and the dielectric relaxation frequency are compared in Table 13.2. The relative changes in each of the two series (dipole rigidly attached to the backbone, and dipole with one component free to rotation in the side group) show that both depend on the same basic type of molecular movement.

Table 13.2

Comparison of kinetic rate constants and dipole-reorientation frequencies

Polymer	Dipole-relaxation frequency (Hz at 25 °C)	Termination rate constant ($dm^3 mol^{-1} s^{-1}$ at 30 °C)

$-CH_2-CH-$
     $\|$
     Cl
	2×10^8	5×10^8

$-CH_2-CH-$
     $\|$
     Br
	3×10^7	$2 \cdot 5 \times 10^8$

$-CH_2-CH-$ (4-chlorophenyl)	3×10^7	$7 \cdot 7 \times 10^7$
$-CH_2-CH-$ (carbazolyl N)	9×10^5	$2 \cdot 5 \times 10^5$

$-CH_2-CH$
     $\|$
     $COOCH_3$
	2×10^9	$2 \cdot 6 \times 10^8$

$-CH_2-CH-$
     $\|$
     $OCOCH_3$
	2×10^9	$2 \cdot 0 \times 10^8$

$-CH_2-C(CH_3)-$
     $\|$
     $COOCH_3$
	$3 \cdot 9 \times 10^7$	$1 \cdot 6 \times 10^7$

$-CH_2-C(CH_3)-\!-CH-\!-C(CH_3)-$ $COOCH_3$ $CO\diagdown_{O}\diagup CO$	$1 \cdot 4 \times 10^7$	$3 \cdot 8 \times 10^6$

(2 per cent citraconic anhydride)

Another structural feature which can affect the segmental motion of polymer chains is the incorporation of comonomer units. It is important to realize that the chain structure will affect the diffusion-controlled and chemically-controlled termination processes differently. In the former case the termination rate will be governed by the composition of the whole chain, or at least the composition extending for several monomer units from the chain end, whereas in the latter case the principal feature affecting reactivity will be the monomer unit containing the site of free-radical activity. As a result the kinetic equations describing the over-all rate of copolymerization will differ in two cases, and it now appears that the conventional equation for simple chemical control determined by the nature of the single end-unit containing the free-radical site is inapplicable.

Consider the copolymerization of sterically hindered monomer with an unhindered monomer. When the copolymer composition is rich in the former, segmental motion will be slow and so the termination rate constant will be low. When the copolymer composition is rich in the 'flexibilizing' monomer, segmental motion will be easier and so the termination rate constant will be higher. The copolymerization of methyl methacrylate and vinyl acetate has been analysed[34] in this fashion and a sensible dependence of rate constant on monomer feed composition obtained. Indeed at 60 °C it appears that the flexibilizing effect almost follows an 'ideal' equation combining the termination constants for homopolymerization and the mole fractions of each unit in the polymer.

Termination may be suppressed not only by the addition of a very hindered comonomer but also by one which exerts dipole–dipole forces that hinder segmental rotation. This has been shown[36] in the effect of citraconic anhydride on methyl methacrylate polymerization (Table 13.2).

13.4.3. The effect of temperature on the free-radical termination reaction

When the termination rate is controlled by segmental rotation, the temperature dependence of the rate will depend upon the energy barriers to rotation. It has already been shown that the activation energy for dielectric relaxation in solution increases with the size of backbone substituent and a comparable effect should be apparent in termination rates. Although this has not yet been observed (due to the difficulty in obtaining accurate activation energies for termination) an even more remarkable phenomenon occurs in chains with bulky steric groups. This is a catastrophic decrease[39] in the termination-rate constants at low temperatures.

In the series vinylbromide, methylmethacrylate, and N-vinylcarbazole, the two latter monomers form hindered chains, and exhibit a marked reduction in termination rate around 220 K. The polyvinylbromide chain is less hindered, however, and as a result the termination rate is higher and exhibits an Arrhenius temperature-dependence over the whole temperature range studied.

The marked reduction in termination rate has been ascribed to an effective cessation of backbone rotation as the energy barriers to rotation become comparable with kT, and the time required for efficient segmental reorientation becomes large.

13.4.4. The effect of polymer concentration on the free-radical termination reaction

All the observations reported in the preceding sections have been made at such low conversion of monomer to polymer that the system viscosity is little affected by inert polymer, and the diffusional approach of the macroradicals is essentially unimpeded by entanglement with preformed chains. However when polymerization is carried to higher conversions, the macroradicals can interact with the preformed polymer, and several very interesting facts can be observed.

One of the stranger phenomena is that in the initial stages of polymerization the termination rate may increase[31] with polymer concentration, and can be described by a virial expression

$$k_t/k_{t,0} = 1 + (A' - [\eta]c + \dots \tag{13.19}$$

where A' is similar to the second virial coefficient and arises from the virial expansion of the diffusive process governing the reaction rate, and $[\eta]$ is the intrinsic viscosity of the polymer formed. Subscript zero refers to zero conversion. Then the termination rate increases with concentration in good solvents when the second virial coefficient term is greater than the intrinsic viscosity, but decreases at higher concentrations when coefficients of the power series describing viscosity become predominant.

At these higher conversions both the enhanced solution viscosity and the restriction of termination arise in the entanglement of polymer chains. Indeed this phenomena leads to a marked auto-acceleration[40,41] in the polymerization which has been called the 'gel' effect. It is sometimes, too, referred to as the 'Tromsdorff' effect, although the work of Norrish[40] and his co-workers predates that of Tromsdorff.[41] Since entanglement occurs when polymer chains are long enough and close enough to interact with their neighbours, the critical concentration at which entanglement effects become evidenced is higher for short chains than for long ones. Consequently the critical conversion at which auto-acceleration commences is higher for those polymerizations leading to low molecular-weight polymer.

The most widely studied[40-8] system has been the free-radical polymerization of methylmethacrylate in which the gel effect is most pronounced. The reduction of the termination rate, with the greatest diminution in systems of high molecular weight, is now well established. By varying the rate of initiation it is possible to alter the average size of the polymerizing radicals independently of the size of preformed polymer. This rather neat little experiment shows[49] that the termination rate in the entanglement region depends on the size of both the reacting chain and the inert polymer.

A detailed kinetic analysis of polymerization at high conversions is difficult because the kinetic orders change with conversion. However Cardenas and O'Driscoll[50] have analysed the data of Balke and Hamielec[47] using a model in which the termination reaction involves unentangled (small) and severely impeded (large) radicals. When the critical degree of polymerization Z_c, and the retarded termination rate constant of entangled chains, $k_{t,e}$, are treated as semi-adjustable parameters, values selected from time-conversion data are not only reasonable, but allow prediction of molecular-weight averages and distributions, as a function of conversion, which are in good agreement with the observations.

At very high conversions the propagation step, too, becomes diffusion-controlled, and polymerization eventually stops when the conversion reaches the composition for which the reaction temperature is the glass-transition temperature. In other words neither propagation nor termination can occur with the very limited molecular motion in the glassy state.

13.5. Luminescence quenching

13.5.1. General comments

The importance of quenching phenomena exhibited by photoexcited states provided a major impetus to the study of diffusion-controlled reactions of non-polymeric species. Effort in this area has taken place in two waves. The first,[5,6] in the 1950s and early-1960s was associated with a general interest in physical photochemistry, and with the use of scintillation and related techniques in radiochemistry and nuclear physics. It was made possible by the development of new sensitive photodetection devices, and the appearance of reliable commercial spectrofluorimeters and phosphorimeters. Despite the considerable advances in instrumentation, time-resolved experiments were limited to the nanosecond region, and the theoretical advances[6] related to short-time behaviour were unable to be tested with any precision. The second is very recent, and is associated with the development of sensitive photon-counting detection techniques and laser excitation sources, particularly those permitting time resolved studies in the picosecond region.

A detailed review of the ways in which theories of diffusion and diffusion-controlled reactions are applied to quenching systems has been given by Alwatter, Lumb, and Birks.[51] Generally eqn (13.4) (or the more exact form allowing for certain encounters being unproductive) is applied, with the observed quenching-rate constant being evaluated either directly from time-resolved measurements or from a Stern–Volmer consideration of quenching efficiency under continuous irradiation,

$$\frac{\Phi_{LO}}{\Phi_{LQ}} = 1 + k_{obs}\tau_0[Q] \tag{13.20}$$

where Φ_{LO}, Φ_{LQ} are the quantum efficiencies of luminescence in the absence and presence of quencher, Q, and τ_0 is the excited-state lifetime in the absence of quencher.

A simplification which is often applied is to equate the hydrodynamic radii of the two reactants to each other, and to halve the encounter radius, so that substitution of a Stokes–Einstein relation for the diffusion coefficient results in cancellation of the radius terms and gives for very efficient quenching

$$k_{obs} = \frac{8kT}{3\eta}\{1 + r_{AB}/(\pi D_{AB} t)^{\frac{1}{2}}\}. \qquad (13.21)$$

For the 'long-time' result when stationary state diffusion has been achieved, the predicted rate constant is simply $8kT/3\eta$ (in molecular units) and contains solvent viscosity as the sole disposable parameter. Methods of improving this relationship by the substitution of more realistic diffusion or microviscosity terms have been suggested,[51] but it can never rigorously be applied to molecules as anisometric, or with such size heterogeneity, as are encountered in polymer systems.

13.5.2. Diffusion-controlled quenching in polymer systems

In addition to the various processes controlled by energy transport discussed in Chapter 6, a number of quenching studies have yielded information on the polymer chain motions in diffusion-controlled quenching.

In fact application of the Stern–Volmer equation (13.20) can be used to yield either the diffusion-controlled rate constant, the excited-state lifetime, or the concentration of quencher so long as independent estimates can be made of any two of these three quantities. Thus the quenching of small fluorescent molecules by a polymer bearing quenching side-groups can be treated[52] assuming a Gaussian distribution function for the density of active segments in the chain. Such a study also shows the polymer concentration at which overlap between polymer coils becomes important. In the same way when the fluorescent groups are bound to the polymer chain,[53] quenching by a small molecule which is a solvent for the polymer can be used to probe the distribution of quencher within the polymer coil.

In a series of experiments using ketones as both fluorescence energy donors (quenched by biacetyl) and acceptors (quenching naphthalene) Heskins and Guillet[54] and Somersall and Guillet[55] showed that polymers with keto side-groups formed less efficient quenchers than small-molecule aliphatic ketones, but that quenching of the polymeric species was just as rapid as that of the small molecules. The experiments probably reflect the reduced penetration of small-molecule donors into the polymer coil, and the migration of donor excitation energy when the polymer is donor.

Attempts[56] to study a reaction analogous to the free-radical termination process using a polymer chain with a single quenching unit to quench a chain of the same polymer with a fluorescent unit have not been successful since the reduced macro-reactant collision rates are too slow to compete effectively with the emission process.

The quenching of phosphorescence in solid polymers by adventitious impurities such as oxygen can be used to monitor oxygen-diffusion rates.[57,58] Particularly interesting are the changes in quenching that occur at polymer transitions,[59] though whether these are caused by changes in oxygen diffusion or radiationless decay is unclear.

References

1. Fowler, R. H. and Slater, N. B. *Discuss. Faraday Soc.* Sept. 1937; *Trans. Faraday Soc.* **34**, 81 (1938).
2. Benson, S. W. and North, A. M. *J. Amer. chem. Soc.* **81**, 1339 (1959).
3. North, A. M. In *Progress in high polymers* (ed. F. W. Peaker and J. C. Robb), p. 95. Iliffe Books, London (1968).
4. North, A. M. *Collision theory of chemical reactions in liquids*. Methuen, London (1964).
5. North, A. M. *Quart. Rev. chem. Soc.* **20**, 421 (1966).
6. Noyes, R. M. In *Progress reaction kinetics* (ed. G. Porter), p. 129. Pergamon, London (1961).
7. Smoluchowski, M. V. *Z. phys. Chem.* **92**, 129 (1917).
8. Frisch, H. L. and Collins, F. C. *J. chem. Phys.* **20**, 1797 (1952).
9. Collins, F. C. *J. colloid Sci.* **5**, 499 (1950).
10. Waite, T. R. *Phys. Rev.* **107**, 463, 471 (1957).
11. Collins, F. C. and Kimball, G. E. *J. colloid Sci.* **4**, 425 (1949).
12. Waite, T. R. *J. chem. Phys.* **28**, 103 (1958).
13. Noyes, R. M. *J. Amer. chem. Soc.* **77**, 2042 (1955); **78**, 5486 (1956).
14. Rabinowitch, E. *Trans. Faraday Soc.* **33**, 1225 (1937).
15. Debye, P. *Trans. electrochem. Soc.* **82**, 265 (1942).
16. Benson, S. W. and North, A. M. *J. Amer. chem. Soc.* **84**, 935 (1962).
17. Allen, P. E. M. and Patrick, C. R. *Makromol. Chem.* **72**, 106 (1964).
18. North, A. M. *Makromol. Chem.* **83**, 15 (1965).
19. Schulz, G. V. and Fischer, J. P. *Makromol. Chem.* **107**, 253 (1967).
20. Moroni, A. F. and Schulz, G. V. *Makromol. Chem.* **118**, 313 (1968).
21. Olaj, O. F. *Makromol. Chem.* **147**, 235, 255 (1971).
22. Burkhart, R. D. *J. polym. Sci. A* **3**, 883 (1965).
23. Kuhn, W. and Kuhn, H. *Helv. Chim. Acta.* **29**, 1533 (1945).
24. Doi, M. *Chem. Phys.* **9**, 455 (1975).
25. Haward, R. N. *Trans. Faraday Soc.* **46**, 204 (1950).
26. North, A. M. and Postlethwaite, D. In *Structure and mechanism in vinyl polymerization* (ed. T. Tsuruta and K. F. O'Driscoll), p. 99. Marcel Dekker, New York (1969).
27. North, A. M. In *Reactivity mechanism and structure in polymer chemistry* (ed. A. D. Jenkins and A. Ledwith), p. 142. John Wiley, London (1974).
28. Bengough, W. I. and Meville, H. W. *Proc. Roy. Soc.* **A249**, 445 (1959).
29. Oster, G. K., Oster, G., and Prati, G. *J. Amer. chem. Soc.* **79**, 595 (1957).
30. Fischer, J. P., Mücke, J. and Schulz, G. V. *Ber. Bunsengesellschaft*, **73**, 154 (1969).
31. North, A. M. and Read, G. A. *Trans. Faraday Soc.* **57**, 859 (1961).
32. Yokota, K. and Itoh, M. *J. polym. Sci. B* **6**, 825 (1968).
33. North, A. M. and Read, G. A. *J. polym. Sci. A* **1**, 1311 (1963).
34. Atherton, J. N. and North, A. M. *Trans. Faraday Soc.* **58**, 2049 (1962).

35. North, A. M. and Postlethwaite, D. *Polymer* **5**, 237 (1964).
36. North, A. M. and Postlethwaite, D. *Trans. Faraday Soc.* **62**, 2843 (1966).
37. O'Driscoll, K. F. and Mahabadi, H. K. *J. polym. Sci.* (In press.)
38. Borgwardt, U., Schnabel, W. and Henglein, A. *Makromol. Chem.* **127**, 176 (1969).
39. Hughes, J. and North, A. M. *Trans. Faraday Soc.* **62**, 523 (1966).
40. Norrish, R. G. W. and Smith, R. R. *Nature* **150**, 566 (1942).
41. Tromsdorff, E., Kohle, H. and Lagally, P. *Makromol. Chem.* **1**, 169 (1947).
42. Fujii, S. *Bull. chem. Soc. Jpn* **27**, 216 (1954).
43. Schulz, G. V. *Z. phys. Chem. N. F.* **8**, 290 (1956).
44. Hayden, P. and Melville, H. W. *J. polym. Sci.* **43**, 215 (1960).
45. Burnett, G. M. and Duncan, G. L. *Makromol. Chem.* **51**, 154, 171, 177 (1962).
46. Horie, K., Mita, I. and Kambe, H. *J. polym. Sci. A-1*, **6**, 2663 (1968).
47. Nalke, S. T. and Hamielec, A. E. *J. appl. polym. Sci.* **17**, 905 (1973).
48. Ito, K. *J. polym. Sci.* **7**, 2995 (1969).
49. Abuin, E. and Lissi, E. A. *J. macromol. Sci. Chem.* **A11**, 282 (1977).
50. Cardenas, J. N. and O'Driscoll, K. F. *J. polym. Sci.* (In press.)
51. Alwatter, A. H., Lumb, M. D. and Birks, J. B. In *Organic molecular photophysics*, Vol. 1 (ed. J. B. Birks), p. 403. Wiley, London (1973).
52. Duportail A., Froelich, D. and Weill, G. *Eur. polym. J.* **7**, 977 (1971).
53. Moldoran, L. and Weill, G. *Eur. polym. J.* **7**, 1023 (1971).
54. Heskins, M. and Guillet, J. E. *Macromolecules* **1**, 97 (1968); **3**, 224 (1970).
55. Somersall, A. C. and Guillet, J. E. *Macromolecules* **5**, 410 (1972).
56. Meagher, J. and North, A. M. unpublished work.
57. Shaw, G. *Trans. Faraday Soc.* **63**, 2181 (1967).
58. Jones, P. F. *J. polym. Sci. B* **6**, 487 (1968).
59. Somersall, A. C., Dan, E. and Guillet, J. E. *Macromolecules* **7**, 233 (1974).

AUTHOR INDEX

Numbers in square brackets refer to the references cited.

SUBJECT INDEX